Practical Mathematics
for Precision Farming

David E. Clay, Sharon A. Clay, and Stephanie A. Bruggeman, editors

American Society of Agronomy
Crop Science Society of America
Soil Science Society of America
5585 Guilford Rd., Madison, WI 53711-5801 USA

agronomy.org • crops.org • soils.org
dl.sciencesocieties.org
SocietyStore.org

ISBN: 978-0-89118-361-7 [print]
ISBN: 978-0-89118-362-4 [electronic]
doi: 10.2134/practicalmath
Library of Congress Control No: 2017954066

Cover design: Karen Brey
Cover photo: Sam Beebe, EcoTrust
Printed in the United States of America

Contents

Contributors

D.E. Clay	South Dakota State University, Brookings, SD 57007-2201
G. Hatfield	South Dakota State University, Brookings, SD 57007-2201
S.A. Clay	South Dakota State University, Brookings, SD 57007-2201
G. Reicks	South Dakota State University, Brookings, SD 57007-2201
A. Varenhorst	South Dakota State University, Brookings, SD 57007-2201
C. Robinson	Illinois State University, Dept. of Agriculture, Normal IL 61790-5020
R.B. Ferguson	University of Nebraska-Lincoln, Dept. of Agronomy, Lincoln, NE 68588
J.D. Luck	University of Nebraska-Lincoln, Biological Systems Engineering Dept. Lincoln, NE 68588
C.G. Carlson	South Dakota State University, Brookings, SD 57007-2201
S.A. Bruggeman	South Dakota State University, Brookings, SD 57007-2201
J. Clay	Washington State University, Pullman, WA
T.M. DeSutter	North Dakota State University, Fargo, ND 58102
A.J. Franzen	South Dakota State University, Brookings, SD 57007-2201
C.L. Reese	South Dakota State University, Brookings, SD 57007-2201
N.R. Kitchen	Division of Plant Sciences, University of Missouri, Columbia, MO 65211
E. Byamukama	South Dakota State University, Brookings, SD 57007-2201
T.P. Trooien	South Dakota State University, Brookings, SD 57007-2201
J. Chang	South Dakota State University, Brookings, SD 57007-2201
B. Arnall	Oklahoma State University, Oklahoma City, OK 74078
S. Fausti	California State Monterey Bay, Seaside, CA 93933
B.J. Erickson	Purdue University, Agronomy Dept., West Lafayette, IN 49707
T. Wang	South Dakota State University, Brookings, SD 57007-2201
C. Graham	South Dakota State University, Brookings, SD 57007-2201

Useful Conversions

Area	Mass	Grain bu weights
Hectare, ha, 10,000 m², 100 by 100 m area, 2.471 acres	1 metric ton, 1000 kg, 1 Mg, 1.102 tons	Corn (shelled) 56 lbs @ 15.5% moisture
Section, 640 acres,	kilogram, kg, 1000 grams, 2.205 pounds,	Wheat 60 lbs @ 13.5% moisture
Acre, ac, area plowed by a oxen in one day, 66 by 660 ft area, 43,560 ft²	gram, g, mass of 1 mL at the melting point of ice, 0.001 kg	Soybean 60 lbs @13.% moisture
Distance	Ton, 2000 lb, 0.90742 metric tons	Sorghum, 56 lbs @13% moisture
Kilometer, km, 1000 meters, 0.6214 miles, 3280.84 ft	Pound, lb, 16 ounces, 0.453592 kg, 453.592 g	Rates
Meter, m, 100 cm, 3.28084 ft, 39.3701 inches	Volume	1 kg ha⁻¹, 0.892 lbs acre⁻¹
Centimeter, cm, 10 mm, 0.393701 inches	Liter, L, 1000 ml, 1000 cm³, 0.264172 US gallons, 1.057 quarts	lb acre⁻¹, 1.121 kg ha⁻¹
Mile, mi, 5280 ft, 1609.34 meters	Milliliter, mL, 0.001 L, 0.033814 US fluid ounces	Temperature
Yard, 36 inches, 0.914 m	U.S. Gallon, gal, 3.785 liters, 8.34 pounds, 4 quarts, 8 pints	F -->C, =0.555 x (F-32)
Foot, ft, 12 inches, 1/3 of a yard, 30.48 cm	Imperial gallon, gal, 4.54609 liters,	C-->F, (1.8 x C)+32
Inch, in, 1/12 of a foot, 25.4 mm, 2.54 cm	U.S. Bushel, 64 pints, 35.2 liters	
	Teaspoon, 0.167 fliud onces, 4.9289 mL	

A Guide to Making Mathematics Practical

1

Jo Clay

Chapter Purpose

This chapter reviews the mathematical skills and approaches needed to solve problems in precision agriculture and to provide a guidance for teaching and understanding mathematics. This chapter describes a common sense approach for solving mathematical story problems, provide examples where the logic leads to correct and incorrect answers, and discusses the solutions using key mathematical ideas.

Key Terms

Problem solving, units and quantities, multiplication and division of fractions, formulas, ratios, proportions, and scale factors, inverse relationships, percentages, unit conversions.

Mathematical Skills

Problem solving, units and quantities, multiplication and division of fractions, formulas, ratios, proportions, and scale factors, inverse relationships, percentages, unit conversions.

The Importance of Practical Mathematics

Implementing precision agriculture requires an ability to identify and solve basic mathematical story problems. However, solving problems is complicated by teaching approaches that emphasize memorization without understanding (Hiebert, 2013). Research indicates, constant practice is needed to remember formulas and procedures that are based on memorization (NCTM, 2000). Once students and adults leave an academic setting, they stop practicing and forget the knowledge they once knew. Teaching for understanding is designed to help people gain a deeper understanding of mathematics, improve problem solving abilities, improve retention of the learned material, and give rise to everyday practical applications of learned mathematical skills. Individual students will gain an enhanced capacity to integrate mathematics into decision making and their everyday lives. In the context of this manual, teaching mathematics is characterized for understanding, as *practical mathematics*. However, in spite of an increased use of teaching for understanding methods and it's corresponding advantages, adoption of practical mathematics has been slow (Dole et al., 2016). A survey of potential employers revealed that Agronomists need better mathematical skills to prepare them for 21st century Agriculture (Erickson et al., 2016).

The goals of this chapter are to discuss why teaching for understanding is very important and to provide a structure for grappling with intuitive ideas. The problems discussed in this chapter are agricultural based and provide multiple solutions to the same problem. One goal is to engage the student in the problem solving process. Embedded in the approaches are models that maybe useful in other problems. You will notice that some of the approaches are based on incorrect mathematics. This is not to confuse you, but to help you

J.A. Clay, Washington State University, Pullman, WA 99614.*Corresponding author (clayjoclay@gmail.com).

doi: 10.2134/practicalmath2016.0115

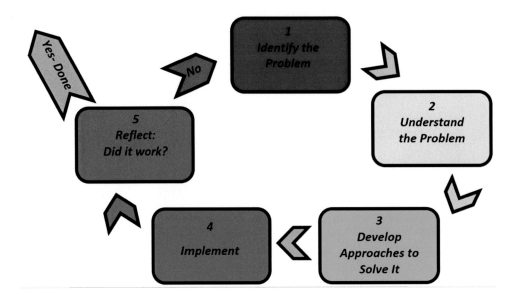

Fig. 1.1. The problem solving process.

become aware of how the mathematics are used to solve problems, and to highlight ways that people incorrectly use mathematical ideas.

In this chapter, problems will be provided and solved using a variety of approaches. The key mathematical ideas will be discussed and you will have the opportunity to deepen your own understanding. Then, you will be asked to identify ways that your enhanced understanding might be useful in your life and career. The goal of this process is to help make connections in your brain to ideas that you have learned, modify misconceptions, and develop new connections so that you can use mathematics to find solutions to a range of problems. Key understandings discussed in the chapter:

· A unit (focus on fractions)

· Operations with fractions (focus on multiplication and division)

· Proportion, Ratio and Scale Factor

· Percentage

· Conversions

Introduction to Problem Solving

The structure of this chapter is different from a typical mathematics book. We ask each of you to engage in thinking about a problem so that you can understand it, identify the intuitive ideas and mathematical knowledge that you have. Then, engage in solving the problem. Figure 1.1 illustrates the five steps of the problem solving process.

Step 1: The first step in solving a problem is to identify it. If you cannot state the problem, you will find it impossible to solve. As you identify the problem, think about the different ways that the problem can be stated. Once you identify the problem, you are ready for step two.

Step 2: Understand the problem by considering the information that is known and what the problem is asking you to find. Many people make a model of the problem in their mind or on paper. The model may be a picture or mathematical in the form of an equation or other notation. In this chapter you will learn some different ways to make models that will help you understand the problem. This is a critical step and one that many people skip.

Step 3: Develop an approach to solve the problem. Sometimes you will jump straight to a mathematical equation or formula because you immediately recognize a way to solve the problem. If you are unable to remember a specific approach, try using mathematical reasoning or simplifying the problem into one that you can solve. Another tool that may help you is to make a table that displays the known information. Remember it is okay to start again.

Step 4: As you implement your approach, remember the fundamental rules that govern operations with fractions and algebra. In this chapter, these fundamental rules will be reviewed using models that are designed to help you understand the rules. We know that if you understand why mathematics operates as it does, you will be more likely to remember how to correctly use the rules or shortcuts.

Step 5: Finally, you need to reflect on the approach. Does the approach make sense in the context of the problem? If it makes sense, is the mathematics correct? Each time you do an operation (or step in solving an equation) you should know why that step is correct. People sometimes use operations incorrectly as you will see in this chapter. If your solution does not make sense, go back to the first step and make sure that you clearly stated the problem, review your model and perhaps make a new model, consider changing the approach that you used, and check the mathematics. Did you perform each operation correctly as you implemented the approach that you used? If you know that your solution is incorrect and you can't figure out what to do, call a friend to talk with, look on the internet for help, or go back to some of your textbooks/notes to review ideas that you may have forgotten.

Some people like bullets to help them remember these key steps. Many people have found the following list helpful in problem solving.

· Restate the problem in your own words.

· Identify what the problem is asking you to find, and clarify the given information.

· Draw a model (picture or other representation) of the problem

· Identify an operation that you would begin with and give a reason for solving the problem.

· Solve the problem using the mathematics that you have learned.

· Reflect on whether the solution makes sense. If it does not make sense, go back and consider where you may have made a mistake.

Section 1: Solving Problems By Separating the Quantity Into Pieces or Units

Solving a problem can often be made easier by separating the known quantity into separate pieces of the same size. For example, when learning to count, you matched a number to each object in a set. Each object represented a number in sequence and you conceptualized as a *unit of one*. For example, you counted the number of sunflower seeds in a tablespoon. Each seed was one unit, and you counted x units (or seeds) in one tablespoon. Later when fractions were introduced, you learned that the unit could be divided into pieces that were the same size. For example, a circle was drawn and represented a unit. It was divided into two pieces and each piece was defined as a half. However, you probably did not investigate when it made sense how to divide a unit in half. For the example preview, it does not make sense to divide a seed in half. The context of the problem and how you define the unit is critical in making sense of a problem and finding a correct solution. The following example illustrates the importance of carefully defining the unit and making sure that it represents the problem context.

A child found that 8 sunflower seeds fit in a tablespoon and was asked to find out how many would fit in a half-tablespoon. Many children would draw on their previous knowledge of dividing circles and draw eight circles. Then they would cut each circle in half and say the answer was 16. An adult might correct the child and say, "Yes, there are 16 half-circles. But, the question is asking you how many sunflower seeds are there if you divide them into two groups"?" The unit, one tablespoon, is split in half. Since there are eight units (seeds) in a tablespoon, a half tablespoon would contain half the units in one tablespoon or (8 seeds × 0.5 = 4). So there are four seeds in a half tablespoon.

The model of splitting a "whole" represented by a circle may lead to an understanding of fractions based on the idea that only a single object can be divided into pieces. An adult may recognize that a group of objects can be divided into subgroups. In the example, the unit changes from one sunflower seed into one tablespoon with eight seeds. Within this problem context, it does not make sense to cut each seed in half. You cannot put two half seed together to make a whole seed. Here we need to think of the unit as the tablespoon of eight seeds. We can divide it in half by making two piles of seeds.

The key understandings are:

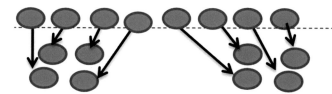

Unit: 8 seeds in a tablespoon

Unit is separated into 2 sets of seeds with the same quantity, 4 seeds in each half.

· *A unit can be divided into same size quantities.*

· *An individual defines a unit to be anything that is suggested by context of the problem.*

The unit may be a liquid with a numerical quantity, a measurement of a field, or a number of discrete objects as in the previous problem. How a unit is represented depends on the context of the problem and individual preference.

PROBLEM 1.1.

A producer wants to use a new weed killer on his/her farm and decide to test it in a small area. The directions tell the producer to dilute the concentrated weed killer with water to make a diluted solution with weed killer making up a fifth of the solution. Producer 1 wants to use 15 ounces of weed killer concentrate. How much water should Producer 1 add?

Before reading ahead, solve the problem or think about how you would go about solving it. The best way to develop your skills is to think about your intuitive reasoning and then take a look at some approaches that others have used. The two approaches yield different answers, only one is correct. See if you can identity the incorrect use of mathematics.

Answer Problem 1.1: Approach A

Producer 1 remembered that to solve problems, first he or she must identify the key information and what was given. Producer 1 also remembers that the word "of" indicates multiplication when you are working with fractions. So Producer 1 notes the phrase "one-fifth of weed killer" and will use 15 ounces. To solve the problem multiply 15 by one fifth.

$$15 \times \frac{1}{5}$$
$$\frac{15}{1} \times \frac{1}{5} = \frac{15}{5} = 3$$

Next, Producer 1 reflects about the reasonableness of his solution and concludes that usually more water is added than concentrate. So, to get a larger number Producer 1 multiplied 3 by 15 to get 45 ounces of water. From past experience, this seems to be a better answer. He/she made the product by mixing 45 ounces of water with 15 ounces of weed killer. However, 45 ounces of water, while seeming more reasonable, is not correct. Why?

Answer Problem 1.1: Approach B

Producer 2 decided to make a model of the problem so that she could figure out how much water to use. To make the model, Producer 2 started with a diagram of the mixed solution because they knew one part of it (the amount of weed killer concentrate) and they wanted to know the other part (amount of water).

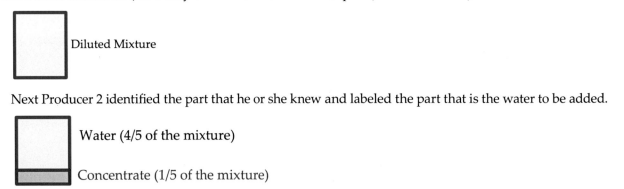

Diluted Mixture

Next Producer 2 identified the part that he or she knew and labeled the part that is the water to be added.

Water (4/5 of the mixture)

Concentrate (1/5 of the mixture)

The diluted mixture can be thought of as the unit and $\frac{1}{5}$ of it is the concentrate. Using their knowledge of fractions, Producer 2 decided that the water is $\frac{4}{5}$ of the diluted mixture. The unit can be broken into 5 equal segments with four of them water. These four segments are $\frac{4}{5}$ of the total. So, the amount of water is 4 times as much as the concentrate. To find the amount of water that is needed, multiply 15 ounces by 4. The correct amount of water needed is 60 ounces.

Discussion of Problem 1.1

Why was approach A incorrect? Approach A is based on memorized knowledge, the word "of" with fractions in a word problem indicates the operation of multiplication. This rule is helpful only in one context, when a given quantity is a unit that needs to be broken into parts. Producer 1's approach would be correct if he/she knew that the quantity was 60 ounces and they wanted to find the number of ounces of weed killer to use. Producer 1 was unable to create a model for the problem and fully understand it. Instead, the producer skipped to step 4 and implemented an approach that was based on memorized information. The producer's approach assumed that the quantity was

60 ounces. They multiplied 60 by $\frac{1}{5}$. By multiplying by $\frac{1}{5}$, Producer 1 in fact was dividing the quantity into 5 equal amounts. However, in this problem you do not know the diluted amount. You only know the amount of the concentrate and multiplying it by $\frac{1}{5}$ will not give you the amount of water to add. In reflecting, Producer 1 made the error of changing the arithmetic to get an answer that made sense. The producer did not have a mathematical reason for changing the arithmetic, so they arrived at a mixture that would not have enough water.

Approach B is based on the key understanding, *a unit can be divided into same size quantities*. Producer 2 recognized the relationship between the unit and its fractional parts. Producer 2 used a model in which the unit is clearly identified and then it is divided into the fractional parts ($\frac{1}{5}$ and $\frac{4}{5}$). The model led Producer 2 to the correct solution to multiply 15 ounces by 4 to find the amount of water to add. The total amount of diluted weed killer is $15 \times 5 = 75$ ounces.

PROBLEM 1.2.

An agronomist has two buckets, a large and smaller one. The smaller bucket holds ⅔ of a gallon of water. He/she can pour the small bucket of water into the larger one and it fill up ⅞ of the large bucket. How many more gallons would it take to fill up the whole bucket? How many gallons does the large bucket hold?

Before reading ahead, solve the problem or think about how you would go about solving it. Only one of the solutions is correct, see if you can find where the error was made.

Answer Problem 1.2: Approach A

Agronomist 1 knows that ⅞ of the bucket is filled. He or she subtracts ⅔ because the problems asks, *How many more gallons are needed?* To find out how many more are needed, Agronomist 1 uses subtraction.

$$\frac{7}{8} \times \frac{3}{3} = \frac{21}{24}$$

$$\frac{2}{3} \times \frac{8}{8} = \frac{16}{24}$$

$$= \frac{21}{24} - \frac{16}{24} = \frac{5}{24}$$

You need to add 5/24 gallons of water. This solution is incorrect.

Answer problem 1.2: Approach B

Agronomist 2 is unsure of where to begin. They knows that $\frac{2}{3}$ gallons fill the small bucket and that $\frac{7}{8}$ is not a measure of gallons. Thus Agronomist 2 knows that they cannot subtract because fractions are not in gallons. Agronomist 2 makes a diagram by placing the small bucket inside the larger bucket.

 ← The part of the large bucket that is not filled.

←The small bucket. It is $\frac{2}{3}$ gallons and fills $\frac{7}{8}$ of the large bucket.

Agronomist 2 recognizes that the relationship between the two buckets. The small bucket is 7/8 of the large one. So, the large bucket is 1/8 larger. Using this relationship, he/she divides the large bucket into 8 pieces (each one is $\frac{1}{8}$ of the large bucket). Agronomist 2 colors in 7 of the pieces ($\frac{7}{8}$). To find the number of gallons in each piece, they divide $\frac{2}{3}$ by 7. Later in the chapter we will discuss division of fractions. For our purpose, Agronomist 2 changes the fraction into a decimal and divides $0.67 \div 7 = 0.095 = 0.10$ (rounded off). They know that each piece is 0.10 gallons. He/she multiplies 0.10 by 8 because there are eight $\frac{1}{8}$s in a whole (unit). Thus, the large bucket is approximately 0.8

gallons, and only 0.10 more gallons are needed to completely fill the large bucket. Working in fractions, the answer is $^{16}/_{21}$ total gallons held by the large bucket. This approach is correct.

Discussion of Problem 1.2

Agronomist 1 did not fully understand the problem. They interpreted the problem to ask how many gallons were needed if the large bucket held $^{7}/_{8}$ gallons. Their approach would be correct for this question. Agronomist 1 found how many more gallons were needed to fill the large bucket if the small one held $^{2}/_{3}$ gallons and the large one held $^{7}/_{8}$ gallons. In this case, Agronomist 1 defined the unit as a gallon, and both fractions referenced this unit. Agronomist 1 also did not go back and reflect on the problem. It does not make sense that the large bucket would have only $^{5}/_{24}$ gallons when the small bucket held $^{2}/_{3}$ gallons. Clearly, $^{5}/_{24}$ is less than $^{2}/_{3}$.

In contrast, Agronomist 2 created a model so that he/she could identify the problem more clearly. In doing so, Agronomist 2 recognized that two units were used in the problem. One unit was a gallon and the second was the large bucket. The problem told them that the small bucket was $^{7}/_{8}$ the size of the large bucket. Recognizing this was an important step to correctly solve the problem and it illustrates that *an individual defines a unit to be anything that is suggested by the problem's context*. Next they used the key understanding, *a unit can be divided into same size quantities*, to find the number of gallons in $^{1}/_{8}$ of the large bucket. After finding the size of $^{1}/_{8}$, Agronomist 2 found the number of gallons in the whole bucket.

Section 1 Practice Exercises

Identify the unit for each problem. You may use a model to help you solve the problem or numbers. Please note, mathematicians use models that they write on paper until they are able to picture them in their minds. Models are useful at times and not useful at other times. In this unit, several models are provided as samples that you can use. You should be able to explain what you did similar to the explanations given in the previous sample problems.

Section 2: Solving Problems That Involve Multiplication and Division of Fractions and Formula

Multiplication

When you were first learned about multiplication, it was mathematically defined *as number of groups of a particular size*. For example: You have 6 bushels of wheat and each bushel weighs 60 pounds. What is the total weight? Here you have 6 groups (a bushel is one group) and the size of each 60 pounds. To find the total you multiply 6 × 60 lb = 360 lb.

Each circle is one bushel

$\quad$ 60 lb $\quad$ 60 lb $\quad$ 60 lb $\quad$ 60 lb $\quad$ 60 lb $\quad$ 60 lb

The problems that follow are illustrations that provide a context for identifying the key understandings of multiplying and dividing fractions. In the discussion of the problems, the key understandings will be presented.

Answer Problem 1.8: Approach A

Producer 1 restated the problem for clarity of thinking, how many bushels are 1/3 of the total number of bushels (6)? Using what he/she learned in the previous section, Producer 1 defined the unit as 6 bushels and drew a model. He/she then divided the unit into 3 groups. Looking at one of the groups, Producer 1 can see that there are 2 bushels in third.

Each bushel soybean is 60 lb. The weight of *two bushels of soybean would be 120 lb of soybean* (2 × 60 lb).

Answer Problem 1.8: Approach B

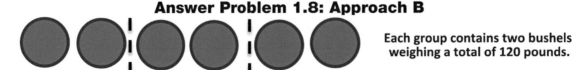

Each group contains two bushels weighing a total of 120 pounds.

Producer 2 used the information from the example, 6 bushels weigh 360 lb. He/she wanted to find 1/3 of this quantity. Producer 2 realized that he/she was finding $^{1}/_{3}$ (part of the group) with a size of 360 lb (size of the group).

1.3. An agronomist has ⅔ of a gallon of water in a bucket, which fills up ⅞ of the bucket. How many gallons would it take to fill up the whole bucket?

1.4. A producer is mixing weed killer. He/she guesses that ⅗ of a bottle of would be the right amount for the sprayer. But, it turns out to fill only about 3/4 of the sprayer. How much of a bottle of concentrate would make a full sprayer?

1.5. You have a farm with a field on the north side of a large hill and a field on the south side of the hill. It takes 40 ears of corn to fill a bushel basket from the field on the north. It takes 24 ears of corn to fill a bushel basket from the field on the south.

   a) Which field produces the biggest ears of corn? Why?

   b) You put 208 ears of corn into a crate. 1/3 of the crate is from the field on the south. How many bushels came from the south field and how many came from the north field?

   c) How many bushels does the crate hold?

   d) In a second crate you fill 3/4 of it with corn from the south field by putting in 108 ears of corn and 1/4 of it with corn from the north field. How many ears of corn are in the crate? How many bushels of corn from the south field do you put in the crate?

1.6. An agronomist wants to use a new weed killer on your farm and decides to test it in a small area. The directions indicate to dilute the concentrated weed killer with water to make a diluted mixture with 1/5 of weed killer. You want to use 15 ounces of weed killer concentrate.

   a) How much water should you add?

   b) Challenge: The active ingredient of Chemical A is 15%, chemical B is 7.3% and chemical C is 5%. Find the number of ounces of each active ingredient. (See Section 3 for percentage)

   15 oz ×(.15) = 2.25 oz Chemical A: 15 oz × 0.073 = 1.095 oz Chemical B: 15 oz × 0.05 = 0.75 oz Chemical C

1.7. Assume the yard that must be sprayed with a chemical measures 44 ft by 75 ft. You plan to use a tank sprayer. The mixing instructions are to mix 2.5 ounces with 1 gallon of water that is applied over a 300 ft² area. How much spray solution (chemical + water) is needed if the spray solution is applied evenly across the area?

1.8. A producer has 6 bushels of grain, some are wheat and some are soybean. One-third of the bushels were soybean? How many bushels are soybean? What is the weight of the bushels of soybean?

In this case, we could think about 1/3 of the bushels are soybean and the problem can easily be solved in two ways. Focus your thinking on the different ways of reasoning. Remember to solve the problem or think about how you would go about solving it before reading ahead. Both solutions are correct, see if you can identify how the models are similar and different.

Mathematically we could write this as ¹/₃ x 360. To solve this, Producer 2 thought about what one-third means: Take the group and put it into 3 same-size chunks.
   The unit is 360 and putting it into 3 chunks means that each chunk is 1/3 and weighs 120 lb (360 ÷ 3).

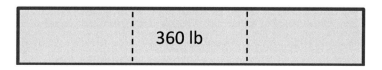

360 lb

Discussion of Problem 1.8

Comparing the two approaches, we first notice that Producer 1 and 2 models are different. Producer 1 thought about each bushel as part of the unit and drew 6 circles. Producer 2 thought about the quantity and divided the total quantity into 3 parts. Second, they used a different order of operation. Producer 1 divided 6 bushels by 3 and then multiplied by 60 lb (written as $6 \times \frac{1}{3} \times 60$). Producer 2 multiplied 6 bushels by 60 lb. and then divided by 3 (written as: $6 \times 60 \times \frac{1}{3}$). Remember that a number $n \div 3$ can be written as $\frac{n}{3}$. Thus $6 \div 3 \times 60 = 6 \times \frac{1}{3} \times 60$.

Key understandings are:

· The unit (total to be broken into pieces) must be defined. It can be a set of objects or a single object as illustrated by Producer 1 and 2.

· The operations used to solve a problem reflect the actions that from a model.

· The order of multiplying and dividing does not matter when working with fractions.

Mathematically, it does not matter if you multiply or divide first. However, you do need to be careful of your notation. The first problem can be written as $6 \div 3 \times 60$ or $6 \times \frac{1}{3} \times 60$. The second can be written as $6 \times 60 \div 3$ or $6 \times 60 \times \frac{1}{3}$. When we change $\div 3$ into a fractional representation for the idea, the two expressions are the same because the order that one multiplies does not matter. Mathematicians call this the commutative property. Producer 1 and 2 thought about the problem differently, but we can see that essentially the work is the same (the operations were done in a different order). The key idea that you need to remember is to make sense of the problem using your mathematical experiences, models that you draw or imagine, and to make sure that the operations you use reflect the actions that you take, as both Producer 1 and 2 demonstrated.

PROBLEM 1.9.

A producer has a honey hive and each fall harvests the honey. The weight of each jar of honey is 1¼ pounds. He/she has 15 jars of honey, how many pounds of honey does the producer harvest?

Remember to solve the problem or think about how you would go about solving it before reading ahead. Both of the solutions are correct, see if you can identify how their reasoning about fractional parts differed.

Problem 1.9: Approach A

Producer 1 has 15 jars of honey and draws a model as shown:

Each circle has the weight of 1 1/4 lb. Counting the jars, Producer 1 knows that he/she has 15 lb and a little more. The little extra is 15 jars (groups) of a size 1/4. He/she knows that 4 fourths make a pound. The amount would be 15 lb and The fractional part (is equivalent to 3 pounds and 3/4 (solve 15 . Total is 15 + 3 3/4 = 18 3/4 lb of honey.

Mathematically Producer 1 could write this as:

$$15 \times 1\ 1/4 =$$
$$(15 \times 1) + (15 \times 1/4) =$$
$$15 + = \textit{Note:} = 15 \div 4 = 3 \text{ remainder } 3 \textit{ or } 3\ 3/4$$
$$15 + 3\ 3/4 = 18\ 3/4$$

Note: $\frac{15}{4} = 15 \div 4 = 3 \text{ or } 3\frac{3}{4}.$

The correct solution is 18 3/4 pounds.

Problem 1.9: Approach B

Producer 2 has 15 jars of honey as shown:

Each jar has 1 1/4 pounds of honey. The circle represents one jar. The quantity 1 1/4 can be thought of as pounds of honey. Producer 2 knows that there are 15 jars. So he or she multiplies 15 by 5/4. Three mathematical methods to solve this are shown. All of the methods are correct. Use the method that makes sense to you.

Each jar can be divided into 5 pieces and each piece is ¼ pound.

Method 1	Method 2	Method 3
15 × 5/4 =	15 × 5 × 1/4	15 × 5 ÷ 4
75/4	75 × 1/4	75 ÷ 4
18 3/4	$^{75}/_4$ = 18 3/4	18 3/4

Notice the use of multiplying the numerators and then "reducing" by dividing. Producer 2 found the correct answer of 18 3/4 pounds of honey.

Discussion of Problem 1.9

Producer 1 and 2 both arrived at the correct answer. Notice that their initial models looked alike (15 circles). However, their thinking was slightly different. Producer 1 took each jar of honey and subtracted out 1 pound. They recognized the need to change the unit from a jar to a pound. They then put the 1 fourth remaining honey into units of 1 pound. Producer 2 took each jar and broke them up into 5 fourths. He/she then converted the jars to pounds by multiplying the number of jars by the conversion ($^5/_4$)– he or she needs 4 fourths to make a pound and he had 5 fourths in each jar.

Division

Division of fractions can be counter intuitive, which is why so many people struggle with this operation. Division is based on "undoing" multiplication or putting a quantity into parts. In essence, you will find either what we call "the number of groups" or "the size of a group", if you know the total.

The problems that follow are illustrations that provide a context for identifying the key understandings of dividing fractions. In the discussion of the problems, the key understandings will be presented.

PROBLEM 1.10.

An agronomist has 38 pounds of phosphorous (P) and 1 ¼ pounds of P will be used in a fertilizer mixture. He wants to divide the P into bags of 1 ¼ pounds. How many bags does he need?

Remember to solve the problem or think about how you would go about solving it before reading ahead. Only one of the solutions is correct, see if you can find where the error was made.

Problem 1.10: Approach A

Agronomist 1 has a big bag of P and he/she needs to find out how many "groups" there are of 1 1/4 pounds. Using logic, he/she knows that there will be less than 38 bags because that is how many bags he/she would need if 1 pound of P fit in one bag. Thinking about this, Agronomist 1 could find the number of fourths in the 38 pounds and then divide the number of fourths into bags that hold 1 1/4 or $^5/_4$ pounds.

Agronomist 1 knows that five of these fourths will fit in one bag. Thus we need to divide 152 fourths by five. 152 ÷ 5 = 30 $^2/_5$ bags will be filled. Agronomist 1's solution is correct.

> *Total of 38 Pounds of P.* This can be divided into fourths. In each pound, there are 4 fourths. Thus there are 38 x 4 fourths = 152 *fourth* pounds of P.

Think of 38 little scoops and each scoop is divided into ¼

Problem 1.10: Approach B

Agronomist 2 knows that there is a total of 38 pounds. This quantity of P needs to be put into bags that holds 1 $^1/_4$ pounds. He or she writes 38 ÷ 1 $^1/_4$. Agronomist 2 has forgotten how to divide mixed numbers. He or she remembers that $1\frac{1}{4}$ can be written as $1 + \frac{1}{4}$ and rewrites the expression as $38 \div (1 + \frac{1}{4})$. Producer 2 rewrites this as $(38 \div 1) + (38 \div \frac{1}{4}) = 38 + (38 \times 4) = 38 + 152 = 190$ bags. What was Agronomist 2's mistake?

Discussion of Problem 1.10

Agronomist 1 used a model to represent the problem and applied mathematical thinking to solve it. Following are symbols showing the steps that were used.

Agronomist 1's Arithmetic	Alternative Arithmetic
$(38 \times 4) \div 5$	$38 \div 1\frac{1}{4}$
$152 \div 5 = 30\frac{2}{5}$	$38 \div \frac{5}{4}$
	$38 \times \frac{4}{5}$
	$\frac{152}{5} = 30\frac{2}{5}$

Notice that when you divide by a fraction you invert and multiply. This is shown by comparing Step 1 in Agronomist 1's arithmetic and Step 3 in the alternative arithmetic.

When you invert the fraction, you are following the steps that you did with the model. You took the total and found the number of fourths (38×4). After you found the number of fourths, you divided by five to find the number of bags. The rule of invert and multiple simply does the thinking for you. Memorizing this rule can be confusing and hard to remember. If you think about the actions of solving the problem, you should remember the rule better.

Key understandings are:

· Division divides a quantity into parts, either the number of groups or the size of a group. First decide on what the total is. The total is always divided into one of the parts.

· Write your equation and make sure that the total comes first (total ÷ part).

· If you have trouble remembering the rule to multiply by the inverse, make a model to show the actions that needs to be taken.

Why was Agronomist 2's solution incorrect? The error was thinking that they could rewrite the quantity $1\frac{1}{4}$ into $1 + \frac{1}{4}$ and then divide the total by each part. In essence, what Agronomist 2 found was the number of bags that could hold 1 lb *and* the number of bags that could hold $\frac{1}{4}$ pound and added these quantities together. Their answer does not make sense. On reflection, if Agronomist 2 started with 38 lb each bag would be partially filled because each bag held $1\frac{1}{4}$ lb, not 1 lb. Thus, there should be fewer than 38 bags filled.

To correct Agronomist 2's approach, imagine 38 pounds of P in a big bag. Note that $\frac{4}{4}$ pounds are in each pound of P. He or she could take five pounds of P and put a pound in four bags, then divide the fifth pound into 1/4 lb quantities and place an additional $\frac{1}{4}$ pound in each of the four bags. The table illustrates the process of taking five pounds and filling four bags so that each bag holds $1\frac{1}{4}$ pounds. The key understanding of *making a model* in for form of a table helps Agronomist 2 correctly solve the problem. He or she can see the relationship of using five pounds to fill four bags.

Number of pounds in big bag	Number of pounds removed	Remaining pounds in the big bag	Number of bags with $1\frac{1}{4}$ lb
38	0	38	0
38	5	33	4
33	5	28	4 + 4 = 8
28	5	23	8 + 4 = 12
23	5	18	12 + 4 = 16
18	5	13	16 + 4 = 20
13	5	8	20 + 4 = 24
8	5	3	24 + 4 = 28
3	3	0	28 + 2 = 30 and remainder of 2

Finally, we have two fourths left over. We need five fourths to fill a bag, so we have two out of the five or two-fifths of a quart. The answer is $30\frac{2}{5}$ bags are filled.

Note: Another way to solve or check the solution is to convert the fraction 1 1/4 to a decimal and divide 38 by 1.25. The focus of this chapter is on fractions, thus converting a fraction to a decimal can simplify finding a solution. The approach of converting to a decimal will lead to a correct solution when the fraction is non-repeating. Fractions like one-third are nonrepeating and converting it to 0.33 can cause serious problems, especially when using large quantities.

Formula

Sometimes precision farmers use formulas and equations to estimate or calculate profit and loss. One example is to estimate harvest crop loss. Sandifolo (2002) created the following formula to find the percent weight loss to tuber crops due to weevil infestation.

CT average weight of clean tuber
#CT number of clean tubers
DT average weight of damaged tubers
#DT number of damaged tubers

$$\% \text{ weight loss} = \frac{(CT \times \#CT) - (DT \times \#DT)}{1} \times \frac{100}{CT(\#CT + \#DT)}$$

Note: This is one way to write the formula. Typically, the denominator 1 is omitted because mathematically it is extraneous. When using a formula, it is helpful to include the denominator 1 for clarity and to ensure that you correctly complete the operation. If you frequently use the same formula with different values, inserting the formula into a spreadsheet is helpful. A spreadsheet can help you track losses over time and in different fields. The spreadsheet also completes the calculations for you without error.

PROBLEM 1.11.

A producer wants to estimate the percent weight loss of the harvest loss using the following data:

Average weight of clean tubers (CT)	32 oz
Number of clean tubers (#CT)	1000
Average weight damaged tuber (DT)	26 oz
Number of damaged tubers (#DT)	500

Answer Problem 1.11: Approach A

Producer 1 begins to solve the problem by writing the formula on paper. Then they substitutes their data in the formula for each variable (CT, #CT, DT, #DT).

$$\% \text{weight loss} = \frac{(CT \times \#CT) - (DT \times \#DT)}{1} \times \frac{100}{CT(\#CT + \#DT)}$$

$$\% \text{weight loss} = \frac{(32 \times 1000) - (26 \times 500)}{1} \times \frac{100}{32(1000 + 500)}$$

Next, Producer 1 computes all of the operations within the parentheses.

$$\% \text{weight loss} = \frac{(32,000) - (13,000)}{1} \times \frac{100}{32(1500)}$$

Producer 1 completes the computations for the numerator and denominator.

$$\% \text{weight loss} = \frac{19,000}{1} \times \frac{100}{48,000}$$

The final steps can be accomplished in several ways. Two methods are displayed below:

Method 1		Method 2	
$\dfrac{19{,}000}{1}\times\dfrac{100}{48{,}000}$		$\dfrac{19{,}000}{1}\times\dfrac{100}{48{,}000}$	
$\dfrac{19{,}000}{1}\times\dfrac{1}{480}$	Write an equivalent fraction by dividing the numerator and denominator by 100. (cancel 00)	$\dfrac{1{,}900{,}000}{48{,}000}$	Multiply the numerators and denominators.
$19{,}000 \div 480$	Rewrite ($\times \frac{1}{480}$) by an equivalent expression.	39.58% loss	Divide the numerator by the denominator.
39.58% loss	Complete the calculation.		

Discussion of Problem 1.11

Producer 1's approach required him or her to remember to substitute the provided data for each variable in the equation. Deriving the formula is beyond the scope of this chapter.

What you need to remember is to:

· Substitute the data

· Follow the rules for order of operation (see Discussion of Problem 1.8)

· Simplify the expression (writing down all of the steps as you simplify can reduce arithmetic errors. Be careful as you "cancel the zero" as shown in Step 2 of Method 1 above)

· Reflect on whether your answer makes sense for the problem

PROBLEM 1.12.

A producer has a field which has both clean and damaged tubers. The field is described as mixed.

Average weight of clean tubers (CT)   32 oz
Number of clean tubers (#CT)   1000
Average weight damaged tuber (DT)   26 oz
Number of damaged tubers (#DT)   500

The above formula is modified to reflect the mixed field.

CT   average weight of clean tuber
#CT   number of clean tubers
DT   average weight of damaged tubers
#DT   number of damaged tubers

$$\%weight\ loss=\frac{(CT-DT)\#DT}{1}\times\frac{100}{CT(\#CT-\#DT)}$$

The data for the mixed field is the same as for Problem 1.11. The difference between the problems is that problem 1.12 has one field that is mixed, and in Problem 1.11 there were two distinct fields, one was damaged and one was undamaged.

Using a slightly different approach, Producer 2 used a 5-step approach to calculate weight loss, as shown.

$\%weight\ loss=\dfrac{(CT-DT)\#DT}{1}\times\dfrac{100}{CT(\#CT+\#DT)}$	Write the formula.
$\dfrac{(32-26)500}{1}\times\dfrac{100}{32(1000+500)}$	Substitute the data into the formula for each variable.
$\dfrac{(6)500}{1}\times\dfrac{100}{32(1500)}$	Calculate the numbers inside the parentheses.
$\dfrac{3000}{1}\times\dfrac{100}{48000}$	Calculate the numerators and denominators.
$\dfrac{3000}{1}\times\dfrac{1}{480}=\dfrac{3000}{480}=6.25\%$	Complete the calculations to find the percentage loss.

SECTION 2: PRACTICE EXERCISES

1.13. A producer was mixing some medicine for a cow. The recipe calls for 5/6 cups of water. As he or she was adding the water, the producer got a phone call and realized that was a third of the amount needed for several sick cows.

   a) How many cups of water does he need to add for several sick cows?

   b) How many 1/2 cup doses of medicine are in a bottle that contains $1^1/_3$ cups?

1.14. An agronomist used a tractor to mow grass. The mileage was 18 $^1/_2$ miles over 4 d. If the agronomist mowed the same distance each day, how many miles were mowed each day?

1.15. A producer has a length of rope that measures 9⅜ yards long. Each farmhand needs a piece $^5/_8$ yards long for lashing. How many pieces of the required length can be cut? What part of a piece for lashing is left over?

1.16. 10 1/2 gallons of water fills up 2 $^1/_3$ buckets. How many gallons are in one bucket?

1.17. An agronomist has two fields of tubers. One is damaged and the other is not. He collects the following data.
Average weight of clean tubers (CT) 18 oz
Number of clean tubers (#CT) 800
Average weight damaged tuber (DT) 16 oz
Number of damaged tubers (#DT) 300

   a) Find the percentage loss.

   b) In another field, he has both damaged and undamaged tubers. Using the same data, find the percentage loss for the mixed field.

Section 3: Solving Problems That Involve Proportions, Ratios, and Scale Factors

Proportion, ratio, or *rate* describes a relationship between quantities that are different. For example, they are used to describe how many miles per gallon a car uses. Precision farmers consider how much herbicides and fertilizer are applied to a field. Note that the two quantities are different, one quantity is measured in gallons and the other is acreage. The proportion of gallons of herbicide or fertilizer to the acreage allows the farmer to apply the proper concentration. A scale factor allows the farmer to quickly determine the correct quantity of chemicals when the acreage increases or decreases.

Scale factors increase or decrease quantities or relationships by doubling or halving; tripling or cutting into thirds, etc. For example, you use 100 gallons of 28–0–0 for 10 acres. How much N do you need for 20 acres? To solve this, the scale factor is two. The number of acres and the quantity of N are doubled. When two quantities increase by the same scale factor, we say this is a *direct proportion*. Two equivalent fractions describe this proportion;

$$\frac{100 \text{ gallons of N}}{10 \text{ acres}} = \frac{200 \text{ gallons of N}}{20 \text{ acres}}$$. Note that the two fractions are equivalent to the same unit ratio, $\frac{10 \text{ gallons of N}}{1 \text{ acres}}$.

Sometimes one quantity increases and the other decreases. For example, it takes one person two days to combine a wheat field. How long would it take two combines? Here you double the combines, so the time should be cut in half. When one quantity increases by a scale factor and the other decreases by the inverse factor, we call it an *inverse relationship*. The key understanding on proportions and rations includes:

   · Direct proportion is a relationship that occurs when the two quantities both increase by the same scale factor f.

   · A unit ratio can be written to express the relationship between two quantities.

   · Two ratios with the same unit ratio form equivalent fractions.

   · An inverse relationship is when one quantity increases by a scale factor f and the other decreases by 1/f.

The following problems are used to develop these key understandings. **Approach A** illustrates the key understanding that *when two quantities have the same ratios they are proportional,* and uses equivalent fraction to solve the problem. **Approach B** illustrates the key understanding of *writing a unit ratio to express the relationship* and uses it to solve the problem. These ideas are extended in the discussion to help you understand why a direct proportion is linear. Problem 1.18 develops the key understanding of *inverse relationships*.

PROBLEM 1.18.

A faucet (A) drips 6 ounces of water in 21 minutes and a second faucet (B) drips 4 ounces in 14 minutes. Do faucet A and B drip at the same rate or is one dripping faster than the other?

Answer Problem 1.18: Approach A

To solve this problem, we need to use the key understanding, *when two quantities have the same ratios they are proportional,* and find the equivalent fraction for the two ratios.

Faucet A
$$\frac{6 \text{ ounces}}{21 \text{ min}} = \frac{6 \text{ ounces}}{21 \text{ min}} \times \frac{14}{14} = \frac{84 \text{ ounces}}{294 \text{ minutes}}$$

Faucet B
$$\frac{4 \text{ ounces}}{14 \text{ minutes}} = \frac{4 \text{ ounces}}{14 \text{ minutes}} \times \frac{21}{21} = \frac{84 \text{ ounces}}{294 \text{ minutes}}$$

Here you can see that we thought of the ratios as equivalent fractions. Changing the ratios so that their denominators are the same, we find that the two ratios that are the same. Because the two equivalent ratios could be found, we say that the ratios are proportional, or equivalent.

Answer Problem 1.18: Approach B

A different way of solving this problem is to "reduce" the two ratios by dividing the numerator and denominator by a form of 1. Notice the arithmetic in the second step.

Faucet A
$$\frac{6 \text{ ounces}}{21 \text{ minutes}} = \frac{6 \div 3}{21 \div 3} = \frac{2}{7}$$

Faucet B
$$\frac{4 \text{ ounces}}{14 \text{ minutes}} = \frac{4 \div 2}{14 \div 2} = \frac{2}{7}$$

The two rates are identical because equivalent fractions can be written for the two rates. This is also called finding the Least Common Denominator and allows us to compare two ratios without conversion to a decimal form to see if they are directly proportional. (If we converted the ratios to a decimal, the decimal expression would be the same.) Again, the key understanding, when two quantities have the same ratios they are proportional.

Extending Approach B to a Strategy Using a Unit Ratio

The solution indicates that in 7 min, 2 ounces of water will drip. This ratio can also be interpreted as the faucet will drip 2/7 ounces in per minute. To perform the calculation, divide the numerator and denominator by seven, resulting in the ratio $\frac{2/7}{1}$. Thus for each minute, the faucets drips $^2/_7$ ounces. This interpretation is called a *unit ratio* and written in ratio format, $\frac{\frac{2}{7} \text{ ounces}}{1 \text{ minute}}$. If we want to know how many ounces we will have in any amount of time, we can simply increase the ratio using the time factor *t*. For example, if we want to know how many ounces there would be after 42 min write the ratio and multiply by 42 as shown below.

$$\frac{2/7 \text{ ounces}}{1 \text{ min}} \times \frac{42}{42} = \frac{2/7 \times 42}{1 \times 42} = \frac{2 \times 6}{42} = \frac{12 \text{ ounces}}{42 \text{ min}}$$

This tells us that in 42 min, we would have 12 ounces of water. Now we ask ourselves, does this make sense? Yes, it should be that amount. If we look at Faucet A, we know that in 21 min it drips 6 ounces of water. We notice that 42 min is twice as long, so the amount of water should be twice as much. The scale factor of 2 can be used to write an equivalent ratio using the ratio $^6/_{21}$ for Faucet A.

The key understandings for find a new ratio with a different value includes:

· Find the unit ratio and then use it to find an equivalent ratio.

· Look at the known ratios and see if there is a relationship between the numbers. If so, use a scale factor to increase or decrease the ratio.

Note that in some cases it is easier to use a ratio that is not in unit format, but sometimes it is easier to use the unit ratio.

A similar process can be used if we want to use ounces as our unit. This enables us to easily find out the number of minutes it would take to get b amount of water.

PROBLEM 1.18. EXTENSION. A faucet drips 2 ounces of water in 7 minutes. How long will it take to get one quart of water?

Answer Problem 1.18. Extension

We could begin by finding the time in terms of 1 ounce and then use the factor 32 because there are 32 ounces in a quart.

$$\frac{2 \text{ ounces}}{7 \text{ min}} = \frac{2 \div 2}{7 \div 2} = \frac{1 \text{ ounces}}{\frac{7}{2} \text{ min}}$$

This tells us that it takes $7/_2$ minutes or 3 1/2 min to drip 1 ounce. To find the number of minutes it takes to get 32 ounces, multiply the time to get 1 ounce by 32.

$$\frac{7}{2} \times 32 = \text{time}$$

$$\frac{7}{2} \times 32 = \frac{7}{2} \times \frac{32}{1} = \frac{7}{2} \times \frac{2 \times 16}{1} = 7 \times 16 = 112 \text{ min}$$

By dividing the number of minutes by 60, we can express the time in hours. 1 h and 52 min. Unit cancellation helps us keep track of unit conversion that is required by problems. Unit cancellation will be further discussed in *Section 5* of this chapter.

Discussion of Problem 1.18. Extension

The key understanding of proportions is that *when two quantities have the same ratios they are proportional.* This means that the two ratios form equivalent fractions that can be written to express this relationship. When you use the unit ratio, the unknown quantity can be found by using the scale factor. When you change a ratio using a scale factor, the relative proportion does not change. This is the definition of a linear relationship. Mathematically, scale factor can be thought of as slope and creates a linear relationship. An algebraic approach can be used to solve Problem 1.18 and a discussion of this approach allows us to further extend our understanding of proportion and ratio.

Answer Problem 1.18 Approach C

The rate of change or slope is another strategy way to determine whether faucet A or B drip at the same rate. If they drip at the same rate, they will form a line going through zero (at time 0, the amount of water should be 0).

A faucet (A) drips 6 ounces of water in 21 min and a second faucet (B) drips 4 ounces in 14 min.

To graph the relationship, make a table and graph the line. We will use time as the independent variable and place it on the horizontal axis (x axis). In math and science, the horizontal axis is usually the independent variable or that variable that is frequently used to measure change. Time is frequently used in this way.

Time (x)	Ounces (y)
0	0
14	4
21	6

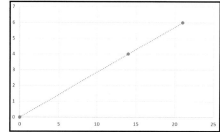

To determine whether the three points are on the same line, find the slope between the three points. If the slopes are the same, then the relationship is linear and the rates are the same. Plotting the three points on a graph will show you that a line connects the three points. This indicates a linear relationship.

$$\text{Slope} = \frac{\Delta y}{\Delta x}$$

Find the slope between (0, 0) and (14, 4).

$$\frac{4-0}{14-0} = \frac{4}{14} = \frac{2}{7}$$

Find the slope between (0, 0) and (21, 6) or between (14, 4) and (21, 6).

$$\frac{6-4}{21-14} = \frac{2}{7}$$

Notice that the two slopes are the same, therefore the relationship is linear. Also note that the slope is the same as the reduced fraction for 4/14 and 6/21. Also notice the slope and unit ratio are equivalent ($\frac{2}{7} = \frac{1}{3\frac{1}{2}}$).
The key understandings includes that:

· Proportions are linear relationships.

· A linear function is formed when two quantity change by the same factor (called slope).

· An inverse relationship is when one quantity increases by a scale factor f and the other decreases by 1/f.

PROBLEM 1.19.

Two people can harvest a field in three days. If one more person joins them, how long will it that the three people to harvest the field? (Assume that the three people will all have productive work to do.)

Remember to solve the problem or think about how you would go about solving it before reading ahead. Only one of the solutions is correct, see if you can find where the error was made. Note: This problem helps you understand inverse relationships.

Answer Problem 1.19: Approach A

Producer 1 writes the ratio 2 people/3 days. They then change the ratio to a unit ratio by dividing the numerator and denominator by 2; $\frac{1 \text{ person}}{\frac{3}{2} \text{ days}}$. The unit ratio indicates that it takes one person $^3/_2$ days to harvest the field. One more person would be three times as long as one person. $\frac{3 \text{ people}}{\frac{3\times3}{2 \text{ days}}} = \frac{3 \text{ people}}{\frac{9}{2} \text{ days}} = \frac{3 \text{ people}}{4\frac{1}{2} \text{ days}}$. It will take 4 1/2 d to cut the field. Can you identify the incorrect thinking?

Answer Problem 1.19: Approach B

Producer 2 reasoned that it should take less time to harvest the field with more people. If two people complete the job in three days, it should take one person twice as long (6 d). To investigate the relationship, a table is constructed. He or she began with the known relationship.

people		2				
days		3				

He/she fills in the relationship that was logically determined.

people	1	2				
days	6	3				

Look at the table and try to write the relationship in words. *When we double the people, it will take half the time.* Use this to fill in the table.

people	1	2		4		
days	6	3		1 $\frac{1}{2}$		

If we triple the people it should take a third of the time. Triple the relationship for one person.

people	1	2	3	4		
days	6	3	2	1 $\frac{1}{2}$		

The solution is it will take two days for three people to harvest the field.

Discussion of Problem 1.19

Producer 1 used his or her knowledge of a unit ratio to calculate a solution. But, they did not fully understand the problem and jumped to an incorrect application of the mathematical ideas. Producer 1 did not recognize that this problem illustrated an inverse proportion. They also did not go back and reflect whether the solution made sense in the problem context. Logically, it makes sense that as you increase the number of people to complete a job, the time that it takes decreases.

In contrast, Producer 2 thought about the context and used their mathematical insights. Producer 2 reasoned that it should take less time if more people were harvesting the field. They created a table to identify a pattern illustrating the inverse relationship. Producer 2 used logic to determine what would happen if one of the variables was increased or decreased. Next Producer 2 wrote the relationship in words and began filling in the table. What other situations can you think of that are inverse relationship?

Section 3: Practice Exercises

1.20. A producer wants to deworm his cattle. He or she decides to use Safe-Guard Dewormer. For beef, dairy cattle, and swine, 0.2 and 0.5 lb of dewormer should be given for body weights of 200 and 500 pounds, respectively.
a) The average weight of 58 young cows is 600 pounds. How many pounds of dewormer does he need?
b) How many pounds of dewormer should he give each animal?

Body weight (lb)	Pounds of dewormer
300	
800	
1900	

1.21. A nozzle (A) wastes 4 ounces of water in 30 min and a second nozzle (B) wastes 3 ounces in 20 min.

a) Do the two nozzles A and B drip at the same rate or is one dripping faster than the other?

b) If you sprayed for 2 h, how many ounces would nozzle A and nozzle B waste?

1.22. A farmer has two tractors, one tractor with a large boom for spraying and a second tractor with a small boom. The small boom is half the size of the large boom. It takes the farmer 4 days to spray a field using the tractor with the large boom. If he uses both tractors to spray the field, how long would it take?

1.23. A farmer uses 220 ounces of herbicide on a 60-acre field. How many ounces would he use on a 45-acre field? How many ounces of herbicide would he use on a 120-acre field? How many on a 150-acre field?

PROBLEM 1.24.

A farmer must control green foxtail that is 9 to 10 cm high. Use the following label to solve each problem.

a) Determine the number of ounces that must be applied per acre (Note: 1 L = 1.06 qt).

b) If they have a 60-acre field to spray, how much Roundup will be needed?

c) If they are operating a sprayer that has a 90-ft boom, traveling at 12 mph, how long in hours and minutes will be needed to spray the 60 acres?

d) If the application rate is 15 gallons of spray solution per acre, how much gallons of spray solution will be needed to spray the field?

e) If the John Deere sprayer is equipped with a 800-gallon stainless steel tank, how often will the spray tank need to be refilled?

7.1	ANNUAL WEED CONTROL WITH ROUNDUP TRANSORB HC LIQUID HERBICIDE		
RATE (L/ha)	GROWTH STAGE	WEEDS CONTROLLED	COMMENTS (Apply in 50-100 L/ha water)
0.5	Weeds up to 8 cm in height	Wild oats, green foxtail, volunteer barley, volunteer wheat Non-Roundup Ready volunteer canola (rapeseed), wild mustard, lady's-thumb, stinkweed	• For wild oats apply at 1-3 leaf stage. • Add 350 mL of a surfactant registered for use such as Agral® 90, Ag Surf®, or Companion™ • For heavy wild oat infestations use 0.67 L/ha rate.
0.67	Weeds 8 cm to 15 cm in height	All annual grasses listed above. All annual broadleaved weeds listed above plus flixweed* and kochia*	• Add 350 mL of surfactant registered for use as listed above. * Suppression only. Refer to higher rates of this table or tank mix table (section 7.2) for control options.

Section 4. Solving Problems with Percentage

Percentage is a special kind of ratio in which one of the ratios is a percent. Remember, we can think of percentage as a part out of 100. A percentage is a ratio in which you are comparing a part to a unit of 100. The ratio can be expressed as a decimal or fraction that is defined as a part out of a whole. With this interpretation, we are changing the fraction to an equivalent fraction with a denominator of 100.

Percentage problems require a different kind of thinking than what we typically use when working with fractions. For example, you know that a herbicide formulation is 47% of chemical A and B. The package tells you that chemical A is 14%. What is the percentage of chemical B. Here you do use subtraction to find the percentage of chemical B because *the total is a percentage and the two parts are percentages of the same total*. You can add and subtract in this context.

A supplier has a sale on seeds. Seeds are on sale for 25% off for the first $100 and 50% off for the second $100. In this context you cannot add the two percentages because they are not part of the same total. In fact, the total quantity changes from $100 to $200. Thus, it is *incorrect* to add 25% + 50% = 75% and then take 75% off $200. If you did, the savings would be $150 and the seeds would cost $50. This should not make sense to you. The supplier would not stay in business. The correct way to solve the problem is to find 25% off the first $100 (which is $25, the cost for the seeds is $75) and 50% off the second $100 (which is $50). The seeds should cost $125 in total. The key understanding is that percentage is a relationship that is based on multiplication. Addition and subtraction can be used in specific situations in which all of the numbers in the equation are percentages and reference the same total.

The key understandings include:

· Percentage is a ratio in which the unit is 100

· Percentage can be expressed in three forms: 15%, $^{15}/_{100}$, 0.15

· Percentage is a relationship based on multiplication

PROBLEM 1.25.
Grapes are 98% water. Manual wants to dry them out to make raisins that are 90% water. If they have 500 pounds of grapes, how many pounds of raisins can they make?

Answer Problem 1.25: Approach A

Producer 1 knows that percentage is based on a ratio of a part out of 100. They know that 98% is water and 2% is solid material. They know that because percentage is a ratio, the relationships are based on multiplication. To better understand the problem, Producer 1 makes a table. Producer 1 finds the pounds of solid by multiplying .02 by 500 lb = 10 lb. This quantity does not change because it is solid matter. To find the pounds of water, subtract 10 (solid matter) from the total pounds. To find the percent water, set up the following set of ratio:

$$\frac{\text{part}}{\text{unit or whole}} = \frac{\text{lb of water}}{\text{total lb}} = \frac{\text{percent}}{100}$$

← Raisins (solid matter, 10 lb)

← Water (490 lbs.)

For the 250 pounds use the following:

Method 1: $\frac{240}{250} = \frac{n}{100}; \frac{240 \div 25}{250 \div 25} = \frac{9.6}{10} = \frac{96}{100} = 96\%$

Total 500 lb

Method 2: $\frac{240}{250} = \frac{n}{100}; 240 \times 100 = n \times 250; 24{,}000 = n \times 250; 24{,}000 \div 250 = n; n = 96\%$

Total lb	500	250 ($\frac{1}{2}$ of 500)	125 ($\frac{1}{4}$ of 500)	100 ($\frac{1}{5}$ of 500)	
Lb of solid	10	10	10	10	The solid does not change.
Lb of water	490	240	115	90	
Percent water	98%	96%	92%	90%	

The first method uses equivalent fractions to find the percentage. The second method uses cross products to find the percentage. Producer 1 continues to fill out the chart until he or she arrives at the correct percentage of water. The answer is 100 pounds of raisins. They reflects back and checks their answer. Does 100 pounds of raisins yield 90% water? The solid part is found by 100 − 10 = 90. 90/100 = 90%. This is the correct solution.

Answer Problem 1.25: Approach B

Producer 2 begins with a sketch and imagines squeezing out all of the water from the raisins. Producer 2 then writes the ratio: $\frac{90}{100} = \frac{n}{500}$ because they want to know what percent of water would need to evaporate from 500 to make a ratio of 90% ($^{90}/_{100}$). Producer 2 solves for n, $n = 450$. They decide that the total weight will be 450 pounds. This approach is not correct. Why?

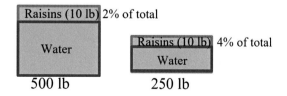

Discussion of Problem 1.25

In the first approach, Producer 1 recognized that *percentage is a ratio* (key understanding) and used a table to help them identify patterns and relationships. The table allowed Producer 1 to keep track of varying amounts. They quickly saw that taking half of the weight did not reduce the percentage very much. They recognized that multiplicative relationships are different from relationships based on addition and subtraction. Looking at the pattern, Producer 1 noticed that when the pounds of grapes were cut in half, the percentage of water decreased by by 2%. He or she continued to cut the pounds in fractional parts and noticed a pattern. Cutting the pounds in half reduced the percentage by 2%; cutting it in $1/4$ reduced the percentage by 4%; cutting it in $1/5$ reduced the percentage by 5%. Since the solid part of the raisins was 2%, it made sense to use a fifth of the total (500 lb) to drop the percentage of water by 10% (5 × 2%). Producer 1 used the key understanding, *percentage can be expressed in three forms: 15%, $^{15}/_{100}$, and 0.15,* as he or she used mathematics to solve the problem.

Producer 2 created a sketch to model the problem and his/her sketch was a great beginning. He/she also recognized the key understanding, *percentage is a ratio in which the unit is 100 and can be expressed as 15%, $^{15}/_{100}$, and 0.15.* His or her error was thinking that 90% of the total (500 lb) would be the weight at 90% water. However, the relationship is inverse. This means that if 10 lb of the solid part of grapes is 2% of a total, then if it doubles the percentage to 4%, the total weight is cut in half. Meaning that for the percentage of solid weight to go up, the amount of water must decrease. In an inverse relationship with percentage, cutting in half means the other increases by two.

Producer 2 needs to have the percentage of solid increase to 10%. This is 5 times the 2% that he/she started with. He or she multiplies the 2% by five and divides the total weight by five. This yields a total weight of 100 pounds and is a correct solution. The key understanding is that *percentage is a relationship based on multiplication.* When one quantity increases (percentage of solid grapes) and the other decreases (percentage of water) an *inverse relationship* exists. It is important to always consider whether the context indicates a direct proportion or an inverse relation-

> ## PROBLEM 1.26.
> An agronomist has three bags of fertilizer. Fertilizer A is 500 lb. of 28-0-0. Fertilizer B is 300 lb of 11-52-0. And fertilizer C is 200 lb of 0-0-60. They decide to mix them together. What is the analysis of the resulting fertilizer blend?

ship. You may want to review these ideas by looking at Problem 1.19 again.

Answer Problem 1.26: Approach A

Agronomist 1 knows 28-0-0 indicates that 28% of fertilizer A is N and there is no P_2O_5 or K_2O in it. To solve the

Total is 500 lb
←28% N

problem, he/she decided to calculate the number of pounds of N in Fertilizer A. He/she knows the total weight of fertilizer A and the percentage that is N.

Agronomist 1 wants to find the number of pounds of N. They write and solve the following equations:

$$\frac{28}{100} = \frac{\text{lb N}}{500}$$

$$\frac{28 \times 5}{100 \times 5} = \frac{\text{lb N}}{500}$$

$28 \times 5 = 140$ lb of N in Fertilizer A

Agronomist 1's approach is correct and they effectively used the diagram to guide his or her computations. Using this method, Agronomist 1 finds the number of pounds of N and P_2O_5 in fertilizer B and K_2O in fertilizer C. Fertilizer B is 300 lb of 11–52–0

To find N: $\frac{11}{100} = \frac{\text{lb N}}{300}$

33 lb of N in Fertilizer B

To find P_2O_5: $\frac{52}{100} = \frac{\text{lb PO}}{300}$

156 lb of P_2O_5
Fertilizer C is 200 lb of 0–0–60

To find K_2O: $\frac{60}{100} = \frac{\text{lb KO}}{200}$

120 lb of K_2O

Agronomist 1 knows that in the mixture, he or she has 1000 lb of fertilizer (500 from Fertilizer A, 300 from Fertilizer B, and 200 from Fertilizer C). Agronomist 1 also knows that there is a total of 173 lb of N (140 from Fertilizer A and 33 from Fertilizer B). To find the percent of N, he or she writes the ratio of the weight of N to the total weight of the mixture. This can be changed to a percentage.

$$\frac{173}{1000} = \frac{17.3}{100}$$ Thus, the percentage of N in the new mixture is 17.3.

Likewise, Agronomist 1 finds the percentage of P_2O_5 and K_2O.

156 lb of P_2O_5

$$\frac{156}{1000} = \frac{15.6}{100}$$ Thus, the percentage of P_2O_5 in the new mixture is 15.6.

120 lb of K_2O

$$\frac{120}{1000} = \frac{12}{100}$$ Thus, the percentage of K_2O in the new mixture is 12.

The new fertilizer mixture is 17.3–15.6–12.

Answer Problem 1.26: Approach B

Agronomist 2 knew that Fertilizer A was 500 lb with 28 N. He or she knew that Fertilizer B was 300 lb with 11 N and 52 P_2O_5. Agronomist 2 also knew that Fertilizer C was 200 lb with 60 K_2O. He or she remembers that the fertilizer is described by the percentage of different components. Agronomist 2 added the two N component together to determine that the N fertilizer grade in the mixture was 39%. He or she noticed that P_2O_5 is only in Fertilizer B and K_2O is only in Fertilizer C. Thus, he/she does not need to do any additional computations. He or she claims the resulting fertilizer is comprised of: N–P_2O_5–K_2O as 39–52–60. Producer 2's solution is incorrect. What was his or her error?

Discussion of Problem 1.26

Agronomist 1 correctly solved the problem. They remembered the key understanding *that percentage is a ratio and percentages cannot be combined when they reference different totals*. In this problem, the number of pounds of each type of fertilizer was different. Thus, he/she began by finding the number of pounds of each component of the fertilizers before he/she could combine them. Agronomist 1 also recognized that the percentage of each component would be different than in the original fertilizers. He or she expected that the percentage of P_2O_5 and K_2O would be much less because the combined fertilizer was so much more (1000 lb). Agronomist 1 used the key understanding of *equivalent fractions* to find the pounds of each component in fertilizers A, B and C and to find the percentage of each component in the new mixture.

In contrast, Agronomist 2 forgot that percentages seldom cannot be combined through addition or subtraction. He or she thought of percentage as a number that indicated a part of a total, but percentage is a ratio with a total of 100. When the total is not 100, the parts cannot be combined. He or she also did not recognize that when the quantity of fertilizer is increased, percentage of a given component decreases unless more of that component is added.

Problem 1.27.

Herbicide classified as acids (e.g., 2,4-D, dicamba) do not readily dissolve in water or petroleum based solutions. Therefore the acid, which is the active ingredient, must be modified into a salt formulation or an ester-type formulation. Depending on the salt or ester type, the molecular weight of the molecule can double. To be able to compare the amount of active ingredient among many different formulations, the acid equivalent is calculated. Acid equivalent calculations are used when salt or ester forms of the herbicide are applied because these forms have variable weight but only the acid is active. The dosage is based on the parent acid of the molecule and the inactive portion of the molecule is excluded. The parent weight of the acid in these formulations is the molecular weight of the acid– 1, because the OH group on the acid is modified by removing the H (molecular weight of 1) and then adding the modifying compound desired. The acid equivalent is the percentage of the parent acid to the molecular weight of the salt or ester. The molecular weight of a parent acid formulation is 542 and the molecular weight of the salt formulation is 630. What is the acid equivalent (ae)?

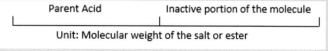

He or she was correct in recognizing that the total amount of fertilizer increased, but Agronomist 2 failed to find the correct percentage of each component using the new weight of 1000 lb.

Answer Problem 1.27: Approach A

Producer 1 knows that the acid equivalent is the ratio of the parent acid to the molecular weight of the salt. Using the definition of a parent acid, he or she knew that it is equal to the molecular weight of the acid formulation minus 1, 542– 1 = 541. The molecular weight of the salt is 630. The ratio is written as $\frac{541}{630}$ = 0.85873. Written as a percentage, the acid equivalent is 85.87% or ae = 85.87%. This solution is correct.

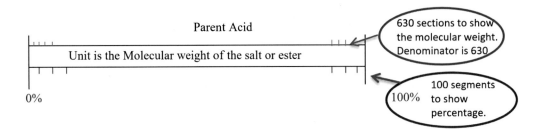

Answer Problem 1.27: Approach B

Producer 2 created a model to show the relationships between the parent acid, molecular weight of the acid formulation, and percentage can be constructed using two number lines.

The unit can be thought of as the molecular weight of the salt and the number of segments that it is divided into is 630. A few of these segments are shown. (There are too many segments and they are too close together to illustrate all of them.) From the double number line, it is easy to see that each of the 630 pieces is smaller than the corresponding 100 pieces. Producer 2 knows by definition that the part that is the parent acid is the molecular weight of the acid formulation minus 1, (542 – 1 = 541). Thus, the fractional part is. To solve the problem we need to convert it to a percentage. Instead of a fraction with a denominator of 630, we want the denominator to be 100. The following equivalent fractions can be used $\frac{541}{630} = \frac{n}{100}$. There are many ways to solve this. One way is to use cross multiplication.

$$\frac{541}{630} = \frac{n}{100}$$
541 × 100 = 630 x n.
54,100 ÷ 630 = 85.873 or 85.87%. ae = 85.87%

Discussion Problem 1.27

Both of these approaches are correct. Both producers began by identifying the unit. The unit was defined as the molecular weight of the salt (630) and neither would be able to solve the problem without this knowledge. They selected an approach; Producer 1 used a mathematical approach that relied on their knowledge of how to write a fraction using a part–whole relationship. They then changed the fraction into a decimal and wrote the decimal as a percentage. Producer 2 began with a model using a double number line. They then wrote a pair of equivalent fractions and used a mathematical procedure of cross products to find the percentage. The key understanding is

PROBLEM 1.28.

Herbicide formulations are often a mixture of 2 or 3 active ingredients (ai). In this problem, the amount of each active ingredient in a mix that is applied to an area will be determined. A herbicide formulation contains a total 5 lb ai per gallon with 5.7% of chemical A and 42% of chemical B.

a) How much of the 5 lb ai is chemical A?

b) How much is chemical B?

Note: 10% ai means that there is 0.1 lb of ai per 1 lb of formulation.

that the parent acid could be written as a fractional part. Using subtraction, it was easy to find the inactive portion of the molecule.

Answer Problem 1.28: Approach A

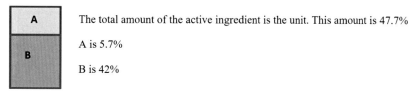

The total amount of the active ingredient is the unit. This amount is 47.7%

A is 5.7%

B is 42%

Producer 1 knew that the total amount of active ingredient is the sum of the active ingredients of chemical A and B. 5.7% + 42% = 47.7%. He or she needed to find out the part of the active ingredient that is chemical A and B. Producer 1 created a model to make sense of the problem.

To find the part of the active ingredient A, Producer 1 wrote it as a fraction . This can be rewritten as the decimal 0.119 (divide the numerator 5.7 by the denominator 47.7). Rounding off it is 0.12.

To find the part of the active ingredient B, Producer 1 wrote it as a fraction. This equals 0.880. For each pound of active ingredient per gallon, Producer 1 used 0.12 lb of A and 0.88 lb of B. (Note: 0.12 + 0.88 = 1.) Our problem states

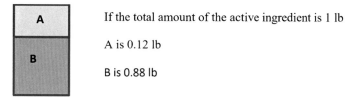

If the total amount of the active ingredient is 1 lb

A is 0.12 lb

B is 0.88 lb

that we want 5 lb per gallon of active ingredient. So, Producer 1 needed five times as much chemical A and B. For chemical A, 5 lb × 0.12 = 0.6 lb and for chemical B, 5 lb × 0.88 = 4.4 lb. This solution is correct.

Answer Problem 1.28: Approach B

Producer 2 wants 5 lb of active ingredient per gallon. It is comprised of chemical A and Chemical B. He or she changed the percentage to a decimal and multiplied it by five to determine the part that is A and B.

For chemical A, Producer 2 changed 5.7% to a decimal (0.057) and multiplied, 0.057 × 5 = 0.285. For chemical B, he or she changed 42% to a decimal (0.42) and multiplied, 0.42 × 5 = 2.1. Rounding off, Producer 2 decided that.29 lb of chemical A and 2.1 lb of chemical B is used. This solution is not correct. Why not?

Discussion Problem 1.28

In approach A, the Producer 1 recognized the importance of the key understanding of *a unit and that an individual can define the unit in a way that is appropriate to the problem context.* Here, we must recognize that the sum of chemical A and B must equal the active ingredient to be used. Producer 1 began by adding the two percentages of A and B to find the total amount of active ingredient. This total represents the unit. Then, the percentage of A and B is found in relation to the unit (47.7%). Finally, the unit is redefined to be 5 lb of active ingredient so the amount of A and B are multiplied by five, the scale factor. To check our reasoning, we add 0.6 lb of A to 4.4 lb of B and get 5 lb of active ingredient.

Why was B incorrect? In approach B, Producer 2 remembered a formula for finding the percentage of something. The problem was that 5 lb was not the unit to be divided. Five pounds represent the total amount of active ingredient that was comprised of two chemicals, A and B. The approach used by Producer 2 indicated that A and B were only part of the total 5 lb. If Producer 2 added the two parts, he or she would have discovered that the compound did not equal 5 lb. (0.29 lb of A + 2.1 lb of B = 2.39 lb ai). The problem called for a total of 5 lb ai not 2.39 lb ai. The problem was thinking that 5 lb was the only unit in the problem. Producer 2 needed to recognize that the first unit was defined in percentage and part of A and B needed to be found using the combined percentage of A and B. The error that Producer 2 made was not identifying the appropriate unit and reflecting back on the problem to see if the answer made sense for the problem.

Section 4 Practice Exercises

1.29. Anhydrous ammonia (AA) is a predominant fertilizer because it is a good source of nitrogen (82%) and is relatively inexpensive. A farmer has 175 lb of AA. How many pounds of N is that?

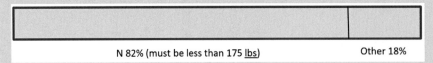

N 82% (must be less than 175 lbs)   Other 18%

1.30. An agronomist wants to buy some anhydrous ammonia. It is 82% N. How many pounds does he need to purchase to have 50 lb of N?

1.31. A producer is following a fertilizer recommendation of 75 lb of Potash per acre. If using 0–0-60, how much of this should be applied?

1.32. If a fertilizer recommendation for corn calls for 120 lb N, 40 lb P_2O_5 and 90 lb K_2O per acre, how much of the following fertilizers should be used per acre?

        82-0-0 _____
        0-20-0 _____
        0–0-60 _____

1.33. A fertilizer dealer will apply 205 lb potassium nitrate (13-0-44) per acre to a field as part of meeting a requirement of 80 lb N and 90 lb. K_2O. How much ammonium nitrate (34-0-0) per acre must also be added to meet the total requirement?
Weights of nutrients in liquid fertilizers:
        10-34-0 = 11.7 lb gallon^{-1}
        9-18-9 = 11.7 lb gallon^{-1}
        28–0-0 = 10.7 lb gallon^{-1}

1.34. What is the amount of nitrogen in 12 gallons of 9–18–9?

1.35. If you need to apply 90 lb of N per acre in the form of 28–0-0, how many gallons should you apply?

Fertilizer Price per Ton
28 (28-0-0)	$287
Anhydrous (82-0-0)	$523
Urea (46-0-0)	$386
DAP (18-46-0)	$463
Potash (0-0-60)	$350

† prices from www.farmfutures.com April 4th 2016 edition

1.36. An agonomist wants to find the most cost effective option to fulfil 300 lb of N, 111 lbs of P_2O_5, and 81 lbof K_2O. Use the table of fertilizer prices below to solve.

1.37. An agronomist has a choice between two insecticides. Insecticide 1 is available in a three-gallon container for $28. Insecticide 2 is available in a five-gallon container for $43. If Insecticide 1 is comprised of 13.1% active ingredient, and Insecticide 2 is comprise of 11.4% active ingredient, which is a better buy?

Section 5. Solving Problems Involving Unit Conversions

Precision farmers sometimes need to convert between units of measurement. Conversion between units of measurement is based on changing the unit discussed in Section 1. If you change 660 feet into miles, you can picture a length that is 660 units long. You know that 1 mile is equal to 5280 feet. Thus, 660 is part of that new measurement. It is precisely $^{660}/_{5280}$. Reduce the fraction and you find out that 660 feet is $^1/_8$ of a mile. However, sometimes changing from one unit of measure to a new unit of measure requires several conversions. For this reason, scientists developed a process called *unit conversion* to change units of measure. Unit conversion is a process that makes use of a ratio that describes the relationship between different measurement units.

To help you understand the connection between your knowledge about fractions and units (as illustrated above), converting 660 feet to miles is solved using unit conversion. The process begins by writing the relationship between feet and miles as a *conversion ratio*. The two following ratios describe the relationship: $\frac{5280 \text{ feet}}{1 \text{ mile}}$ or $\frac{1 \text{ mile}}{5280 \text{ feet}}$. Both of these ratios are called *conversion ratios* because they describe the relationship between two measurements. The correct conversion ratio for a particular problem is selected by paying attention to the units of measure. Selecting the correct one allows you to "cancel" the unit that you are changing when you multiply the measure that you want to change by the conversion ratio. If you want to change 660 feet to miles, you begin by writing a ratio to describe 660 feet.

Step 1: $\frac{660 \text{ feet}}{1}$

Step 2: Select the conversion ratio $\frac{1 \text{ mile}}{5280 \text{ feet}}$ because it allows you to cancel "feet". This is the unit that you are changing.

Step 3: Write the expression to convert the unit of measure. $\frac{660 \text{ feet}}{1} \times \frac{1 \text{ mile}}{5280 \text{ feet}}$

Step 4: Cancel the unit of measure that is found in the numerator and denominator. Then complete the arithmetic.

$$\frac{660 \text{ feet}}{1} \times \frac{1 \text{ mile}}{5280 \text{ feet}} = \frac{660}{5280} \text{ miles} = 0.125 \text{ miles or } \tfrac{1}{8} \text{ miles}$$

Step 5: Does this solution make sense? Would you expect 660 feet to be a fraction of a mile? Yes, it should be less than a mile. It should also be less than half of a mile. You could estimate that half of a mile is approximately half of 5000 feet which is 2500 feet. Clearly 660 feet is much smaller. If you use the wrong conversion factor, your answer will not make sense because it will be either too big or too small.

Discussion

In section 1, you learned that a unit can be divided into same size quantities, and that the unit is defined by the context of the problem. The problem of converting feet to miles requires you to change one measurement unit to another. The given unit of measure (feet) is smaller than the new unit of measure (miles). This tells you that the given unit of measure is part of the larger one. Thus, you divide by the larger unit. Unit conversion changes this thinking into a process. It requires you to select a conversion ratio that allows you to "cancel" the given unit of measure (feet). Previously, this was accomplished by thinking about how the units change and then dividing.

The key understandings to remember when using unit conversion include:

· The relationship between two units of measure is a ratio. This ratio is called a conversion ratio and can be used to change units of measure.

· Write the initial measurement as a ratio and then select the conversion ratio that allows you to cancel the unit of measure that you want to change.

· Reflect on the solution to make sure it "makes sense." This is critical because occasionally the wrong conversion ratio is used.

PROBLEM 1.38.

Convert 5 miles per hour to feet per second.

Answer Problem 1.38: Approach A

Begin by converting miles to feet. This can be accomplished by one of two methods.

Step 1:

Method 1: Section 1 changing units		Method 2: Unit cancelation	
	Picture five miles in your mind and notice that there are lots of feet in each mile. One mile contains 5280 feet.	$\dfrac{5 \text{ miles}}{1}$	Write the measurement as a ratio.
$5 \times 5280 = 26{,}400$	Complete the calculation.	$\dfrac{5280 \text{ feet}}{1 \text{ mile}}$ or $\dfrac{1 \text{ mile}}{5280 \text{ feet}}$	Select the first conversion ratio. The miles will cancel.
26,400 ft		$\dfrac{5 \text{ miles}}{1} \times \dfrac{5280 \text{ ft}}{1 \text{ mile}} = 26{,}000 \text{ ft}$	Cancel miles and complete the calculation.

Step 2: Next, convert hours to seconds.

Method 1: Section 1 changing units		Method 2: Unit cancelation	
	Picture 1 h. There are 60 min in 1 h.	$\dfrac{1 \text{ hour}}{1}$	Write the measurement as a ratio.
Picture 60 s in each minute. Multiply 60 by the number of minutes.	There are 60 s in 1 min. 60×60 is the number of seconds in 1 h.	$\dfrac{60}{\text{min}}$ or $\dfrac{1 \text{ hr}}{60 \text{ min}}$	Select the first conversion ratio. The hours will cancel.
$3600 \text{ s} = 1 \text{ h}$		$\dfrac{1 \text{ hr}}{1} \times \dfrac{60 \text{ min}}{1 \text{ hr}} = \dfrac{60 \text{ min}}{1}$	Cancel hours and complete the calculation.
		$\dfrac{60 \text{ s}}{1 \text{ min}} = \dfrac{1 \text{ min}}{60 \text{ s}}$	Convert minutes to seconds.
		$\dfrac{60 \text{ min}}{1} \times \dfrac{60 \text{ s}}{1 \text{ min}} = 3600 \text{ s}$	Select the conversion ratio, cancel and calculate.

Step 3: Finally, find the speed in feet per second. The original speed was five miles per hour. Write it as a ratio and then substitute the equivalent measurements using feet and seconds.

$$\dfrac{5 \text{ miles}}{1 \text{ hr}} = \dfrac{26{,}400 \text{ ft}}{3600 \text{ s}}$$

Simplify the ratio by dividing the numerator by the denominator ($26{,}400 \div 3600$). The result is 7.33 ft s^{-1} or $7\,^1/_3$ ft per second. Does this make sense? Use some estimation to consider the magnitude of the answer. If you go 5 miles in 1 h, you should go 1 mile in about 10 min or $^1/_{10}$ of a mile in 1 min. One tenth of a mile is 528.0 feet, thus you should go about 528 in 1 min. One minute is 60 seconds, thus your distance traveled in 1 s should be about $^1/^{60}$ of the distance in 1 min. Rounding 528 to 600 lets you quickly estimate a distance of 10 ft s^{-1}. This is close to your answer and confirms that your calculations are correct. Do not expect an estimation based on reasoning to exactly match your calculation because rounding off was used.

Answer Problem 1.38: Approach B

Step 1: Begin by writing the speed as a ratio and a second ratio using the required units of measure. Note that f and s are used to indicate the unknown number of feet per second. You will find these two unknowns.

$$\frac{5 \text{ miles}}{1 \text{ hour}} = \frac{f \text{ feet}}{s \text{ seconds}}$$

Step 2: Write all of the conversion ratios that are needed to describe the relationship between feet and miles and between hours and seconds. Note that there is not conversion ratio for hours to seconds so two conversion ratios are needed, one for hours and minutes and one for minutes and seconds.

Miles and feet $\qquad \dfrac{1 \text{ miles}}{5280 \text{ feet}} \text{ or } \dfrac{5280 \text{ feet}}{1 \text{ mile}}$

Hours and minutes $\qquad \dfrac{1 \text{ hour}}{60 \text{ minutes}} \text{ or } \dfrac{60 \text{ minutes}}{1 \text{ hour}}$

Minutes and seconds $\qquad \dfrac{1 \text{ minute}}{60 \text{ seconds}} \text{ or } \dfrac{60 \text{ seconds}}{1 \text{ minute}}$

Step 3: Select the conversion ratios that cancel the units of measure that you want to change. Begin with either miles or hours. Both methods are illustrated below.

Method 1: Begin with distance measurements		Method 2: Begin with time measurements	
$\dfrac{5 \text{ mi}}{1 \text{ hr}} \times \dfrac{5280 \text{ ft}}{1 \text{ mi}}$	Notice that the unit "miles" cancels	$\dfrac{5 \text{ mi}}{1 \text{ hr}} \times \dfrac{1 \text{ hr}}{60 \text{ min}}$	To cancel hours
$\dfrac{5 \text{ mi}}{1 \text{ hr}} \times \dfrac{5280 \text{ ft}}{1 \text{ mi}} \times \dfrac{1 \text{ hr}}{60 \text{ min}}$	The unit "hours" cancels	$\dfrac{5 \text{ mi}}{1 \text{ hr}} \times \dfrac{1 \text{ hr}}{60 \text{ min}} \times \dfrac{1 \text{ min}}{60 \text{ s}}$	To cancel minutes
$\dfrac{5 \text{ mi}}{1 \text{ hr}} \times \dfrac{5280 \text{ ft}}{1 \text{ mi}} \times \dfrac{1 \text{ hr}}{60 \text{ min}} \times \dfrac{1 \text{ min}}{60 \text{ s}}$	The unit "mins" cancels	$\dfrac{5 \text{ mi}}{1 \text{ hr}} \times \dfrac{1 \text{ hr}}{60 \text{ min}} \times \dfrac{1 \text{ min}}{60 \text{ s}} \times \dfrac{5280 \text{ ft}}{1 \text{ mi}}$	To cancel miles
	Cancel intermediate units	$\dfrac{5 \text{ mi}}{1 \text{ hr}} \times \dfrac{1 \text{ hr}}{60 \text{ min}} \times \dfrac{1 \text{ min}}{60 \text{ s}} \times \dfrac{5280 \text{ ft}}{1 \text{ mi}}$	Cancel intermediate units
$\dfrac{5 \times 5280 \text{ ft}}{60 \times 60 \text{ s}} = \dfrac{26,400 \text{ ft}}{3,600 \text{ s}} = 7\frac{1}{3} \text{ ft s}^{-1}$	Complete calculation	$\dfrac{5 \times 5280 \text{ ft}}{60 \times 60 \text{ s}} = \dfrac{26,400 \text{ ft}}{3,600 \text{ s}} = 7\frac{1}{3} \text{ ft s}^{-1}$	Complete calculation

Consider whether the answer makes sense. Use the reasoning in Approach A to estimate whether the result is close to an estimate or your experience for working in the field.

Discussion Problem 1.38

Both approaches yield the same results. That is because the methods are mathematically the same. Once you practice Approach B, you will find that you can convert measurements into different ones quickly and accurately because the process helps you keep track of each unit that is converted. It is critical that you write the two conversion ratios that can be used before you begin. Remember to select the one that allows you to cancel the unit that is being changed. The measurement that you want should not be cancelled, if it is, then you made an error selecting the conversion ratio.

Steps to follow:

1. Write the measurement as a ratio
2. Write the relevant conversion ratios

3. Select the ratio that allows you to cancel the unit of measure that is being changed. You may need to use several ratios to end up at the desired unit of measure (i.e., changing mile per hour to feet per second).

4. Write an expression using the initial ratio and the conversion ratios selected.

5. Cancel the intermediate units when they are in the numerator and denominator and keep the desired units.

6. Complete the calculation and ask yourself if your answer makes sense.

Reflect on these steps and combine them to make three or four steps that make sense to you. Reflecting and shortening the steps will help you avoid errors. Practice will allow you to develop your confidence. Always ask yourself if the conversion makes sense using your knowledge from the field or estimation.

Section 5 Practice Exercises

1.39. Convert 15 ft s^{-1} to miles per hour.

1.40. A tractor travels 0.25 miles in 45 s. Find the equivalent speed in miles per hour.

1.41. One square mile is covered with water 3 in deep. Find the water quantity in

a) Gallons

b) Gallons per acre

1.42. Convert 10 mi hr^{-1} to ft s^{-1}

1.43. Convert 70 mi hr^{-1} to ft s^{-1}

1.44. Convert 7.5 ft s^{-1} to mi hr^{-1}

1.45. A field with sides, width = 1.5 mi, length = 0.75 mi. Find the area enclosed in acres.

1.46. A field with sides, width = 2000 m, length = 5000 m. Find the enclosed area in hectares.

1.47. Change 5 gal to cubic inches.

1.48. A tank has the following dimensions: 18 in × 20 in × 30 in. Find the volume in gallons.

1.49. Change 15 in^3 s^{-1} to gal min^{-1}.

1.50. Change 400 acres to hectares.

Challenge Problems

1.51. You expect that the maximum output from a 90-foot boom with nozzles spaced 20 in apart is 1 gallon per minute. Note: The rule of thumb in the industry is to oversize the pump on the sprayer by 50% so that there is enough capacity in the pump to agitate the solution properly when using a formulation that requires agitation. This agitation is needed so that a homogenous solution that can be sprayed on a crop.

a) You have a 90 foot boom with nozzles spaced 20 in apart. How many nozzles do you have? (90 ft × 12 = 1080 inches) (1080/20 = 54 nozzles)

b) What size pump must be on the sprayer? (54 nozzles @ 1 gal min^{-1} × 1.5 = 81 gal min^{-1})

c) If the tank holds 1000 gallons, how long before the tank needs to be refilled? (1000/54 = 18.5 min)

d) If the sprayer is spraying 1.818 acre min^{-1}, how many acres have been sprayed? (18.5 × 1.818 acre min^{-1} = 33.66 acre)

e) If it takes 10 minutes to fill the tank, what is your field efficiency? (18.5/(18.5+10) = 64.5%)

1.52. A combine is harvesting grain with a ground speed of 5 mi hr^{-1}. The header has a width, w = 30 ft. The grain yield is 50 bushels per acre and the combined gran tank volume is 200 bushels.

a) Find the harvest rate in acre hr^{-1}.

b) Find the running time to fill the grain tank.

1.53. The producer owns a Hagie sprayer similar to the one pictured below. This Hagie has 15 nozzles that are spaced 30 inches apart. They will use this sprayer, traveling at 9 mph, to apply liquid fertilizer with the aid of his or her Ag Leader technology. The goal: top dress N on 160 acres of corn that has a row spacing of 30 inches. Assume the Hagie is equipped with a 1000 gallon tank.

   a) How long is the tool bar for this sprayer?

   b) How many hours and minutes will be needed to apply 28-0-0 fertilizer if the operator is 100% efficient?

   c) If the application rate is 30 lb acre^{-1} and the density of the fertilizer is 10.67 lb gallon^{-1}, how many times will he or she need to refill the tank?

Note: http://www.fluidfertilizer.com/pdf/Fluid%20Characteristics.pdf.
(Image Source: John Fulton.)

APPENDIX A: ANSWERS TO PRACTICE QUESTIONS

1.3. Three fifths of a bottle fills 3/4 of the sprayer. A fifth of the bottle is the same quantity as $^{1}/_{4}$ of the sprayer. You need $^{4}/_{4}$ of the sprayer to fill it. Thus, multiply $^{1}/_{5}$ of the bottle by 4. To fill the sprayer, use $^{4}/_{5}$ of the bottle.

1.4. $(2/3) + \dfrac{\frac{2}{3}}{8} = 0.7499$ gal

1.5. $2x + 40y = 208$, $2x = y$, $x = 2$, $y = 4$

   a. South has larger ears, because fewer ears are needed to fill a bushel.

   b. Let S = the number of bushels from the south field

   Let N = number of bushels from the north field

   2 S = N, because there are twice as many bushels from the N field (N field is 2/3 of the crate and the south is 1/3).

   24 S = number of ears of corn in the South field

   40 N = number o fears of corn in the North field

   40 × 2S = 80 S, number of ears of corn in the North field

   24 S + 80 S = 104, total number of ears of corn

   104 S = 208, S = 2. There are 2 bushels of corn in the South field.

N = 2 S, 2 × 2 = 4. There are 4 bushels of corn in the North field.

c. 6 bushels

d. 3/4 of the crate is 108 ears of corn from the south field. To find the number of bushels, divide 108 by 24 to find 4.5 bushels of corn from the south field. 4.5 bushels fill 3/4 of the crate. Divide by 3 to find the number of bushels in 1/4 of the crate. 4.5/3 = 1.5 There are 1.5 bushels of corn from the north field. 1.5 × 40 = 60 ears from the north field. 60 + 108 = 168 ears of corn.

1.6. $(15)(x + 15) = 0.2x = 60$ ounces of water

Chemical A: 15% of 15 ounces = 2.25 ounces of A

Chemical B: 7.3% of 15 ounces = 1.095 ounces of B

Chemical C: 5% of the 15 ounces = 0.75 ounces of C

1.7. Find the area of the field by 40 × 75 = 3,300 ft². The directions tell you the quantity to use the 300 ft². Divide 3,300 ft² by 300 to find 11. You need 5 tablespoons of concentrate and 1 gal of water for each 300 ft². Multiply 5 tablespoons by 11 to find 55 tablespoons of concentrate. You will use 1 gal × 11 = 55 gal of water.

1.13.

   a. $(^5/_6)$ is the $^1/_3$ of the amount needed. $^5/_6$ × 3 = 2.5 cups for two cows

   b. (1.33 cups) × (1 dose/0.5 cups) = 2.66 doses

1.14. 18.5/4 days; 4/625 miles per day

1.15. (9 × 3/8)/(5/8) = 15 pieces

1.16. 10.5 gallons are in 2.5 buckets. There are five half buckets. Divide 10.5 by 5 to find the number of gallons in a half bucket. 10 ÷ 5 = 2.1 gallons in a half bucket. A whole bucket is twice as much. So, the answer is 4.2 gallons.

1.17.

   a. = [(18-16) × 300][100/18 × (800 + 300)] = 3.03% loss

   b. same as a., [(18-16) × 300][100/18 × (800 + 300)] = 3.03% loss

1.20. Each pound of cattle requires 0.001 lb dewormer. So, 300 lb cattle require 0.3 lb dewormer, 800 lb cattle require 0.8 lb dewormer, and 1900 lb cattle require 1.91 lb dewormer.

1.21.

   a. Faucet A: 4 ounces/30 s or 1 ounce/7.5 s

     Faucet B drips 3 ounces/20 s or 1 ounce/6.67 s

     So, Faucet B drips faster

   b. There are 3.600 s in 1 hr. There are 7,200 s in 2 hr.

     Faucet A drips 4 ounces/30 s, or 960 ounces

     Faucet B drips 3 ounces/20 s, or 1,080 ounces

1.22. When both tractors work, they can spray 1.5 times as much as when one tractor sprays the field. Let x represent the number of days that both tractors will take. $1.5x = 4$, $x = 2.66$ days

1.23. (220/60) × 45 = 165 oz; (220/60)(120) = 807 oz.

1.24.

   a. (0.67 L/h)(1 h/ 2.471 acres)(1.06 qt/L) = 56 ounces per acre

   b. 60 × 56 = 3360 ounces

   c. 60 acres × 43560 ft² per acre × 1 mile/5280 ft × 1 h/12 m × spray/90 ft = 0.458 hr

d. $15 \times 60 = 3360$ ounces

e. $800/900 \times 0.458 = 0.407$ hours. It needs to be filled with 100 gal at 0.407 hours.

1.29. 175 lb AA × 0.82 A/ lb AA = 143 lb N

1.30. 50 lb N × 1 lb AA/0.82 lb N = 61 lb AA

1.31. (75 lb Potash) × (1/160) = 125 lb of 0-0-60

1.32.
120 lb × 1/0.82 = 146 lb AA
40 lb × 1 lb/0.2 lb = 200 lb of 0-20-0
80 lb × fert/0.6 lb = 150 lb 0-0-60

1.33.
205 × 0.13/lb fert. = 26.65 lb N
80 - 26.65 = 56.35 lb N needed
56.35 lb N × 1 lb fert./0.34 lb N = 165 lb 34-0-0

1.34. 11.7 lb per gal × 0.09 lb N/lb fert. × 12 gal = 12.64 lb N

1.35. 90 lb N per acre × 1 lb fert/0.28 lb N × gal/10.7 lb = 380 gal per acre

1.36. 241 lb DAP, 323.89 lb AA, 135 lb 0-0-60

1.37.

	Gallon	Cost	Active ingredient	Amount of active ingredient	Cost of active ingredient per gallon
Insecticide 1	3	$28	13.1%	0.393 gal	$71.25
Insecticide 2	5	$43	11.4%	0.57 gal	$75.43

1.39. 15 ft s^{-1} × mile/5280 ft × 3600 s hr^{-1} = 10.2 mi hr^{-1}

1.40. 0.25 mile/45 s × 60 s/min × 60 min/hr = 20 mi hr^{-1}

1.41. mile^2 × 640 acre mile^{-2} × 43560 ft^2 acre^{-1} × 0.25 ft × 7.48 gal ft^{-3} = 52,132,608 gal, 81,457 gal per acre

1.42. 10 mile/hr × 1 hr/3600 s × 5280 ft/1 mile = 14.7 ft per second

1.43. 70 mile/hr × 1 hr/3600 s × 5280 ft/1 mile = 102.7 ft per second

1.44. 7.5 ft/s × 3600 s/mile × 1 mile/5280 ft = 5.11 miles per hour

1.45. 1.5 m × 0.75 m × 640 acre/mile^2 = 720 acres

1.46. 2000 × 5000 × ha/10000 m^2 = 1000 ha

1.47. 5 gal × 231 in^3/gal = 1155 in^3

1.48. 18 × 20 × 30 × gal/231 in^3 = 46.6 gal

1.49. 15 in^3/s × 1 gal/231 in^3 × 60 s/min = 3.89 gal/min

1.50. 400 acres × ha/2.471 acres = 162 acres

1.52.

   a. mile/hr × 30 ft × 5280 ft/1 mile × acre/43560 ft^2 = 18 acres

   b. 200 bushels × hr/18 acres × 1 acre/50 bushels = 0.22 hr

1.53.

   a. 15 rows × 30 in/row = 450 in or 37.5 ft

   b. 160 × 1 hr/9 mile × 1 mile/5280 ft × 37.5 ft × 43560 ft^2/acre = 3.911 hr or 3 hr and 54.7 min

ACKNOWLEDGMENTS

Support for this document was provided by Washington State University, Precision Farming Systems community in the American Society of Agronomy, International Society of Precision Agriculture, and the USDA-AFRI Higher Education program.

REFERENCES

Dole, S., L. Bloom, and K. Kowalske. 2016. Transforming pedagogy: Changing perspectives from teacher-centered to learner-centered. Interdisciplinary Journal of Problem-Based Learning. 10(1). doi:10.7771/1541-5015.1538

Erickson, B., D.E. Clay, S.A. Clay, and S. Fausti. 2016. Knowledge, skills and ability needed in the precision ag workforce: An industry survey. 13th International conference of precision agriculture. St. Louis, MO. 31 July–3 Aug. 2016.

Hiebert, J. 2013. Conceptual and procedural knowledge: The case of mathematics. Routledge, New York.

NCTM. 2000. Principles and standards for school mathematics. National Council of Teachers of Mathematics (NCTM), Reston, VA.

Sandifolo, V.S. 2002. Paper 15: Estimation of crop losses due to different causes in root and tuber crops: The case of Malawi. FAO Corporate Document Depository. Food and Agriculture Organization, Rome, Italy. http://www.fao.org/docrep/005/Y9422E/y9422e0f.htm (verified 11 Oct 2017).

Simple Programming for Automating Precision Farming Calculations

2

Aaron J. Franzen,* C. Gregg Carlson, Cheryl L. Reese, and David E. Clay

Chapter Purpose

Agronomists need to make routine calculations to determine seeding and fertilizers rates. Errors associated with these calculations can be reduced by writing simple programs for custom calculations. Software available on most computers is suitable for this purpose. The purpose of this chapter is to demonstrate how to develop macros and functions within Microsoft Excel with detailed instructions for a PC. Directions for an Apple Macintosh are provided in the Appendix. The example provides instructions for developing a program that will convert corn grain at a known weight and moisture content to marketable bushels at 15.5% moisture.

Key Terms

Programming, Microsoft Excel (Microsoft Corporation, Redmond, WA), macros, functions, visual basics, VBA.

Mathematical Skills

Using regression analysis, fitting equations using iteration that minimizes bias.

Computer programs can help reduce human errors associated with routine calculations, allowing for quick entry, formatting, and calculation for a wide range of problems. Such custom programs can range from simple to complex. For example, custom programs can be used to assess changes in soil health, determine herbicide residuals, fertilizer rates, and yield predictions. Many computers have software that can be used to create custom scripts, functions, or macros to automate tasks that are done repeatedly. In the Microsoft Windows system for the PC, Visual Basic for Applications (VBA) is supported by Excel and other Microsoft Office products. Visual basic was derived from BASIC (**B**eginner's **A**ll-purpose **S**ymbolic **I**nstruction **C**ode), a family of programming languages that has been used for over fifty years. The examples in this chapter are designed to be compatible with Microsoft Excel 2013, and provide a starting point for both older and newer software versions.

Software program are used for a multitude of problems that range from controlling the action sequences in computer games to controlling complex environmental simulation models. Even through software programs have very different applications they all follow a specific set of rules. However, different software languages have different rules. By understanding the rules, computer programs can be created that automate tasks that are routinely conducted. These rules are integrated into the programming language.

South Dakota State University, Brookings, SD 57006. *Corresponding author (aaron.franzen@sdstate.edu)
doi:10.2134/practicalmath2017.0023

Copyright © 2017. American Society of Agronomy, Crop Science Society of America, and Soil Science Society of America, 5585 Guilford Rd. Madison, WI 53711, USA. *Practical Mathematics for Precision Farming.* David E. Clay, Sharon A. Clay, and Stephanie A. Bruggeman, (eds.).

The ability to create and run custom programs, while very powerful, could also present computer security issues. This is because a custom program might access the file system on your computer, potentially causing data loss or corruption, personal information being compromised, or even create a way for the program's authors to access your computer remotely after running the program. Since there is a security issue when running custom programs, the default setting for Microsoft Excel is that all macros or custom functions are disabled. When using custom programs in Excel 2013, you should first check the *trust center* to ensure that Macros are enabled. This is accomplished by clicking the *file* tab, followed by clicking *options*. Click *Trust Center* on the left pane, followed by clicking *Trust Center Settings* and *Macro Settings*. Check *Enable all macros*. When saving your program you must select *save as* and under *save as type* select *macro-enabled workbook*.

Programing can get very complicated quickly, especially if there is no predefined set of tasks or calculations that the program should execute. The "best practice" for writing good programs is to outline the list of tasks and calculations that should be completed using comments in the code prior to writing any of the actual code. If you are adding on to a previous program, it is also a good idea to start in the same way: writing an outline in code comments, then writing the code. In BASIC, any text that comes after a single quote, ', is considered a comment. Comments can be whole-line comments to describe what a multi-line set of commands should do, or end-of-line comments that explain what the individual line is intended to do. The comment outline created prior to writing the code is very useful in helping you remember what you wanted the code to do, especially when you are working on the code for multiple sessions where you have to leave for a time and return later. Comments are also critical if you are working on a program with a partner, as part of a group, or if you are going to share the code with others later on.

After writing the code, you may run into errors when you attempt to run the program. The most probable error for beginners to encounter in VBA code is the syntax error. A syntax error occurs when the code is written in a way that doesn't conform to the programming language syntax. Syntax errors might be simple typographic errors or misspelled words. Before the program can be run, the code must be "compiled." Compiling code means that the text of the code is converted into executable form that will then be run by the software package. If you click "Compile VBAProject" on the VBA debug menu, any syntax errors will be highlighted so that you can address them. Other errors may be more difficult to diagnose, such as a **run-time error** or **logic error**. In either of these cases, the code will compile successfully, but will return an error when the program runs. An example of a run-time error might be a program that is meant to multiply two numbers from the Excel spreadsheet together. If one of the spreadsheet cells contains text instead of a number, the program would fail do to an incompatible data type.

Introduction to VBA Statements

Data types and variables are a key concept in computer programming. The data type defines what type of value a variable can store. A variable is a named container that the program uses to store data in memory. In VBA you create variables to store information using the Dim (short for dimension) statement. The VBA statement Dim bananas As Integer creates a variable named "bananas" that is of the Integer type. Variable and program names cannot contain spaces or symbols, only underscores, numbers, and letters. The first character in a name must be a letter. Variables can have a range of characteristics and can represent words or numbers. The most commonly used data types in VBA include *Integer* (whole numbers from -32,768 to 32,767), *Boolean* (True or False), *String* (multiple text characters to form a word or sentence that are surrounded by quotations), *Single* (decimal numbers), and *Double* (higher precision decimal numbers that require more memory than *Single*). The VBA compiler can also create arrays of variables, which are multiple variables with the same name using the Dim statement. To create an array variable to hold 10 fertilizer application rates, you would use the statement `Dim rates(9) As Single`. To access each member of the array from your VBA program, you would use its index. The index is an integer number from zero to the number of the last member of the array. The Dim statement creates an array where the number used is the last element the array. This means that `Dim fertilizerRates(1) As Single` creates an array with two members, 0 and 1. The program would access the data to get or set the value of the two members with `fertilizerRates(0)` and `fertilizerRates(1)`.

When naming variables, use variable names that make sense. Long and informative names are better than single letter names, except in the case of numbers that are only used temporarily in the program. For example, do not name a calculated Nitrogen recommendation as P or K as this may be confused with a phosphorus or potassium recommendation. A common way to create informative variable names is to use multiple words with the first letter of each capitalized. With that in mind, it is often better to use a name like NitrogenRateCalculated rather than just N.

Programs generally execute one line at a time, from the top to bottom of the text. Programs can be written so that they complete a single calculation or many calculations. When starting, keep the program small and use descriptive names that make sense.

In many situations you may want a different calculation to be done depending on some condition. This can be accomplished using a `For` (which creates a loop to execute the code a fixed number of times) followed by a `Next` (to end the loop) or `If` (which returns one value if the condition is true and another value if the condition is false)

PROBLEM 2.2.

What is the meaning of the following statements?

```
Dim num As Integer=0
num=5 + 7
```

ANSWER:

In the first line an integer called "num" is assigned a value of 0.
In the second line, the num is reassigned a value of 12.

PROBLEM 2.3A.

What is the value of sum after the program exists the **For** loop?

```
Dim i As Integer = 1
Dim sum as Double = 0
For i=1 to 3
     sum = sum +i
Next i
```

ANSWER:

In this calculation, the first time through the loop i is defined as 1 and sum is 0+1. In the second time through the loop i is 2 and sum + i is 3 (2+1=3). In the third time through the loop i is three and sum+i is 6 (3+3). At this point the statement after next i is run.

PROBLEM 2.3B.

If the **For** statement is changed to $i = 1$ to 4 what is the next value?

ANSWER:

At loop 3 the value was 6, so for the 4th loop the value is (6+4) = 10

PROBLEM 2.4.

What will the following double For loop do?

```
Dim i as integer, j as integer
For i=1 to 5
    For j=1 to 2
        Cells(i, j).Value = 10
    Next j
Next i
```

ANSWER:

It will produce a data set where all the values are 10.

PROBLEM 2.5.

Conduct the following operation using VBA rules.

    a. 3+ 6/2

    b. (3+6)/2

    c. (3+6)/2×5

    d. ((3+6)/(2×5))

ANSWER:

    a. 6

    b. 4.5

    c. 22.5

    d. 0.9

statement. The For statement allows the programmer to repeat a calculation a number of times. All of the commands between For and Next are executed once for each specified value in the For statement. For example:

```
Dim i As Integer = 0
For i = 0 to 10
    Repeated Calculations
Next i
```

In this For loop, i is the counter that is initially set at zero. Each time through the For statement, i is increased by 1, followed by completing the assigned task between the For and Next statements. When the Next i statement increments to $i = 11$, the program exits this loop without performing the "Repeated Calculations."

The If statement is used to specify what calculations are completed based on a specific condition being achieved. The If…then statement allows you to do a specified activity if the conditional statement goal is met. For example,

```
If i = 6 then "Saturday"
    End If
```

The If…then…else statement allows you specify an alternative activity in the case that the condition is not true. For example,

```
If i = 6 then "Saturday"
else "not Saturday"
End If
```

VBA Mathematics

The first step in writing VBA programs is understanding and mathematically defining the problem. The mathematics supported by VBA is slightly different than the mathematics supported by Microsoft Excel. Selected mathematical operators are shown in Table 2.1. In VBA, the program conducts tasks in a specified order. First, multiplication and division are conducted followed by addition and subtraction. Second, placing brackets around an operation specifies the order

Table 2.1. Selected VBA functions.

<	less than	+	addition
>	great than	−	subtraction
=	equal to	*	multiplication
< =	less than but equal to	/	division
abs(x)	absolute value	exp(x)	e^x
log (x)	natural log (x)	log(x)/log(10#)	log base 10
sqrt(x)	square root	cos(x)	returns the cosine of a value

that the mathematics is conducted. Third, when the operation contains both multiplication and division, the program starts on the left. To avoid confusion we recommend that you use brackets to define the order of the mathematics.

VBA Programming: A Problem Solving Tool

Computer programming is a powerful tool for solving problems and automating tasks that are done repeatedly. As with most tools, there are often many approaches to get to the desired outcome. The examples presented in this chapter are intended to provide an introduction to programming with VBA, and it is not intended to serve as a reference manual. Additional information for building programs is available in Walkenbach (2015) or Carney (2016) and Kiong (2009). The examples that follow provide possible solutions to determine the marketable weight of grain so that a grower can calculate the value of the grain when sold to a grain elevator.

Grain Moisture Content

Problem solving starts with defining the problem. In this problem, grain at one moisture content will be converted to the weight at another moisture content. Additional information on grain moisture contents are available in Carlson and Reese (2016), and on drying grain is available in Carlson (2016). Most grains produced as commodities are bought and sold based on either their dry weight equivalent, or more commonly, on an equivalent basis moisture content. In either the dry equivalent or wet basis equivalence cases, the offered price is normally dollars per bushel, with the test weight of the standard bushel given. For instance, the marketable basis for corn is 15.5% moisture content, wet basis, and the test weight is 56 lb bu^{-1}. The equation for wet basis moisture content is:

$$\text{Moisture Content} = 100\% \times \frac{\text{Weight}_\text{Water}}{\text{Weight}_\text{WetGrain}},$$

Where:

$$\text{Weight}_\text{WetGrain} = \text{Weight}_\text{Water} + \text{Weight}_\text{DryCorn}$$

When the crop is harvested, the moisture content is commonly different than the marketable basis. In fact, the moisture content can vary significantly in yield monitor data taken from a single field. We want to develop a program that can take grain weight at the harvested moisture content and convert it to the basis level. Preferably, we should be able to use the program many times to justify taking the time to write it. Since there are thousands of lines of yield data with different weight and moisture contents in a yield monitor file for a single field, it is safe to say that this program can be used many times.

If we rearrange the first equation above to solve for the water weight in the sample, we get the equation:

$$\text{Weight}_\text{Water} = (\text{MoistureContent} / 100\%) \times \text{Weight}_\text{WetGrain}$$

If the yield data includes the weight and the moisture content of the wet sample the dry weight can calculated with the equation:

$$\text{Weight}_\text{DryGrain} = \text{Weight}_\text{WetGrain} - \text{Weight}_\text{Water}$$

Now that we have the dry weight of the corn, we can calculate the equivalent weight of water at 15.5%:

$$\text{Weight}_\text{Water@15.5\%MC} = \frac{0.155 \times \text{Weight}_\text{DryGrain}}{(1 - 0.155)}$$

The total equivalent weight of our corn sample adjusted to the 15.5% MC basis is:

$$Weight_{WetGrain@15.5\%MC} = Weight_{Water@15.5\%} + Weight_{DryGrain}$$

Using this series of equations, we would calculate the marketable weight of 56 pounds of corn at 26% moisture content as follows:

```
(General)                                          basisWeight

Function basisWeight(WetWeight As Single, MC As Single) As Single
'Calculate the water weight as MC*WetGrainWeight
'Calculate the dry grain weight
'Calculate the water weight at the basis MC percent
'Calculate the grain weight at the basis MC percent
End Function
```

While this isn't too many steps, it is a process that could be generalized and turned into a VBA **function**. A function is a set of commands that are run to perform a specific calculation. In VBA, functions can be called from other functions, from macro programs, and even from the spreadsheet itself. What follows is a multistep process to (i) prepare your installation of Excel to allow VBA programming, and (ii) write a function to convert corn at a given moisture content to its equivalent weight at 15.5% marketable moisture content in Excel for Microsoft Windows. The appendix shows the process to prepare Excel on Apple macOS.

Step 1: Before you can develop with VBA within Excel, the Developer tab must appear on your Microsoft Excel ribbon. (Note below that *"File Home Insert Page Layout ...."* is the ribbon). The steps described below are designed for Excel within the Microsoft Office Professional Plus 2013 package. This approach may be different for different Microsoft Excel versions. Note that in the ribbon below, Developer is not available. This is normal. To make the Developer tab available on your ribbon, begin by clicking on *File* as found below.

Developer will now appear on your Ribbon as seen below.

Step 2: Click *DEVELOPER* and then *Visual Basic.*
After clicking *Visual Basic* the following screen will appear.
This is the VBA workspace, and is where you will begin to write both functions and macros.

Step 3: A blank Excel file will not have any custom functions or macros defined. Before they can be defined, we need to add a "module" to the file. To add the module, right click on the Insert menu at the top of the workspace and select "Module." Upon inserting, you should have a text editor in the right pane of the VBA window, and a "Module1" in the outline in the upper left frame. The text editing area is where you write the actual code.

When saving your program you must select *file, save as, My Documents,* and under *save as type* select macro-enabled workbook, type in your file name, and save.

Step 4: Declare the function. Click your mouse in the white text editor in the VBA window and begin by typing `Function basisWeight(WetWeight As Single, MC As Single)` into the editor, and hitting return. When you hit return, you should have `End Function` appear in the text editor as in the image below.

The first line beginning with Function is your function declaration. The declaration is the line that tells VBA that you are creating a function named `basisWeight`. The text between the parentheses is telling VBA that your function requires two input variables, `WetWeight` and `MC`, both as decimal numbers. All of the commands that you enter between the declaration and the `End Function` line make up the function.

Step 5: Plan out your functions calculations. Earlier in this chapter it was suggested that the best practice is to use comments in the VBA code to spell out the sequence of calculations before writing the statements to carry out those calculations. Although this example problem is straightforward, let's still follow that best practice. Add comments in the function to spell out the four steps to calculate the marketable weight. Hopefully you come up with something similar to:

Step 6: Write the code. The ultimate goal of this example is to calculate the marketable weight of grain given the wet weight and moisture content, and it is possible to do this in a single function in VBA. But since it might be useful to know the dry weight and other intermediate steps on their own, it makes sense to use multiple functions. In this case, we'll create a function for each of the steps indicated in the comments. The first step is to calculate the water weight, so we'll create a function called waterWeight with the same input variables as our basisWeight function. Since this is a wet-basis moisture content, it can never be more than 100% water. Recognizing this, we can put in checks to see if the user has used the percent or the fraction for their moisture content. See the function text below.

```
Function waterWeight(WetWeight As Single, MC As Single) As Single
If MC <= 1# And MC >= 0# Then 'MC is a fraction
    waterWeight = MC * WetWeight
Else
If MC <= 100# And MC >= 0# Then 'MC is a percentage
    waterWeight = MC * WetWeight / 100
    End If
End If
End Function
```

This function checks to see if the MC value is between zero and one. If so, it assumes the user has entered a fractional moisture content. If the MC is between zero and 100, it assumes it was entered as a percentage and divides the calculation by 100. Note, there is no handling of cases where the user accidentally enters a negative number or a number greater than 100. It is possible to handle such cases, but error handling is beyond the scope of this chapter.

Two other functions for dry weight and water weight at the basis MC are shown in the figure below. The final output of the basisWeight function is the marketable grain weight, which is the sum of the dry weight and the water at the marketable MC. This calculation was done in the main basisWeight function. The final set of functions for converting to marketable grain weight is:

```
eneral)                                                      basisWeight
Function basisWeight(WetWeight As Single, MC As Single) As Single
Dim water As Single, dryGrain As Single, waterBasis As Single
'Calculate the water weight as MC*WetGrainWeight
water = waterWeight(WetWeight, MC)
'Calculate the dry grain weight
dryGrain = DryWeight(WetWeight, water)
'Calculate the water weight at the basis MC percent
waterBasis = water155(dryGrain)
'Calculate the grain weight at the basis MC percent
basisWeight = dryGrain + waterBasis
End Function


Function waterWeight(WetWeight As Single, MC As Single) As Single
If MC <= 1# And MC >= 0# Then 'MC is a fraction
    waterWeight = MC * WetWeight
Else
If MC <= 100# And MC >= 0# Then 'MC is a percentage
    waterWeight = MC * WetWeight / 100
    End If
End If
End Function


Function DryWeight(WetWeight As Single, waterWeight As Single) As Single
DryWeight = WetWeight - waterWeight
End Function


Function water155(DryWeight As Single) As Single
water155 = (0.155 * DryWeight) / (1 - 0.155)
End Function
```

Table 2.2. The weight and moisture content basis of standard bushel of different annual crops.

Crop	Weight bushel	Specified moisture	lb at 0% moisture
Corn	56	15.50%	47.32
Soybean	60	13%	52.2
Wheat	60	13.50%	51.9
Barley	48	14.5%	41.04
Oat	32	14%	27.52
Rye	56	14%	48.16

Notice that each of the functions is declared with an "As Single" at the end of the declaration. This tells VBA that the result of the function is a decimal number, not an integer or Boolean.

The functions that we have created are specific functions for corn, but could be written for any grain or other agricultural product that has a marketable basis moisture content. Table 2.2 shows a list of common commodities marketed this way. You could also write a generic function that has a third input variable: specified basis moisture. This would allow you to use the same function to do the conversion on any crop with a specified basis moisture. In fact, this could be generalized even more if the conversion used the equation:

$$Weight_{Marketable} = \frac{Weight_{Initial} \times (1 - MC_{Initial})}{(1 - MC_{Marketable})}$$

To practice, write the code for this solution. How different is this code than the code used in the program. What other factors need to be defined as variables?

An advantage of VBA functions is that they are available to use in other functions or even directly in the spreadsheet. After creating the set of functions in VBA, Excel should recognize the function names as you type in spreadsheet cells. In the figure below, Excel has recognized that I started typing = basisW in cell F3 since it is a VBA function and is giving me the option to autocomplete. This figure is with the *Show Formulas* option selected in the Formulas tab.

SUM	▾ : ✕ ✓ *fx*	=basisW			

	A	B	C	D	E	F
1	Moisture Content	Weight	Water Weight	Dry Weight	basis water	basis weight
2	0.26	56	=WaterWeight(B2,A2)	=DryWeight(B2,C2)	=water155(D2)	=basisWeight(B2,A2)
3	26	56	=WaterWeight(B3,A3)	=DryWeight(B3,C3)	=water155(D3)	=basisW
4						*fx* basisWeight

In the figure below, *Show Formulas* is deactivated and the results of the formulas are shown. Note that the If...then statement in the VBA function allows the calculations to work for decimal or percentage versions of moisture content.

	A	B	C	D	E	F
1	Moisture Content	Weight	Water Weight	Dry Weight	basis water	basis weight
2	0.26	56	14.5600	41.4400	7.6014	49.0414
3	26%	56	14.5600	41.4400	7.6014	49.0414
4	26	56	14.5600	41.4400	7.6014	49.0414

So far in this chapter we have been using VBA functions to perform calculations that are useful performed repeatedly. The remainder of the chapter introduces VBA macros. A macro is similar to a function, allowing common tasks to be automated so they can be done repeatedly. While functions are generally good for working with numbers to perform calculations, macros are better suited to creating custom formatting or reports that might be used often. In the rest of the chapter we will develop two macros. The first will prepopulate a worksheet with text and some formatting, while the second will take user-supplied numbers and compute the resulting moisture content on the worksheet.

Step 1: Click *DEVELOPER* and then *Macros*.

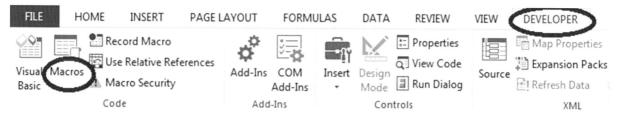

After clicking *Macros* the following screen will appear. Now type a *c* in Macro name.

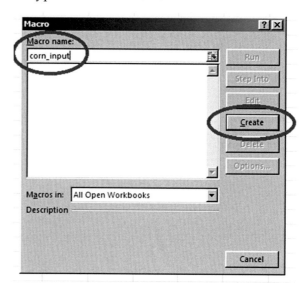

Note that *Create* button is now available. Finish typing the name "corn_input" and click *Create*. Macro names are picky. Use underline "_" instead of a blank space (if you typed in a blank space *Create* becomes unavailable).

Step 2: After you click *Create,* the screen will change and have a box in a gray background, Sub corn_ input() and *End Sub* will appear in the new box. See below.

```
Book1 - Module1 (Code)
(General)

    Sub corn_input()
    |
    End Sub
```

Type in the lines below between Sub corn_input()and End Sub. Sub is shorthand for subroutine. A Subroutine is used to accomplish a specific task within a program. You can activate a subroutine calling the name of the subroutine. For example, the statement call corn_input() will activate the commands within the corn_input() subroutine. Remember to not use spaces in the commands, using spaces within the parentheses for text strings (i.e. "Initial MC") is ok.

The first four lines after Sub corn_ input() create prompts in the spreadsheet. Sheet1.Cells(3,2) is one of the ways for VBA to address the Excel cell B3. Yes, this appears to be backward. B3 is Column B and Row 3. Sheet1.Cells(3,2) is Row 3, Column 2 (which is Column B). The next 2 lines will add color where the input should occur. Sheet1.Cells(4,2).Interior.Colorindex = 6 changes the cell (4,2), which is also known as cell B4 (yes note again that it is backward), to yellow.

```
Sub corn_input()
Sheet1.Cells(2, 2) = "Initial MC"    'Put label in B2
Sheet1.Cells(3, 2) = "decimal"       'Put units in B3
Sheet1.Cells(2, 5) = "Initial Weight"   'Put label in E2
Sheet1.Cells(3, 5) = "Pounds"            'Put units in E3

Sheet1.Cells(4, 2).Interior.ColorIndex = 6   'Format B4 yellow background
Sheet1.Cells(4, 5).Interior.ColorIndex = 6   'Format E4 yellow background
End Sub
```

Step 3: To get back to the initial Excel spread sheet click on the *Excel* icon in the upper left hand corner as seen below. You will always use this icon to get back to the original spreadsheet.

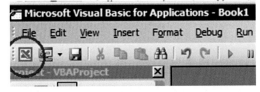

In the Microsoft Excel, with *Developer* selected click on *Macros*. The following box will appear.

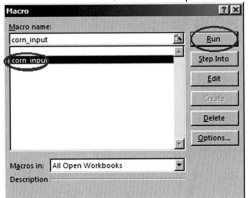

Select the "corn_input" macro and click on *Run*. The following spreadsheet should appear.

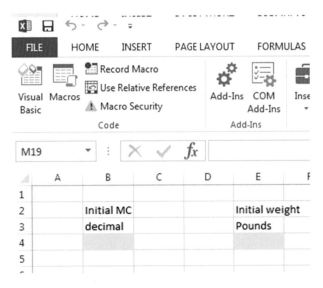

Step 4: Into cells B4 and E4 type 0.24 and 56. (56 pounds of corn at 24% moisture content).

Step 5: Now create a second macro named `calculate _ bushel`. This is done by clicking on *Macros* as before. Select *macro name*. The cell will go empty and you will type a new name `calculate _ bushel`. The Create button will become active. Click Create and the framework (top and bottom lines) for the new subroutine will appear. Type in the new code as seen below:

```
Sub calculate_bushels()
MC = Sheet1.Cells(4, 2)              'Get the moisture content from B4
wtcorn = Sheet1.Cells(4, 5)          'Get the wet weight from E4
bushels = basisWeight(CSng(wtcorn), CSng(MC))    'compute bushels with previous function
Sheet1.Cells(2, 8) = "Bu 15.5%"      'Put a lable in H2
Sheet1.Cells(3, 8) = bushels         'Put the computed bushels in H3
Sheet1.Cells(3, 8).Interior.ColorIndex = 8 'Format H3 aqua background
End Sub
```

Step 6: Click the Excel button at the top of the page. From the developer ribbon select *Macro* again. The following box will appear:

PROBLEM 2.6.
Modify and use the example functions to calculate 1 bushel of wheat (60 lbs) at 24% moisture to wheat at the marketable MC of wheat at 13.5% moisture.

ANSWER:

$$\text{Weight}_{\text{Marketable}} = \frac{60 \text{ lb} \times (1-0.24)}{(1-0.135)} = 52.7 \text{ lb}$$

$$\text{Bushels}_{\text{Marketable}} = 52.7 \text{ lb} \div 60 \frac{\text{lb}}{\text{bu}} = 0.88 \text{ bu}$$

PROBLEM 2.7.
How would you write code in VBA that determines 4^5?

ANSWER:

```
Sub Raise_to_power()
Sheet1.Cells(2, 2) = "Number to"
Sheet1.Cells(3, 2) = "raise to power"
Sheet1.Cells(3, 5) = "power"
Sheet1.Cells(2, 5) = "Number of"
Sheet1.Cells(1, 1) = "Power"
Sheet1.Cells(4, 2).Interior.ColorIndex = 6
Sheet1.Cells(4, 5).Interior.ColorIndex = 6
End Sub
Sub calculate_Power()
x = Sheet1.Cells(4, 2)
y = Sheet1.Cells(4, 5)
Z = x ^ y
Sheet1.Cells(6, 3) = "The Number Is"
Sheet1.Cells(7, 3).Interior.ColorIndex = 8
Sheet1.Cells(7, 3) = Z

End Sub
```

PROBLEM 2.8.
Develop a VBA program that will determine the corn yield at 15.5% moisture for this sample yield monitor file.

Coordinates		Yield (bu acre^{-1})	Moisture %	Yield (bu acre^{-1} 15.5%)
x	y			
1	1	150	22	138.5
1	2	140	21	130.9
1	3	135	20	127.8
1	4	120	19	115.0
2	1	130	20	123.1
2	2	140	22	129.2
2	3	180	24	161.9
2	4	200	25	177.5

ANSWER 2.8
Set up the macro as described above. The program is below:

```
Sub Macro _ example _ 5()
Dim Yield(2, 4) As Single
Dim Moist(2, 4) As Single

'data input
Yield(1, 1) = 150
Moist(1, 1) = 22
Yield(1, 2) = 140
Moist(1, 2) = 21
Yield(1, 3) = 135
Moist(1, 3) = 20
Yield(1, 4) = 120
Moist(1, 4) = 19
Yield(2, 1) = 130
Moist(2, 1) = 20
Yield(2, 2) = 140
Moist(2, 2) = 22
Yield(2, 3) = 180
Moist(2, 3) = 24
Yield(2, 4) = 200
Moist(2, 4) = 25

'Text to appear on spread sheet
Sheet1.Cells(2, 3) = "Coordinates"
Sheet1.Cells(3, 3) = "X"
Sheet1.Cells(3, 4) = "y"
Sheet1.Cells(3, 5) = _
"Yield (bu/acre)"
Sheet1.Cells(3, 6) = _
"Moisture %"
Sheet1.Cells(3, 7) = _
"Yield (bu/acre 15.5%)"
```

Additional Problems

2.9. Develop a program that will calculate soybeans that weigh 60 lb with 13.5% moisture.

2.10. How would you write code in VBA that determines the value $\log_{10}(880.63)$?

Check to see if you have entered 0.24 and 56 into cells B4 and E4 as instructed above.

Select `calculate _ bushel` from the Macro box (above) and Click *Run*.

This example shows that that 56 lb of corn at 24% will be 0.899 bushels corn at 15.5% moisture. Or, 56 lb of corn at 24% moisture dried to 15.5% moisture is 50.4 lb of corn. This value can be checked with the equation,

The following will appear. Your results indicate that 56 pounds of 24% moisture corn is 0.8994 bushels of 15.5% corn.

	A	B	C	D	E	F	G	H
1								
2		Initial MC			Initial weight			bu 15.5%
3		decimal			Pounds			0.899408
4		0.24			56			
5								
6								

$$\text{grain weight}_{\text{final}} = \frac{\text{grain weight}_{\text{initial}} \times (100\text{-}\% \text{ Initial moisture})}{(100\text{-}\% \text{ final moisture})}$$

The resulting amount of grain is 50.4 lbs. To save your program, select File, Save As, select My Documents, type your program name in file name, and in Save as Type select Excel macro-enabled work book.

ACKNOWLEDGMENTS

Support for this document was provided by South Dakota State University, South Dakota Soybean Research and Promotion Council, Precision Farming Systems community in the American Society of Agronomy, International Society of Precision Agriculture, and the USDA-AFRI Higher Education program (2014-04572).

This chapter used a number of screenshots from Microsoft Excel 2013. Because we followed the instruction from Microsoft in the use of these products, the screen shot are used with permission from Microsoft. The use of commercial products does not endorse one product over another, it is for the convenience of the reader.

REFERENCES AND ADDITIONAL INFORMATION

Carlson, C.G., and C.L. Reese. 2016. Chapter 35: Grain marketing-Understanding corn moisture content, shrinkage and drying. In: D.E. Clay, C.G. Carlson, S.A. Clay, and E. Byamukama, editors, iGrow corn: Best management practices. South Dakota State University, Brookings, SD.

Carlson, C.G. 2016. Chapter 53: Corn storage and drying. In: D.E. Clay, C.G. Carlson, S.A. Clay, and E. Byamukama, editors, iGrow corn: Best management practices. South Dakota State University, Brookings, SD.

Carney, K. 2016. Excel VBA for complete beginners. Home and Learn. http://www.homeandlearn.org/ (accessed September 30, 2016). [2016 is year accessed].

Kiong, L.V. 2009. Excel VBA made easy: A concise guide for beginners. ExcelVBATutor.com http://www.excelvbatutor.com/vba_book/vbabook_ed2.pdf (verified 2 Aug. 2017).

Walkenbach, J. 2015. Excel VBA programing for dummies, 4th edition. John Wiley and Sons, New York.

An Introduction to Experimental Design and Models

3

David E. Clay,* Gary Hatfield, and Sharon A. Clay

Chapter Purpose

There is a perception that precision farming is the use of state-of-the-art technologies to apply variable rate treatments across fields. However, precision farming is also the use of these tools to apply locally-based knowledge to improve management decisions. This chapter provides an overview on how experimentation and mathematical models can lead to improved yield or net return from differing agricultural systems in a site specific manner. This information can then be used to determine if decisions should be varied across the field. The goals of this chapter is to improve your understanding how to conduct strip trials and the associated analysis, interpretation of laboratory results, and how use of zero- and first-order mathematical models in problem solving.

Experimental Designs

In many on-farm experiments, producers often want to know if one treatment is better than another or if a specific treatment compared with another will increase yields. However, in order to make valid comparisons, experiments must be designed appropriately. Because responses may be landscape dependent, multiple areas with the same treatment need to be established to estimate the range of response.

The arrangement in space is called the experimental design. The goal of the experimental design is to allow for valid statistical comparisons between the various treatments. There are many different experimental designs, ranging from simple to complex. This chapter discusses how to design and analyze simple experiments. In precision farming, many experiments use randomized complete block, strip trial, or matched pair designs. For information on other types of experimental designs, refer to Steel et al. (1997), and information for

Key Terms

Hypothesis testing, type I error, type II error, experimental design, data collection, sample analysis, interpretation, mathematical model, and rate constants.

Mathematical Skills

Learn how to conduct a statistical analysis of strip-trial experiment, conduct t tests, calculate confidence intervals, means, and medians, and use zero and first order kinetics in problem solving.

D. Clay, G. Hatfield, and S.A. Clay, South Dakota State University, Brookings, SD 57007-2201. *Corresponding author (david.clay@sdstate.edu).

doi: 10.2134/practicalmath2017.0104

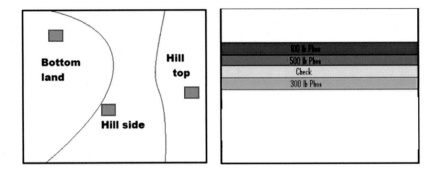

Fig. 3.1. Example of block design (left) and fertilizer strip design (right) applied across a landscape (Carlson and Reicks, 2013).

Grid	Random	Management zones

Fig 3.2 sampling diagrams:

Grid

X X X

X X X

X X X

X X X

Random

X X X
X

X

X X X

Management zones

X X X X

X X X X
X

X X X

X X X

Fig. 3.2. Examples of different sampling approaches. The management zones may be defined based on landscape position, management zones, or soil type information.

analyzing on-farm experiments are available in Knighton (1998), Wittig and Wicks (1998), Nielsen (2008), Rempel (2002), and Tangren (2002).

In a randomized block design, all treatments are contained within an area that has similar characteristics. Each block contains all the treatments. A blocked experiment is used to minimize the impact of site on the results. In Fig. 3.1, the three blue squares represent blocks located in the bottom land, hill side, and on the hill top. The plot size for each treatment should be similar. An experiment should contain at least 3 replications or blocks. Often, the equipment used to plant, fertilize, spray, or harvest a field is too big to easily manage in small plot trials. Therefore, treatments may be applied in strips across the entire field (strip trial) (Fig. 3.1).

The experimental design tells where to put the treatments and identifies which comparisons are valid. During the experimental planning, additional information such how and when to collect information should also be identified. In the strip trial experiment shown in Fig 3.1, each treatment is replicated once within this block. Additional blocks will be located in adjacent areas or fields (Fig 3.2). To define boundary conditions of the treatments, experiments should be repeated in time and space. In agronomic field experiments, treatments are usually replicated at least four times, whereas in greenhouse experiments, treatments may be replicated eight or more times. The number of replications will depend on the size of the area, the expense of the trial, and other factors.

Once the design is developed and treatments applied, observations (measurements of a pertinent factor) are collected. Observations may include the number of germinated seeds in a 30 m² area, or the amount of soil soluble P contained in a soil sample. Observations within a single treatment can be analyzed to estimate the mean, median, or the rate of change, if the observations are a time function. The mean is the average value of a population, whereas the median is defined as the value where 50% of the population is above the value and 50% is below the value. In field studies, these mean and median values are often different (Clay et al., 2002).

Precision vs. Accuracy

Accuracy describes how close the measurements are to the true value. Correct answers are identified by using appropriate laboratory quality assurance (QA) protocols. Precision is a term that describes the variation (or scatter) of the values in a data set; high precision has limited scatter in the data set. High precision has limited scatter and

PROBLEM 3.1A.

In a given area, the yield monitor reports that the yield is 9, 9.5, 8.5, 8.3, 10.2, 11.5, and 11 Mg ha^{-1}. What are the mean and median values?

ANSWER:

The mean is calculated with the equation, $\overline{X} = \dfrac{\sum_{i=1}^{i=n} X_i}{n}$.

The mean ($\overline{X}$) is calculated by adding up each observation (X_i) and dividing by the number of values (n) that were summed. In this example,

$$\text{mean} = \frac{9+9.5+8.5+8.3+10.2+11.5+11.0}{7} = 9.71$$

The median is determined by organizing the observations from low to high values, and then determining the value where 1/2 of the values are above and 1/2 are below the median.
Median determination

8.3, 8.5, 9, 9.5, 10.2, 11, 12.5

In this case, the median value is 9.5, as three values are lower than 9.5 and 3 values are greater than 9.5.

PROBLEM 3.1B.

In another portion of the field the yield monitor values are 3.5, 13.6, 4.7, 9.5, 18, 20, 5.1. Determine the mean and median values.

ANSWER:

Mean = 3.5+13.6+4.7+9.5+18+20+5.1/7 = 10.6
Median = 3.5, 4.7, 5.1, 9.5, 13.6, 18, 20 = 9.5

Note that the median of both data sets is 9.5; however, the mean values are quite different.

PROBLEM 3.1C.

Calculate the mean and median of the dataset with the values 2, 4, 6.

ANSWER:

Mean = 4; Median = 4

PROBLEM 3.1D

Calculate the mean and median of the dataset with the values 3.9, 4.0, 4.1.

ANSWER:

Mean = 4; Median = 4.

Note that in problems 3.1c and 3.1d the mean and median are the same even though the numbers are very different.

it does not indicate if the value is correct (Fig. 3.3). In Fig. 3.3 a high variance is indicative of imprecise information, although low variance does not imply accurate data. High variance in the data complicates the development of accurate recommendations.

To mathematically describe precision, the variance (s^2) or standard deviation (s) is calculated. Low variance indicates that the data is tightly grouped. The variance (s^2) is determined with the equation,

$$s^2 = \frac{\sum_{i=1}^{n}(X_i - \overline{X})^2}{n-1}$$

where, $\overline{X}$ is the mean and X_i are the values for each observation., and n is the number of observations. The standard deviation (s) is the square root of the variance (s^2). Examples of high and low standard deviation and variance are provided in Problem 3.2. The mean (calculated in Problem 3.1) is calculated with the equation: $\overline{X} = \dfrac{\sum_{i=1}^{n}(X_i)}{n}$

Many data management software programs calculate the mean, median, variance, and standard deviation. In Excel, the mean, median, variance, and standard deviation are calculated with the commands ' = average (first data point, last data point)', ' = median (first data point, last data point)', ' = var(first data point, last data point)' and ' = stdev (first data point, last data point)', respectively.

PROBLEM 3.2A.

Calculate the variance and standard deviation for the values, 9, 9.5, 8.5, 8.3, 10.2, 11.5, and 11.

ANSWER:

1. The mean is calculated in Excel using the command, = average (start.end). In this case the average or mean in 9.71.

2. Variance = $(9-9.71)^2 + \ldots + (11-9.71)^2/(7-1) = 9.1/6 = 1.5$. In Excel this is calculated using the commend = var(start,end).

3. Standard deviation = square root of 1.5 = 1.23. In Excel this is calculated with the command, = stdev(start,end).

PROBLEM 3.2B.

Determine the mean, variance, and standard deviation of the data set: 3.5, 13.6, 4.7, 9.5, 18, 20, 5.1.

ANSWER:

Mean 10.6.
Variance = $(3.5-10.6)^2 + \ldots + (5.1-10.6)^2/(7-1) = 268/6 = 44.7$
Standard deviation = square root of 44.7 = 6.69
Note that in problems 3.2a and 3.2b both datasets had a median of 9.5, however in dataset 1 the variance is low. This indicates precision, whereas in dataset 2, there is high variance, indicating that the mean value is imprecise.

PROBLEM 3.2C.

Determine the mean, variance, and standard deviation for the data set, 2, 6, and 4.
Dataset 3. 2, 6, 4.

ANSWER:

Mean is 4.
Variance = $(2-4)^2 + (4-4)^2 + (6-4)^2/(3-1) = 8/2 = 4$
Standard deviation = square root of 4 = 2

PROBLEM 3.2D.

Determine the mean, variance, and standard deviation of the following dataset, 3.9, 4.0, 4.1

ANSWER:

Mean is 4
Variance = $(3.9-4)^2 + (4-4)^2 + (4.1-4)^2/(3-1) = 0.02/2 = 0.01$
Standard deviation = square root of 0.01 = 0.1
Note that both dataset 3.2c and 3.2d have the same mean, however, the variance of 3.2c is 4, indicating imprecision and no accuracy.

Hypothesis Testing

When thinking about comparing treatments, an idea or hypothesis is proposed. This hypothesis is then tested through experimentation. There are two basic outcomes from the study, that the response to the selected treatments is the same, indicating no effect (also called the null hypothesis or H_o), or that the response differs among the treatments (called the alternative hypothesis, H_A). The H_A can be simple, i.e., there is a response difference, or more complex, one treatment response will be 20% greater than another treatment. Additional details on hypothesis testing are available in Clay et al. (2011) and Freund and Wilson (1997). The null or alternative hypotheses provide insight into if a one-tailed or two-tailed test should be selected. If the null and alternative hypotheses indicates that the treatment can increase or decrease the measured parameter, then a two-tailed test should be used, whereas a one-tailed test is used if the treatment can only increase or decrease the parameter, but not both.

An example of an alternative hypothesis is that, relative to a 0 P rate, applying 60 kg P ha^{-1} will increase the corn yield. The null hypothesis is that applying 60 kg P ha^{-1} does not increase the corn yield. This is a one-tailed test as written as the emphasis is that differences occur only if the treatment increases yield. A strip trial could be used to test this hypothesis. The treatments in the strip trial might be, 0 and 60 kg P ha^{-1}, with each treatment being replicated three or more times.

	Accurate	Inaccurate
Precise	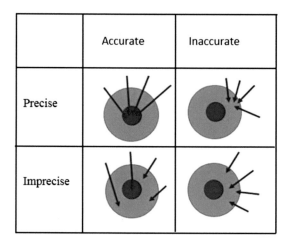	
Imprecise		

Fig. 3.3. A visualization of precise and accurate measurements.

During the experiment, information is collected and analyzed to determine if the null hypothesis should be accepted or rejected. Essentially, the strip or each plot in a randomized block experiment provides data which can be used to produce evidence if the null hypothesis should be accepted or rejected. The means of the treatments across replications and variance of each mean are calculated and compared based on the stated statistical criteria. If yield statistically increased due to the addition of 60 kg P ha^{-1}, then the alternative hypothesis is supported, whereas if the response to 60 kg P ha^{-1} is deemed similar to the no P treatment, the null hypothesis is supported.

In hypothesis testing, two types of errors are possible. The first error occurs when you reject the null hypothesis when it is true (i.e., saying there is a response difference when none occurred). This type of error is called a false positive or type I error. The second type of error occurs when you accept the null hypothesis when it is false (i.e., saying there is not a response when a response occurred). This type of error is called a false negative or type II error. The amount of acceptable error is called the alpha (α) value. The alpha value is the probability of rejecting the null hypothesis when it is true. The α value is set by the researcher before the experiment is conducted and is the standard on how extremely different the data must be before the null hypothesis is rejected. The smaller the α value, the more stringent the test and the greater the probability of concluding that there is no difference in response. The p-value is the probability of obtaining similar results if the null hypothesis is true. The test criterion, or t-value is calculated based on treatments means and their standard deviations and is dependent on the total number of replications – 1 (degrees of freedom).

The test statistic is determined with the equation: $\dfrac{\left(\overline{X_1}-\overline{X_2}\right)}{s\sqrt{\dfrac{1}{n_1}+\dfrac{1}{n_2}}}=$ where, $\overline{X}$ are the means of the two populations,

n is the number of replications, and s is the standard deviation.

This equation states that the t statistic (t-value) is the mean of one treatment subtracted from the other and divided by the common standard deviation times the square root of 1/number of replications. After the t-value is calculated, it is compared with tables containing 'critical t-values' for the p-value of interest [see Table 3.1 for different p-values and degrees of freedom]. The critical t-values differ with degrees of freedom and p-value selected, and if it is a one- (there is only a difference one way) or two-tailed (there could be a positive OR negative difference) test. If the resulting t-value is greater than the critical t-value in the table, there is a significant difference between the two treatments (Table 3.1).

The t-values in Table 3.1 are the critical values used to determine if two means are similar or different. The tabled t-value when combined with the standard deviation can be used to calculate the confidence interval for the mean. The confidence interval represents the range of values for a given level of significance. The degrees of freedom are the number of independent pieces of information that go into the estimate of a parameter. For example,

PROBLEM 3.3.
For each set of hypotheses how many tailed test should be used?
H$_o$: Variety A and B have similar yields.
H$_a$: The two yields are different.

ANSWER:
two-tailed test is used because two possible outcomes are possible.

PROBLEM 3.4
H$_o$: N fertilizer does not increase yield.
H$_a$: N fertilizer increases yield.

ANSWER:
One-tailed test because only one possible outcome is possible.

Table 3.1. The critical *t*-values for different probability levels and degrees of freedom (df). The *t*-values at 0.10, 0.05, and 0.025 probability levels represent the probability value exceeding a specific value for a one sided test. One sided and two sided comparisons have different critical *t*-values for identical probability levels. For example, the *t* value for a two-sided test with a 95% confidence level would be under the column $t_{0.025}$. The two tailed test uses $t_{0.025}$ as opposed to $t_{0.05}$ because the 5% is split between both sides of the curve. Degrees of freedom for each *t*-values is *n*-1. The diagram below shows a one-tailed test and the associated critical *t*-values.

	One-tailed values						
Degrees freedom	$t_{0.10}$	$t_{0.05}$	$t_{0.025}$	Degrees freedom	$t_{0.10}$	$t_{0.05}$	$t_{0.025}$
1	3.0078	6.314	12.706	18	1.33	1.734	2.101
2	1.886	2.92	4.303	19	1.328	1.729	2.093
3	1.638	2.353	3.182	20	1.325	1.725	2.086
4	1.533	2.132	2.776	21	1.323	1.721	2.08
5	1.476	2.015	2.571	22	1.321	1.717	2.074
6	1.44	1.943	2.447	23	1.319	1.714	2.069
7	1.415	1.895	2.365	24	1.318	1.711	2.064
8	1.397	1.86	2.306	25	1.316	1.708	2.06
9	1.383	1.833	2.262	30	1.31	1.697	2.042
10	1.372	1.812	2.228	40	1.303	1.684	2.021
11	1.363	1.796	2.201	60	1.296	1.671	2.00
12	1.356	1.782	2.179	80	1.292	1.664	1.99
13	1.35	1.771	2.16	100	1.29	1.66	1.984
14	1.345	1.761	2.145	1000	1.282	1.646	1.962
15	1.341	1.753	2.131	Inf.	1.282	1.645	1.96
16	1.337	1.746	2.12				
17	1.333	1.74	2.11				

the degrees of freedom for a CI for mean containing n measurements is n-1. The degrees of freedom is defined as *n-1*, where *n* is the number of observations. Most computers have software programs for conducting these types of analysis.

Confidence Intervals

The two sided (±) confidence interval is calculated using the equation, $\pm t \times \frac{s}{\sqrt{n}}$, where the critical *t*-value is obtained from Table 3.1. In this calculation, use the critical *t*-value for the two sided test at an appropriate probability level. Because the data in Table 3.1, provide the *t*-values for a one sided test ($t_{0.05}$), values for 95% confidence interval are obtained under the column $t_{0.025}$. This value is used because 2.5% of the population is contained in each of the two tails of the population distribution. For example, for the 95% confidence interval, the critical *t*-value for 25 degrees of freedom for a two sided test is 2.06.

Population Distributions

In agriculture, it is often assumed that populations follow a bell-shaped curve (Fig. 3.4). This means that the population distribution has about the same numbers of very high and very low values (tails of the curve). However, in precision farming, many populations have an uneven population distribution (skewed distribution). The description of the skewness is left-skewed (i.e., there is a tail on the left side of the mean) or right-skewed (there is a small number of points on to the right side of the mean) (Clay et al., 2002; Fig. 3.4). If the population is skewed to the right, the median (point where 50% of the values are greater than and less than) is less than the average (Fig. 3.4). Soil nutrient analysis, and weed densities in a field are often skewed to the right. When the data is shewed to the left, the data set contains more small values than large values.

PROBLEM 3.5.

A farmer measures the yield in 12 fields. Six of the fields are hybrid A and six are hybrid B. The grower wants to know if the two hybrids have different yields at the 95% level. First state the null and alternative hypotheses and if the test is one-tailed or two-tailed. Hybrid A has a mean yield of 12 Mg ha^{-1} (215 bu acre^{-1}). Hybrid B has a yield of 10 Mg ha^{-1} (178 bu acre^{-1}). The variance (s^2) of each mean is 1.488 Mg ha^{-1}. Calculate the t-value and determine if the t-value justifies concluding that there is a difference between hybrid yields.

ANSWER:

H$_o$: The two means are the same.
H$_a$: The two means are different.
This is a two-tailed test as one hybrid could have higher or lower yield than the other. The alternative hypothesis does not specify how the two are different. In this case the critical t-value for a 0.05 α value (t$_{0.025}$) with 10 (6+6–2) degrees of freedom is 2.228 (see Table 3.1). There is a loss of two degrees of freedom because two means were calculated. In this example, the critical t-value is associated with t$_{0.025}$ because a two-tailed test was specified in the hypothesis statement. (If H$_A$ stated A > B, then a one-tailed test (t$_{0.05}$) would be used

$$\frac{(\overline{X}_1 - \overline{X}_2)}{s\sqrt{\frac{1}{n_1}+\frac{1}{n_2}}} = \frac{(12-10)}{1.22\sqrt{\frac{1}{6}+\frac{1}{6}}} = 2.82$$

and the critical t-value would be 1.812). The calculation for the test value is . In this calculation, the standard deviation (s) is the positive square root of the variance (s^2) or 1.488. The test statistic is The test statistic is then compared with the appropriate critical t-value. The critical t-values for a 95% probability level is 2.228 (Table 3.1). Because 2.82 is greater than 2.228, the null hypothesis is rejected and the alternative hypothesis, that the hybrid means differ, is accepted.

When the soil analysis of a composite soil sample is completed, the reported value represents the average or mean, not the median. A consequence of this distribution is that the soil test result may overestimate the amount of nutrients in the soil, thereby underestimating the amount of fertilizer needed to optimize yields. Potential solutions to this problem are to collect grid soil samples or use prior information to identify areas that should and should not be included in the composite soil sample (Clay et al., 2002). For example, if the field previously contained a livestock confinement area, that area should be analyzed separately from the rest of the field (Fig. 3.5). Research has shown that old animal confinement areas can impact soil test P for over 50 yr. Once the population distribution is understood, models can be used to process this information into better decisions.

Using Zero and First Order Models to Solve Problems

In agriculture, mathematical models can be used to improve science based predications and enhance understanding. The zero and first order models are commonly used to define the rate of change. In zero order kinetics, the parameter of interest (y) is calculated with the equation, $y = mt + b$, where t is time, m is the rate of change and b is the y-intercept. In zero-order kinetics, the rate of change is independent of the concentration of the reactants and

the half-life is the initial concentration (b) divided by the slope (m) multiplied by two: $t_{\frac{1}{2}} = \frac{b}{2 \times m}$. Silber et al. (2010) reported that the dissolution of minerals from biochar may follow zero-order kinetics.

In first order kinetics, the rate of change is dependent on the concentrations of the reactants. First order kinetics is described by the equation: $\frac{dy}{dt} = -ky$. In this equation, the rate of change is a function of the amount of reactants (y) in the system and the first order rate constant (k). The rate constant represents the portion of the reactants or substrate that reacts in during a given time period. Integration of the function, $\frac{dy}{dt} = -ky$, results in the equation, $y_t = y_0 e^{-k}$ which can be transformed into ln(yt) = -kt + ln y_0. The half-life is calculated with the equation $t_{half-life} = \frac{\ln 2}{k}$. First order or modified first order rate equations have been used to calculate herbicide degradation (Di et al., 1998; Charnay et al., 2005) (cont. turnover of carbon compounds in soil (Adair et al., 2008; Boyle and Paul 1989; Charnay et al., 2005, Chang et al., 2016; Clay et al., 2010, 2015, Kemanian and Stöckle, 2010; Lehmann and Kleber, 2015; Parton et al., 1993; Walker, 1974).

Problem 3.6.

You conduct a strip trial experiment in five fields. The experimental design is a randomized block. Each block is a portion of land that was split in half. Treatment A was applied to one half and treatment B was applied to the other half. Each block is a replication where it is assumed that the soil and environmental conditions are similar across the block. The data represents hypothetical data from two treatments, and could represent yields, soil test results, or weed populations. Within a block, the treatments are randomly assigned. The question is, are the two treatments different?

Blocks	Treatment A	Treatment B
1	100	110
2	120	130
3	140	150
4	135	150
5	125	135
Mean	124	135

A B Block 1

B A Block 2

B A Block 3

B A Block 4

A B Block 5

Answer:

A randomized block experiment can be analyzed within the Excel program. To do this select data analysis, Anova, Two-Factor Without Replication. Fill in the input range and select OK. In this analysis, rows represent blocks and columns represent the two treatments.

Anova: Two-Factor Without Replication

SUMMARY	Count	Sum	Average	Variance
Row 1	2	210	105	50
Row 2	2	250	125	50
Row 3	2	290	145	50
Row 4	2	285	142.5	112.5
Row 5	2	260	130	50
Column 1	5	620	124	242.5
Column 2	5	675	135	275

ANOVA Source of Variation	SS	df	MS	F	P-value	F crit
Rows	2060	4	515	206	6.98E-05	6.388233
Columns	302.5	1	302.5	121	0.000388	7.708647
Error	10	4	2.5			
Total	2372.5	9				

†The rows represent the blocks and the columns represent the treatment. The p-value 0.000388, is less than 0.05. Based on this analysis we conclude the two treatments are different.

‡This analysis approach will not calculate interactions between the treatments.

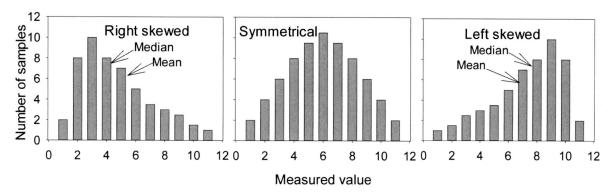

Fig. 3.4. Hypothetical histogram that are skewed to the right, bell-shaped curve, and skewed to the left. Soil nutrient distributions and weed densities are often skewed right, whereas yield monitor data may be skewed left (Clay et al., 2002).

PROBLEM 3.7.

Ten soil samples are collected from a production field, each sample is analyzed and the mean is 15 ppm (15 mg kg^{-1}) and the variance is 5. What is the 95% confidence interval (CI) of the mean? The α value is (100-confidence)/100, and for confidence intervals we use a two-sided test. This calculation assumes a normal distribution.

ANSWER:

The 95% confidence is calculated with the equation,

$$CI = t \times \frac{s}{\sqrt{n}} = t_{0.025,9} \times \frac{\sqrt{5}}{\sqrt{10}} = 1.6$$

The critical *t*-value for a two-sided test with nine degrees of freedom is 2.262. Confidence intervals always use two sided tests. Therefore, for an α value of 0.05, use the $t_{0.025}$ value. Based on this value, the 95% confidence interval is 15 ± 1.6.

PROBLEM 3.8.

If you know that the soil test P distribution is skewed to the right, and that the average value is 12 ppm (12 mg kg^{-1}) and the median is 9 ppm (9 mg kg^{-1}), how would you vary your P recommendation?

ANSWER:

Fifty percent of this field has a P concentration < 9 ppm, your recommendation needs to be adjusted accordingly. Without additional information it is difficult to identify how to adjust the values. One approach to obtain this information is to grid soil sample.

PROBLEM 3.9.

The soil test P values from a field that was grid soil sampled are: 2, 6, 4, 8, 7, 10, 22, and 15 ppm. What are the mean and median of these samples?

ANSWER:

mean = 9.25 ppm
median = 7.5 ppm

Based on the mean and median, would the histogram be normal, or skewed right or left?
Skewed distribution to the right (the median is less than the mean). Calculate the 95% confidence interval for the mean.

Variance = 41.9; std deviation = 6.47; $t_{0.025,7}$ = 2.365
CI = 2.365×6.47/sqrt(8) = 2.365×2.29 = 5.41

This calculation can be done in Excel using the confidence function ' = confidence(α, standard_dev, size)'
   ' = confidence(0.025,6.47,8)'

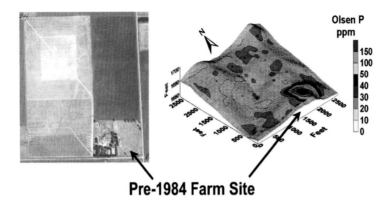

Pre-1984 Farm Site

Fig. 3.5. Black and white photograph of a quarter section taken in 1956, and a soil P grid map collected in 2001. Note: In 2001, there was no visible trace of a farm site on this quarter section of land (Courtesy D.E. Clay). The soil test Olsen P-values indicate that livestock were confined at this homestead. When grid soil sampled, samples taken from the old homestead result in skewed histograms as shown in Fig. 3.4.

Herbicides often are applied for their residual activity. This means that the herbicide stays in the soil and kills emerging weeds for days or weeks after application. Other herbicides are only applied after the weed has emerged (postemergence) and these have little or no effect after application. For example, atrazine is a photosynthetic inhibitor used as a preemergence or postemergence herbicide to control a wide range of weeds including kochia (*kochia spp.*), common cocklebur (*Xanthium spinosum* L.), smooth pigweed (*Amaranthus hybridus* L.), Palmer pigweed (*Amaranthus Palmeri*), tall waterhemp [*Amaranthus tuberculatus* (Moq.) J.D. Sauer], common waterhemp (*Amaranthus rudis* L.), and wild sunflower (*Grindela camporum* Greene), whereas 2,4-D is a postemergence herbicide that is used to control broadleaf plants. As residual herbicides are degraded, they become less effective. For example, planting a broadleaf cover crop shortly after applying atrazine would have limited success. However, if the cover crop is planted one year after application, it most likely would be successful. Additional information on herbicide persistence and carryover is provided in Colquhoun (2006).

Summary

In summary, this chapter provides an introduction into experimentation and use of mathematical models. Examples are provided on how to conduct relatively simple on-farm studies. This chapter also provides a discussion on the types of data that are collected from production fields, and provides possible approaches on how to make the information more useful. Zero- and first-order kinetics and where these models can be used in problem solving are discussed.

Additional problems

3.14. You collect 20 soil samples from a production field, the mean (average) is 15 ppm (15 mg kg^{-1}) and the variance is 3, what is the 95% confidence interval of the population mean? What is the 90% confidence interval of the population mean?

3.15. You collect 10 samples from a production field. The soil test P concentration of those samples are 9, 10, 11, 8, 12, 10, 15, 14, 22, and 7. What is the mean, median, variance, and standard deviation of this set of numbers? How would you fertilize this field?

3.16. Soil organic matter has first order rate constant of 0.015 lb (lb×year)$^{-1}$. What is the half-life of 10,000 kg of residue? How much of this residue will remain in the soil after 40 yr?

3.17. You conduct an on farm study that contained two N rates and 4 replicates. The yield for the 0 lb N ha^{-1} rate is 90 bu acre^{-1} and the yield for the 100 lbs N ha^{-1} rate is 140 bu acre^{-1}. The variance is 20. At the 0.05 level, did N significantly increase the crop yield?

ACKNOWLEDGMENTS

Support for this document was provided by South Dakota State University, Precision Farming Systems community in the American Society of Agronomy, International Society of Precision Agriculture, and the USDA-AFRI Higher Education program (2014-04572).

REFERENCES

Adair, E.C., W.J. Parton, S.J. Del Grosso, W.L. Silver, M.E. Harmon, S.A. Hall, I.C. Burke, and S.C. Hart. 2008. Simple three pool model accurately described patterns of long-term litter decomposition in diverse climates. Glob. Change Biol. 14:2636–2660.

Boyle, M., and E.A. Paul. 1989. Carbon and nitrogen mineralization kinetics in soil previously amended with sewage sludge. Soil Sci. Soc. Am. J. 53:99–103. doi:10.2136/sssaj1989.03615995005300010018x

Carlson, C.G., and G. Reicks. 2013. Chapter 54: Developing on-farm research protocols. In: D.E. Clay, C.G. Carlson, S.A. Clay, L. Wagner, D. Deneke, and C. Hay, editors, iGrow Soybeans: Best management practices for soybean production. South Dakota State University, SDSU Extension, Brookings, SD.

Charnay, M.P., S. Tuis, Y. Coquet, and E. Barriuso. 2005. Spatial variability in 14C-herbicide degradation in surface and subsurface soils. Pest Manag. Sci. 61:845–855. doi:10.1002/ps.1092

Chang, J., D.E. Clay, A.J. Smart, and S.A. Clay. 2016. Estimating annual root decomposition in grassland systems. Rangeland Ecol. Manag. 69:288–291. doi:10.1016/j.rama.2016.02.002

Clay, D.E., S.A. Clay, C.G. Carlson, and S. Murrell. 2011. Mathematics and calculations for agronomists and soil scientists. International Plant Nutrition Institute, Peachtree Corners, GA. http://www.ipni.net/ipniweb/portal.nsf/0/C69CD7F53F A540D80625785B0068FEF8/$FILE/Math%20_METRIC_TOC_2015.pdf (verified 30 June 2017)

Clay, D.E., C.G. Carlson, T. Schumacher, V. Owens, and F. Mamani Pati. 2010. Biomass estimation approach impacts on calculated SOC maintenance requirements and associated mineralization rate constants. J. Environ. Qual. 39:784–790. doi:10.2134/jeq2009.0321

Clay, D.E., N. Kitchen, C.G. Carlson, J.L. Kleinjan, and W.A. Tjentland. 2002. Collecting representative soil samples for N and P fertilizer recommendations. Crop Manage. doi:10.1094/cm-2002-12xx-01-MA.

Clay, D.E., G. Reicks, C.G. Carlson, J. Moriles-Miller, J.J. Stone, and S.A. Clay. 2015. Tillage and corn residue harvesting impacts surface and subsurface carbon sequestration. J. Environ. Qual. 44:803-809 doi:102134/jeq2014.07.0322.

Colquhoun, J. 2006. Herbicide persistence and carryover. A3819. University of Wisconsin Extension, Madison, WI. http://corn.agronomy.wisc.edu/Management/pdfs/A3819.pdf (verified 30 June 2017).

Di, H.J., L.A.G. Aylmore, and R.S. Kookana. 1998. Degradation rates of eight pesticides in surface and subsurface soil under laboratory and field conditions. Soil Sci. 163:404–411. doi:10.1097/00010694-199805000-00008

Freund, R.J., and W.J. Wilson. 1997. Statistical methods. rev. eds. Academic Press, New York.

Kemanian, A.R., and C.O. Stöckle. 2010. C-Farm: A simple model to estimate the carbon balance of soil profiles. Eur. J. Agron. 32:22–29. doi:10.1016/j.eja.2009.08.003

Knighton, R. 1998. Setting up on-farm experiments, SSMG 17. In: D.E. Clay and S.A. Clay, editors, Site-specific management guidelines. International Plant Nutrition Institute, Peachtree Corners, GA.

Lehmann, J., and M. Kleber. 2015. The contentious nature of soil organic matter. Nature 528:60–68. doi:10.1038/nature16069

Nielsen, R.L. 2008. A practical guide to on-farm research. Purdue University, Lafayette, IN.

Parton, W.J., J.M.O. Scurlock, D.S. Ojima, T.G. Gilmanov, R.J. Scholes, D.S. Schimel, T. Kirchner, J.C. Menaut, T. Seastedt, E. Garcia Meya, A. Kamnalrut, and J.I. Kinyamario. 1993. Observations and modeling of biomass and soil organic matter dynamics for the grassland biome worldwide. Global Biogeochem. Cycles 7:785–810.

Rempel, S. 2002. On farm research guide. The Garden Institute of Alberta, Edmonton, AB.

Silber, A., I. Levkovitch, and E.R. Graber. 2010. pH-dependent mineral release and suface properties of cornstraw biochar: Agricultural implication. Environ. Sci. Technol. 44:9318–9323. doi:10.1021/es101283d

Steel, R.G.D., J.H. Torrie, and D.A. Dickey. 1997. Principles and procedures of statistics a biometrical approach. 3rd Edition, McGraw-Hill, New York.

Tangren, J.A. 2002. Field guide to experimental designs, Washington State University, Pullman, WA.

Walker, A. 1974. Simulation model for predicting herbicide persistence. J. Environ. Qual. 3:396–402. doi:10.2134/jeq1974.00472425000300040021x

Wittig, T.A., and Z.W. Wicks, III. 1998. Simple on-farm comparisons. In: D.E. Clay and S.A. Clay, editors, Site-specific management guidelines. International Plant Nutrition Institute, Peachtree Corners, GA.

Mathematics of Longitude and Latitude

4

David E. Clay,* Terry A. Brase, and Graig Reicks

Chapter Purpose

In precision farming, latitude and longitude values are routinely used to identify locations, create application maps, and assess spatial variability. However, historically legal land descriptions were not based on satellite-based location systems (GPS or GNSS). To identify land ownership, an understanding of all location systems are needed. This chapter discusses the various techniques used to identify land ownership based on location and to calculate latitude and longitude values.

Historic Land Classification and Locations

Historically, location was needed to navigate the seas as well as construct buildings. Early on, all locations were relative. For example, a pile of rocks was relative to the location of a stream. It is believed that the peg and rope technique was used to build complex structures such as Stonehenge (Johnson and Pimpinelli, 2008). In this basic building method, a peg, at point A, is pounded into the soil, and a straight base line that does not intersect the peg is drawn on the soil surface. A rope that is slightly longer than the distance between the peg (A) and base line is attached to the peg. This rope, attached to point A (peg) is used to draw a circle around point A. The intersection between the base line and the circle are called points C and D. The point half way between C and D is called E. The angles between the A, E, and C is 90 degrees and the angle between A, B, and E is 90 degrees. Based on these basic measurements, complex structures were created.

However, the peg rope approach does a poor job at identifying ownership boundaries that may follow a stream, river, or mountain, which were written down or verbally passed down from one generation to another. For example, in medieval Europe, a communal memory of the boundaries was created by walking around the village borders.

In the United States, land surveys were historically conducted using the metes and bounds system and the rectangular survey system. The English system, metes and bounds, was brought to North America when the land was settled. In this system, metes refers to the distance between two points,

Key Terms

Latitude and longitude, unit conversions, rectangular system, metes and bounds, land survey, township, range, triangulation, dead-reckoning, North Star, South Celestial Pole, Global Positioning System (GPS), Global Navigation Satellite System (GNSS), data projection, datum, Universal Mercator Coordinate System (UTM).

Mathematical Skills

Converting latitude and longitude values from one scale to another, understanding global positioning systems, and understanding how locations were determined historically.

D.E. Clay and G. Reicks, South Dakota State University, Department of Agronomy, Horticulture and Plant Science, Brookings, SD 57007-2201; T.A. Brase, West Hills Community College, Coalinga, CA 93210. *Corresponding author (david. clay@sdstate.edu).

doi: 10.2134/practicalmath2016.0111

whereas the direction might be provided by the bearing of a compass. For example, "take 10 steps to the north of the oak tree." Bounds is the general physical description of the boundary. For example, "along a stream." The primary problem with the metes and bounds system is that landmarks can move or disappear over time.

In the United States, as settlement moved from the Atlantic Coast to the Great Plains, land classification switched to the rectangular system. This system was based on three reference points, a longitude running through Greenwich England (0 longitude), the equator (0 latitude), and the center of the Earth. Associated with this system are 36 principal meridians, established by the surveyor general of the U.S., running North and South, and used as a reference for subdividing public lands in a large region. The principal meridians are arbitrarily located, do not coincide with the latitude and longitude system (Fig. 4.1), but are used to divide townships between east and west. Baselines running east and west were also established.

In the rectangular system, lines parallel to the baselines were surveyed (Fig. 4.2). These lines were called towns and they were located every 6 miles. Strips of land running North and South, parallel to the principal meridians were called ranges. The intersection of a town and range is called a township with the dimensions of six miles by six miles. Each township contains 36 sections, each being one mile by one mile (Fig. 4.3). The sections are numbered boustrophedonically on a map, meaning that the first range starting at the north-east corner, is numbered from one to six right to left, and numbers in the next range are 7 to 12, left to right, (i.e., Section 7 is directly below (to the south of) Section 6, and Section 12 is directly below section 1) (in a snake-like fashion starting in the northeast corner). Each section has the approximate dimensions of one mile by one mile and contains 640 acres. Sections are split in quarter sections, with the dimensions of 1/2 mile by 1/2 mile containing approximately 160 acres. Based on the land description, the size of a field or land parcel can be estimated. For example, the size of a land parcel located at S1/2, SE1/4, S5, R1W, and T1N is 80 acres [640/(4 × 2)]. (640 represents the acres within a section; 4 is the denominator of the SE 1/4; and 2 is the denominator of the S 1/2) The historical definition of an acre is the amount of land that could be plowed in one day by an oxen. The acre dimensions were designated as one furlong (660 ft) by one chain (66 ft) or 43,560 ft2. However, because the distance between the longitudes decrease as you travel from the equator to the pole, discontinuities exist in the township range system. To account for these discontinuities, quarter sections along the northern and western township borders were adjusted.

World Exploration

During the 13th, 14th, 15th, and 16th centuries, the Age of Exploration was limited by the ability of sea captains to identify their location. In response to numerous navigation errors (i.e., ship wrecks), Great Britain created the

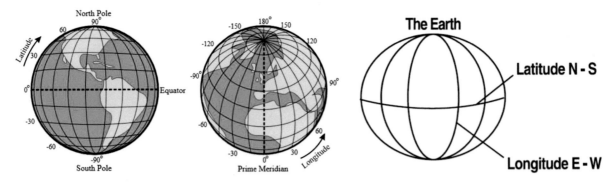

Fig. 4.1. A spheroid Earth with the locations of the latitudes and longitudes (Courtesy of Wikimedia Commons).

Problem 4.1
How many acres are located in N1/4, SE1/4, S3, T2N, R1W? Could you show the approximate location of this in Fig. 4.2?

Answer:
The answer is 40 acres. This calculation is based on a section containing 640 acres, and a section containing four quarter sections.

$$\frac{40 \text{ acres}}{\text{quarter quarter section}} = 640 \text{ acres} \times \frac{1 \text{ section}}{4 \text{ quarter sections}} \times \frac{1 \text{ quarter section}}{4 \text{ quarter quarter section}}$$

Longitude Prize, which would be awarded to a person or persons who discovered a practical method for determining longitude for ships at sea. France, Spain, and Holland offered similar prizes.

Prior to an accurate technique to identify longitude, sea captains used dead reckoning and generally stayed within the sight of land. In dead reckoning, the location is based on the previous location and takes into account speed and direction of travel. For example, if you are at point A and you travel one mile north, your new location is one mile north of point A. However, at sea, currents make measurement of actual speed difficult. Relative speed was determined by dropping a weighted line containing knots overboard. Based on the change in the number of knots that could be observed, the relative speed was calculated.

During this period of time, people discovered how to accurately locate their position. Similar to identifying boundaries in medieval Europe, early explorations stayed within sight of land. With time, new techniques that involved the locations of the sun and stars, along with measurements of speed, and dead reckoning were created to identify locations.

Early explorers did not have the capacity to accurately measure longitude and latitude. Latitude was generally measured using celestial navigation. In the Northern Hemisphere, the latitude was determined by measuring the altitude of the North Star (Polaris) above the horizon. The North Star is located at a fixed point in space from which a location can be calculated. In the Southern Hemisphere there is not a single star that marks the southern celestial pole. However, a fixed point was identified by using the Southern Cross, a star constellation easily observed in the Southern Hemisphere. The process of locating the southern celestial pole involved drawing an imaginary line between the two of the four stars in the Southern Cross (Gacrux (the northern most star, or 'top' of the cross) and Acrux (the southern most star of the constellation, or 'bottom' of the cross) (Fig. 4.4). The Southern Celestial pole was determined by extending this line 4.5 times the length (from Gacrux to Acrux) of the Southern Cross downward to the south. The invention and use of the sexton improved the measurements of these angles.

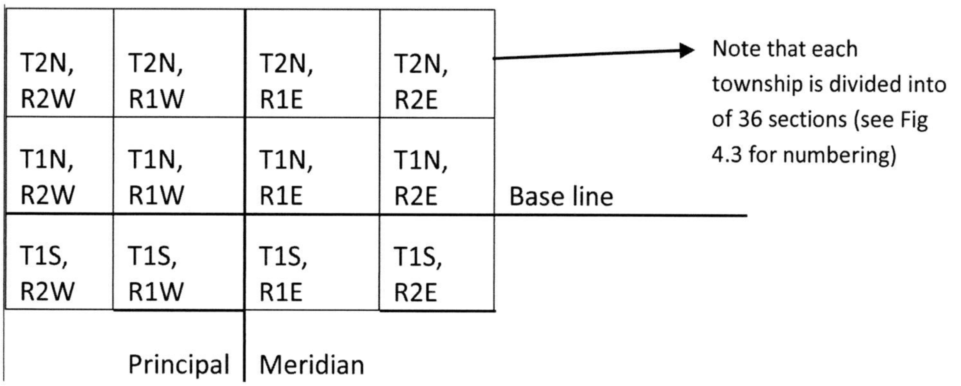

T2N, R2W	T2N, R1W	T2N, R1E	T2N, R2E	
T1N, R2W	T1N, R1W	T1N, R1E	T1N, R2E	Base line
T1S, R2W	T1S, R1W	T1S, R1E	T1S, R2E	

Note that each township is divided into of 36 sections (see Fig 4.3 for numbering)

Principal | Meridian

Fig. 4.2. The locations of townships. Each township is 6 miles by 6 miles. A township is shown in Fig. 4.3.

6	5	4	3	2	1
7	8	9	10	11	12
18	17	16	15	14	13
19	20	21	22	23	24
30	29	28	27	26	25
31	32	33	34	35	36

Fig. 4.3. A map showing the location of sections within township (as shown in Fig. 4.2).

PROBLEM 4.2.

What is the longitude of Lincoln, Nebraska if the time difference (between Lincoln and Greenwich England) when the sun is highest in the sky is 6.45 h?

ANSWER:

Because each day is 24 h, the Earth spins 15° each hour (360/24 = 15). In 6.45 h, the Earth has revolved 96.7 degrees (6.45 × 15 = 96.7 degees). To identify the location, the latitude, longitude, and altitude must be known. For ocean exploration, the altitude was known (sea level). Latitude was determined by measuring the sun's angle when it reached its highest point in the sky. To identify the longitude, explorers (sailors) needed a standard (time) from which they could measure distance from a reference point. The reference point was Greenwich England, and the clock was called a marine chronometer.

Longitude was harder to measure, and eventually it was determined that if you measured the time when the sun reached the highest point in the sky relative to the time at fixed point (Greenwich, England) the difference in time between the fixed point and your point was related to longitude. The relationship between time and longitude was that the Earth was spinning at the rate of 15° each hour (360/24).

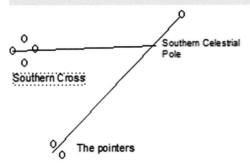

Fig. 4.4. Locating the southern celestial pole from the southern cross. The stars in the southern cross are Mimonsa, Delta Crucis, Gacrux, and Acriux. The Southern Celestrial pole is the point directly above the southern pole. The Southern Celestrial pole provides a fixed point for reference, and the southern sky turns around this point.

Global Navigation Satellite Systems

The wide-scale availability of Global Navigation Satellite Systems (GNSS) has changed how land is located and defined. The number of operational satellites varies over time and is dependent on the number of spare satellites in orbit. Originally, the minimum number of satellites for a full constellation was 24, which were located in six earth-centered orbital planes. The number of currently operating satellites is obtained at http://tycho.usno.navy.mil/gpscurr.html. Each satellite broadcasts its location (ephemeris) and a timing code. GPS receivers on Earth use this information to calculate the distance between the satellite and the receiver. Because there are multiple satellite systems, the term that describes the collective group of satellites used for positioning is GNSS.

The ability of GNSS or the subset GPS to calculate a location is based on a very accurate clock. For example, an error of 0.001 s can produce an error in the calculated distance of 300 km (186 miles). By combining data from multiple satellites, the location of the receiver is determined. The constellation of the satellites influences the accuracy of the calculated location, which can be improved by evenly distributing the satellites in space. The accuracy of the GNSS calculation resulting from the satellite constellation is quantified with the dilution of precision (DOP) value. Some receivers can calculate and report the DOP. Accuracy increases with lower DOP values; values less than 2 are considered good. Objects that block the ability of a GNSS unit to receive satellite transmissions increase error. Signals can be blocked by trees, buildings, and landscape features.

In three-dimensional space, information from four satellites are required to calculate a location. The current satellite constellation has been designed to ensure that a receiver located anywhere on the Earth can receive a signal from at least four satellites simultaneously.

Global Nagivation Satellite System Errors

GNSS position errors can be caused by: (i) clock errors, (ii) the Earth's gravitational field impacting satellite orbits, (iii) poor satellite configuration; (iv) atmospheric interference, and (v) multipath errors. To improve accuracy, a differential correction signal (DGNSS) can be used. In differential correction, the location of a known point (base station) is used to correct the accuracy on the moving receiver. In differential correction, the accuracy decreases with increasing distance from the base station. This accuracy may be defined as 4 cm + 3 ppm. Based on this definition, this receiver would have a positional accuracy of 4 cm at the base station. However, this accuracy would decrease by 3 cm with every million cm from the base station. For example, a rover located 2 million cm (12.4 miles, or 20 kilometers) from a base station would have positional accuracy of 10 cm [4cm +2,000,000 cm ×(3cm/1,000,000 cm)] . Real time kinematic GNSS systems provide the best accuracy. (Continued on p. 70)

Exercise to Understand GNSS

Step 1. Use a compass to draw a circle. The satellite is located in the center of the circle. The distance between the center of the circle and the line represents the distance that a signal from the satellite can travel in a specific period of time. By measuring the time, you can then calculate the distance. In 3-D you would be located on a surface of a sphere, however we will represent this 3-D problem as a 2-D problem.

Based on the distance, you are somewhere on the surface of the circle. If a second satellite sends a signal you can then calculate your distance from the second satellite. In 2-D, this is represented by drawing a second circle.

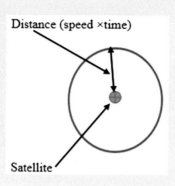

Step 2. Draw a second circle.

Based on this exercise, you are located somewhere at the intersection of these two circles, which shows two possible solutions. Which point you are located at is resolved by adding a third circle. The accuracy of the location increases with the number of satellites used to identify your location. In three dimensions, at least 4 satellites are needed to calculate a location. The process used to identify your location is loosely characterized as triangulation.

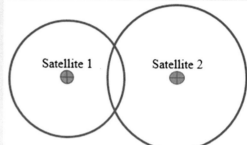

Step 3. Add another circle to identify a single intersection of the three circles. This intersection is the correct location.

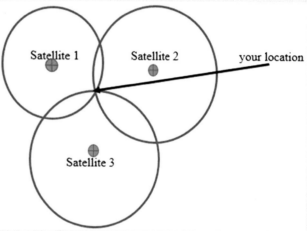

PROBLEM 4.2A

What is the radius of the center of a circle with the following equation, $(x-5)^2 + (y-4)^2 = 100$. In this system x in east-west and y is north-south.

ANSWER:

For the equation of a circle with the equation, $(x-h)^2 + (y-k)^2 = r^2$, the h and k terms are the center of the circle.

PROBLEM 4.2B

If x is 5, what is y?

ANSWER:

$(y-4)^2 = 100$, $y = 14$

PROBLEM 4.3.

Convert 96.83556 to degrees, minutes, and second.

ANSWER:

Minutes = 0.83556 × 60 = 50.1336
Seconds = 0.1336 × 60 = 8.016

ANSWER:

96 degrees, 50 min, and 8.01 s (96° 50' 8.016")

PROBLEM 4.4.

Convert 96 degrees, 49 min and 40 s into decimal degrees.

ANSWER:

Divide the minutes by 60 and seconds by 3600. Add minutes and seconds together.

$$\text{Decimal degrees} = 96\,\text{deg.} + 49\,\text{min}\frac{1\,\text{deg.}}{60\,\text{min}} + 40\,\text{s}\frac{1\,\text{deg.}}{3600}$$

Decimal degrees = 96+0.81667+0.01111 = 96.82778

In the future, GNSS receivers may have the capacity to use corrections from multiple base stations. Additional information on GNSS errors are available in Stombaugh (2018).

Latitude and Longitude

The reference points for latitude and longitude values are the equator and the zero degrees longitude, which is the longitude that travels through the Royal Observatory in Greenwich England. Latitude is defined as the polar coordinate, degrees north or south from the equator (up to 90 degrees north or south) (Fig. 4.1). In the latitude and longitude system, location can be reported in minutes and seconds or decimal degrees. Examples of this, using a position in Gettysburg, SD, is Decimal Degrees (DD) is 44.988283, -99.953828. The same position in Degrees, Minutes, Seconds (Latitude DMS) is 44°59' 17.8188", -99°57' 13.7808. A degree contains 60 min or 3600 s.

To convert a value in degree, minutes, and seconds to a value expressed as degree minutes, the seconds must be divided by 60 and added to the minutes. To convert minutes to decimal degree, minutes must be divided by 60, and to convert seconds to decimal degrees, seconds must be divided by 3600. This conversion is based on 1 degree containing 3600 s. In the example above, 44° 59 min and 17.8188 s, the 59 min is 0.9833333 decimal degrees (= 59/60) and 17.8188 s is 0.004949 (17.8188/3600) decimal degrees. Adding together the decimal degrees for the degrees + minutes + seconds gives us 44.988283 (= 44+0.983333+0.004949).

To convert decimal degrees to degrees, minutes, and seconds, the degree decimal must be multiplied by 60 to get minutes. The minute decimal degrees are then multiplied by 60 to get seconds. Using the example above, the decimal degrees of 0.988283 is multiplied by 60 to produce 59.29698 min (= 0.988283 × 60). The decimal portion of 59.29698, is then selected (0.29698) which is then multiplied by 60. The minutes portion is 17.8188 (= 0.29698 × 60). Putting them together gives us the Latitude DMS of 44°59' 17.8188".

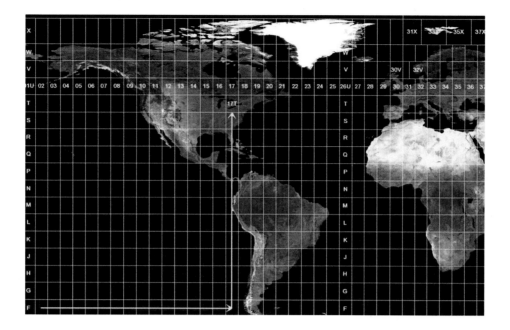

Fig. 4.5. UTM latitudinal zones overlaid over the United States (Courtesy of Public Domain, available here: https://commons.wikimedia.org/w/index.php?curid=1601744).

Additional Rules to Understand About Latitude and Longitude

Distance Between Longitudes Depend on the Latitude

Because longitude lines converge at the poles, the distance between longitude lines are not constant and depend on where on earth you are located. At the equator, the distance between longitude lines is 69.172 miles, whereas at the north and south poles the distance is zero. At 45 degrees latitude, half-way between the equator and the north or South Pole, 0.0001 degrees latitude from north to south is about 36.46 feet. These distances are relatively constant from north to south. However, longitude values vary greatly and decrease as you travel to the poles. At 45 degrees latitude, the distance of 0.0001 degrees longitude from east to west is about 25.87 feet (Fig. 4.1), whereas at the pole this value approaches zero. Further discussion on determining the distance between two GPS points is available in Carlson and Clay (1999).

Data Projections and Datum

Precision farming practitioners need to understand the difference between projections and datum (reference point). Data projections are used to display 3-dimensional body (earth or field surface) in a 2-dimensional coordinate system. The transformation from 3-D to 2-D is a not trivial process, and many different types of projections can be found. Common projection types include the Mercator, Gall-Peters, Miller Cylindrical, Mollweide, Goode, Homolosine Equal-area, Sinusoidal equal area, and Robinson. Each projection type contains substantial errors.

The datum is the specific mathematical calculation used to fit the earth to a 3-dimensional ellipsoid. The shape of the Earth is not a perfect sphere. Isaac Newton hypothesized that with time, centrifugal force would cause the Earth to bulge in the center. The resulting shape is called an ellipsoid. Different datums produce different latitude and longitude values for the same location. A datum commonly used in North America is NAD27. This datum was developed in 1927 and it is commonly used in USGS maps. Other common datums include NAD83, WGS84, and ITRF00. The center of the earth is the center point for the WGS84 and NAD83 ellipsoids, and as more is learned, the models are adjusted. In the United States, when a GPS system is first purchased, the factory setting is typically WGS84.

Universal Transverse Mercator coordinate system

The Universal Transverse Mercator (UTM) system between 80 degrees south and 84 degrees north into 60 zones. Each zone has a width of 6 degrees longitude (Fig. 4.5). Universal Transverse Mercator coordinates are known as "Northings" and "Eastings" and are actually linear measures (meters) from each grid's starting reference line. By convention the easting is followed by the northing. Latitude and longitude and UTM values have a different

Additional Questions

4.6. How does GPS differ from UTM?

4.7. Do you need more than one satellite to identify a location?

4.8. How many acres are located in E1/2, SW1/4, S2, T55N R3W?

4.9. Convert 90.678321 to degrees, minutes, and seconds.

4.10. Convert 98 degrees 32 min and 29 s to decimal degrees.

4.11. If the sun reaches its peak 20 h after peaking in Greenwich England, what is the longitude?

4.12. What is the longitude if the time difference between Greenwich England and where you are located is 10 h?

4.13. If DGNSS has an accuracy of 2 cm + 2 ppm, what is the accuracy 1, 2, and 6 miles from the base station?

organizational structure. For example latitude and longitude values look like 67.5 degrees, whereas UTM has values such as D14 706832 mE, 4344683 mN. In this coordinate: "D" references the UTM latitudinal zone; "14" references the UTM longitudinal zone. Thus, D14 identifies the specific grid zone for which the coordinate is provided. Every grid zone on earth could have this same coordinate so it is important to include this. The easting is the measurement in meters in the east-west direction within the zone, whereas the northing is the distance in meters the north-south direction within the zone. A northing of 0 is on the equator if the location is north of the equator. If the location is south of the equator, the equator is often identified as 10,000,000 m. If a northern of 8500,000 mN is provided, the location would be 1500,000 south (10,000,000-8,500,000) of the equator. To avoid confusion, the full coordinate system should specify if the location is north or south of the equator. One of the advantage of the UTM system is that the location is provided in meters from a reference point.

Summary

For numerous purposed we have needed to identify our location. Since the advent of civilization, this has been accomplished using a wide variety of approached ranging from the location relative to a stream or river to more recently using GPS. In many locations multiple systems are being used, and therefore it is important to understand the local customs. Additional information about GPS is available in Johansen et al. (1999), Krishna (2016), Pfrost et al. (1999), and Zhang (2015).

ACKNOWLEDGMENTS

Support for this document was provided by South Dakota State University, South Dakota Soybean Research and Promotion Council, the Precision Farming Systems community in the American Society of Agronomy, International Society of Precision Agriculture, and the USDA-AFRI Higher Education Grant (2014-04572).

REFERENCES

Carlson, C.G., and D.E. Clay. 1999. The earth model- calculating field size and distance between points using GPS coordinates #11. In: D.E. Clay, editor, Site specific management guidelines. Potash and Phosphate Institute, Norcross, GA.

Johnson, A., and A. Pimpinelli. 2008. Pegs and ropes: Geometry at Stonehenge. Nature Precedings. hdl:10101. npre.2008.2153.1

Johansen, D.P., D.E. Clay, C.G. Carlson, K.W. Stange, S.A. Clay, and K. Dalsted. 1999. Selecting a DGPS for making topography maps. SSMG 14. In: D.E. Clay, editor, Site specific management guidelines. Potash and Phosphate Institute, Norcross, GA.

Krishna, K.R. 2016. Push button agriculture: Robotics, drones, satellite-guided soil and crop management. Apple Academic

Press and CRC Press, Boca Raton, FL. doi:10.1201/b19940

Pfrost, D., W. Cassidy, and K. Shannon. 1999. Global positioning satellite receivers. SSMG-6. In: D.E. Clay, editor, Site specific management guidelines. Potash and Phosphate Institute, Norcross, GA.

Stombaugh, T. 2018. Satellite-based systems (GPS and GNSS) for precision agriculture. In: K. Shannon, D.E. Clay, and K. Kitchen, editors, Precision Farming Basics, ASA, CSSA, SSSA, Madison, WI.

Zhang, Q. 2015. Precision agriculture technology for crop farming. CRC Press. Boca Raton, FL. doi:10.1201/b19336

Spatial Statistics

Gary Hatfield*

5

Spatial Data

Spatial data is comprised of variables of interest collected at different documented locations. A variable of interest may be elevation, slope, plant population, weed counts, soil nutrient level, crop reflectance, or yields that are measured using a yield monitor during harvest. The source of the information may be derived from sensors or laboratory analysis. In all cases, the location of the sampling points is known; it may be reported latitude and longitude or Northings and Eastings. Table 5.1 lists yield monitor data for nine different spatial locations.

Mapping data can help reveal relationships or the lack of relationships between spatial locations (Kolaczyk et al., 2005). As an example, corn yield per harvested acre is mapped in Figure 5.1. A visual inspection reveals that the highest yields are concentrated in the states of Iowa, Minnesota, Illinois, Nebraska, and South Dakota. This map also reveals there is little corn production west of the Rocky Mountains. This map was created using the United States Department of Agriculture National Agriculture Statistics Service website at https://www.nass.usda.gov/Charts_and_Maps/Crops_County/.

Key Terms

summary statistics, spatial data, spatial autocorrelation, semi-variogram, interpolation, inverse distance weighting, kriging

Mathematical Skills

Analysis of spatial data using spatial statistics and geostatistical methods.

Chapter Purpose

Many precision agriculture activities rely on the analysis of spatial information to control equipment for optimal applications, determine economic optimization, and assess ecological sustainability. This chapter provides an overview of spatial data, means, medians, kurtosis, variance, autocorrelation, semi-variance, and spatial interpolation methods.

Table 5.1. A sample data set containing yield monitor, latitude, and longitude information.

Site	Latitude	Longitude	Dry yield
			bu ac^{-1}
1	43.2350	-96.9000	168.5
2	43.2350	-96.8985	226.8
3	43.2350	-96.8970	242.3
4	43.2325	-96.9000	164.8
5	43.2325	-96.8985	219.4
6	43.2325	-96.8970	233.2
7	43.2300	-96.9000	177.8
8	43.2300	-96.8985	245.4
9	43.2300	-96.8970	204.7

South Dakota State University, Department of Mathematics and Statistics, Box 2225, AME 256, Brookings, SD 57007-2201.*Corresponding author (gary.hatfield@sdstate.edu).

doi: 10.2134/practicalmath2016.0102

Why Spatial Data is Different from Other Types of Data

Datasets are often comprised of columns and rows, where columns are variables and rows are observations. In Table 5.1, there are four columns (site, latitude, longitude, and dry yield) and nine observations. Each site's dry yield is georeferenced to allow the spatial properties among and between the observations to be investigated. It is this georeferencing that makes spatial data different from other types of data. This allows spatial autocorrelation to be investigated and incorporated into models that may also include explanatory variables.

An important property of spatial data is that measurements made at one location may be correlated and not independent from information collected at other locations. For example, the deposition of sediments from erosion at point B is dependent on the slope at point A. In addition, there is a tendency for measurements at nearby locations to be more similar than measurements at distant locations. This concept is captured in Tobler's First Law of Geography "Near things are more related than distant things" (Fotheringham and Rogerson, 2009).

Spatial information is represented by four types of data models. These are:

1. Point: a single point location,
2. Line: a set of ordered points,
3. Polygon: an area, marked by one or more closing lines, and
4. Grid: a collection of points or rectangular cells organized as a regular lattice. The grid is a special form to represent spatial data, such as measurements from an image sensor (Bivand et. al., 2008).

Data that is not associated with a geographic location is referred to as nonspatial, or aspatial, data and usually consist of attribute data. This type of data can often be considered as measurements on random samples from the same population. Properties of the population can be inferred from the properties of the sample. These properties may include average, dispersion, and shape. An example of non-spatial data are the chemical analysis for protein content of grain samples from a wheat shipment delivered to an elevator for storage or sale.

Mean, Median, Kurtosis, Variance, and Skewness for Nonspatial Data

Spatial and nonspatial data can be summarized using various statistical techniques that measure spread and dispersion. Summary statistics include the mean and median as measures of the average and variance as a measure of dispersion of the data (Table 5.2). Values such as skewness and kurtosis provide information about the shape of the distribution.

The mean is the arithmetic average of observations. It is calculated by summing up the values of a variable for a set of observations and then dividing by the number of observations that were summed. One must be careful when calculating the mean as it will be impacted by extreme observations. The median is a value that divides the ordered observations in half. It is a robust measure of the parameter as it is not influenced by extreme observations. The variance is a measure of

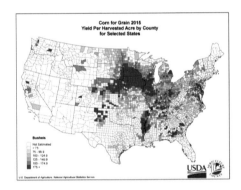

Figure 5.1. Corn yield per harvested acre by county.

PROBLEM 5.1

Create a map of yield per harvested acre by county for soybeans using the United States Department of Agriculture National Agriculture Statistics Service website at https://www.nass.usda.gov/Charts_and_Maps/Crops_County/.

ANSWER:

Go to https://www.nass.usda.gov/Charts_and_Maps/Crops_County/, select soybeans, and then select PNG or PDF for yield per harvested acre by county.

PROBLEM 5.2

Calculate the mean, median, variance, skewness, and kurtosis for the following data obtained by measuring Foreign Material (FM), expressed in percent total, on nine (n = 9) random samples from a shipment of corn: 2.47, 1.34, 0.44, 1.45, 0.14, 1.51, 0.90, 2.47, 1.66.

ANSWER: (Detailed calculations below and R code in Appendix B.)

Mean:

$$\bar{x}=\frac{1}{9}\sum_{i=1}^{9}x_i=\frac{1}{9}(2.47+1.34+0.44+1.45+0.14+1.51+0.90+2.47+1.66)=1.376$$

Median:

Order data: 0.14, 0.44, 0.90, 1.34, 1.45, 1.51, 1.66, 2.47, 2.47
Since n = 9 is odd, the median is the middle ordered observation, median

Variance:

$$s^2=\frac{1}{9-1}\sum_{i=1}^{9}(x_i-\bar{x})^2=\frac{1}{8}\begin{bmatrix}(2.47-1.376)^2+(1.34-1.376)^2+(0.44-1.376)^2+\\(1.45-1.376)^2+(0.14-1.376)^2+(1.51-1.376)^2+\\(0.90-1.376)^2+(2.47-1.376)^2+(1.66-1.376)^2+\end{bmatrix}=0.641$$

Skewness:

$$\frac{1}{9}\left(\sum_{i=1}^{9}(x_i-\bar{x})^3\right)\left[\left(\sqrt{\frac{9}{\sum_{i=1}^{n}(x_i-\bar{x})^2}}\right)^3\right]$$

$$=\frac{1}{9}\begin{bmatrix}(2.47-1.376)^3+(1.34-1.376)^3+(0.44-1.376)^3+\\(1.45-1.376)^3+(0.14-1.376)^3+(1.51-1.376)^3+\\(0.90-1.376)^3+(2.47-1.376)^3+(1.66-1.376)^3+\end{bmatrix}\left[\left(\sqrt{\frac{9}{5.128}}\right)^3\right]$$

= -0.043

Kurtosis:

$$\frac{1}{9}\left(\sum_{i=1}^{9}(x_i-\bar{x})^4\right)\left[\left(\sqrt{\frac{9}{\sum_{i=1}^{n}(x_i-\bar{x})^2}}\right)^4\right]$$

$$=\frac{1}{9}\begin{bmatrix}(2.47-1.376)^4+(1.34-1.376)^4+(0.44-1.376)^4+\\(1.45-1.376)^4+(0.14-1.376)^4+(1.51-1.376)^4+\\(0.90-1.376)^4+(2.47-1.376)^4+(1.66-1.376)^4+\end{bmatrix}\cdot\left[\left(\sqrt{\frac{9}{5.128}}\right)^4\right]$$

= 2.061

Excess Kurtosis = Kurtosis − 3.0 = 2.061 − 3.0 = -0.939

dispersion as it measures how compact or spread out the data is. Skewness is a measure of the shape of a distribution and specifically measures the symmetry of the distribution of the data. Values equal to zero or near zero indicate symmetry. Positive values indicate a distribution that is skewed to the right and negative values indicate skewed to the left. Kurtosis also measures the shape of a distribution with respect to the tails of a distribution (Westfall, 2014). A distribution with a kurtosis value less than three is called platykurtic and indicates more of the variance is the result of infrequent extreme observations in the tails of the distribution. A distribution with a value of kurtosis greater than three is called leptokurtic and its tails are longer and fatter. A kurtosis value of three indicates tails of a distribution that are similar to a normal distribution.

Table 5.2. Summary and descriptive statistics used to describe the data set. To use these formulas, is a data value and is the number of observations.

Statistic	Measure	Formula
Mean	Average	$$\bar{x}=\frac{1}{n}\sum_{i=1}^{n}x_i$$
Median	Average and the point where 50% of the data is greater than and less than.	Order data from smallest to largest n odd: Middle observation n even: Average of two middle observations
Variance	Dispersion	$$s^2=\frac{1}{n-1}\sum_{i=1}^{n}\left(x_i-\bar{x}\right)^2$$
Skewness	Symmetry Positive: Skewed right Negative: Skewed left	$$\frac{1}{n}\left[\sum_{i=1}^{n}\left(x_i-\bar{x}\right)^3\right]\left[\left(\sqrt{\frac{n}{\sum_{i=1}^{n}\left(x_i-\bar{x}\right)^2}}\right)^3\right]$$
Kurtosis	Tails of a distribution If > 3: Heavy tailed 3: Normal distribution If < 3: Light tailed	$$\frac{1}{n}\left[\sum_{i=1}^{n}\left(x_i-\bar{x}\right)^4\right]\left[\left(\sqrt{\frac{n}{\sum_{i=1}^{n}\left(x_i-\bar{x}\right)^2}}\right)^4\right]$$

Summary statistics are used to assess the shape of the population distribution. Comparing the mean and the median provides information about the direction of skewness. If the mean is larger than the median, this indicates potential skewness to the right. If the mean is less than the median, this indicates potential skewness to the left. If the mean and median are similar, this indicates a potential symmetric distribution.

Interpretation of Statistics for Problem 5.2

This distribution is symmetric since the mean = 1.376 and median = 1.45 are close together and skewness = -0.043 is very close to zero. There is an indication the distribution is platykurtic since the observed value of 2.06 is less than three; however, this could be due to the small sample size with only nine observations.

Differences Between Software Programs

If software is used to calculate summary statistics, then one must be mindful of the imbedded formulas. Some software programs use the normal distribution, which has a kurtosis of three, as the base shape and this is called excess kurtosis. Another modification incorporates the sample standard deviation, , in the calculation. The formula used by Excel in the KURT function is:

$$\text{Excess Kurtosis}=\left[\frac{n(n+1)}{(n-1)(n-2)(n-3)}\left(\sum_{i=1}^{n}\left(\frac{x_i-\bar{x}}{s}\right)^4\right)\right]-\frac{3(n-1)^2}{(n-2)(n-3)}$$

Using Excel to analyze the data in Problem 5.2 results in an excess kurtosis value of -0.65. This formula is also used in other software programs. An open source program called R (R Core Team, 2016) provides three options for calculating kurtosis in the Package e1071 (Meyer et al., 2015). The three equations found in the Help documentation for this package are given in Table 5.3.

The R code to load this package and run all three options (with results presented as comments) is given below. This code is also provided in Appendix B.

```
library(e1071)
kurtosis(x, na.rm=TRUE, type=1)  # -0.9393707
```

```
kurtosis(x, na.rm=TRUE, type=2)  # -0.6464204
kurtosis(x, na.rm=TRUE, type=3)  # -1.371848
```

High Density Data Versus Low Density Data

Data density is a measure of the number of observations per spatial unit. An example of high density data is yield monitor data that is collected at prescribed intervals as a combines harvests a field. Low density data is collected infrequently over a spatial area. Examples include soil tests or scouting reports from a few locations in a field. A way to measure data density is to examine the number of observations per acre.

Spatial Scale and Autocorrelation

Precision agriculture uses data at various spatial scales. The most common is a sub-field scale. Spatial scale is used to describe the size of an area for which measurements are applicable. For example, in Figure 5.1, the corn yield per harvested acre is summarized by counties. The data for each county could be further subdivided into yield for individual farms in a county. And this could be subdivided into fields within farms, and even further into individual yield monitor observations within fields.

Spatial resolution requirements for point samples have three major considerations. First, the strength of the relationships across space. Second, the representation of the spatial information in the existing databases. Third, the spatial resolution needed to implement proposed management practices (Sadler et al., 1998).

Spatial autocorrelation measures the correlation of a single variable with itself at its various locations, that is, as it varies through a landscape. More specifically, spatial autocorrelation introduces a deviation from the assumption of independent observations of classical statistics (Griffith, 2003). Autocorrelation can result from at least two different mechanisms. First, autocorrelation is the result of the scale used to describe the data set. For example, within a soil type or management zone, the yields of each are independent of each other, whereas across soil types the yields exhibit spatial autocorrelation with each other. In this example, yields in the two soil types may be different due to differences in soil texture or water holding capacity. The second type of correlation results from the plant at point A being directly influenced by the plant at point B. For example, if a soybean plant is too close to its neighbor it may grow taller to increase its ability to harvest light. The presence of spatial autocorrelation violates the assumption of independence required by many statistical procedures.

Bolker (2008) shows that spatial autocorrelation reduces the amount of information in the data because measured values at data points are more similar to each other. The total amount of information in the data is smaller than if the data were independent. However, it is this similarity that is exploited by the use of spatial statistics.

Measuring Spatial Autocorrelation with Moran's I Statistic

The most common measure of spatial autocorrelation is Moran's I. It tests for global spatial autocorrelation over a specified study area. Before spatial autocorrelation can be measured, a definition is needed for observations that are close together, or considered neighbors. This can be based on adjacency or distance (Shekhar et al., 2009).

Two common definitions for adjacency are the Rook's and Queen's cases. For the Rook's case, locations are considered adjacent if they share a boundary with positive length [Figure 5.3 (a), (b), (c)]. The Queen's case considers locations adjacent if they not only share a boundary with positive length, but also share a vertex, or corner

Table 5.3. Three definitions of kurtosis used in various software packages.

Option	Equation
Type=1	$g_2 = \frac{1}{n}\left(\sum_{i=1}^{n}(x_i-\bar{x})^4\right)\left[\left(\sqrt{\frac{n}{\sum_{i=1}^{n}(x_i-\bar{x})^2}}\right)^4\right]-3$
Type=2	$G_2 = \left[(n+1)g_2+6\right]\frac{n-1}{(n-2)(n-3)}$
Type=3	$(g_2+3)\left(1-\frac{1}{n}\right)^2-3$

1	2	3
4	5	6
7	8	9

Figure 5.2. Study area consisting of nine locations and their identification numbers.

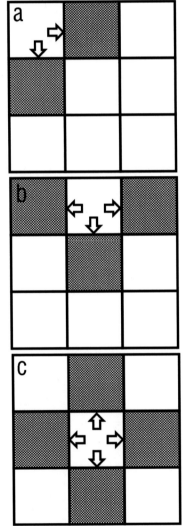

Figure 5.3. Neighbors defined using Rook's Case. (a) Corner location (b) edge location (c) interior location.

[Figure 5.4 (a), (b), (c)]. Neighbors for a given location are identified by a chessboard pattern. Each of these methods is illustrated using a study area that has nine locations arranged in a regular lattice pattern, or grid, as depicted in Figure 5.2. Entries are unique identification (ID) numbers for each location. For example, the northwest location has ID = 1.

It is seen that for the Rook's Case, each corner in a regular grid has two neighbors, each edge has three neighbors, and all interior points have four neighbors.

It is seen that for the Queen's Case, each corner in a regular grid has three neighbors, each edge has five neighbors, and all interior points have eight neighbors.

Neighbors can also be defined based on distance between locations, where the common measure is Euclidian distance. Neighbors are usually based on distance bands defined by the analyst.

By definition, a location cannot be a neighbor with itself. The weights for all pairs of observations may be presented in a weight matrix, denoted by W, with elements w_{ij} that define the relationship between locations where measurements were made. The weight matrix is usually symmetrical with zeroes on the main diagonal. Other elements in an adjacency matrix can be defined as w_{ij} = 0 if locations i and j are neighbors; otherwise, w_{ij} = 0.

The formula for calculating Moran's I is:

$$\text{Moran's } I = \frac{n \sum_{i=1}^{n} \sum_{j=1}^{n} w_{ij} \left(x_i - \bar{x} \right) \left(x_j - \bar{x} \right)}{\left(\sum_{i=1}^{n} \sum_{j=1}^{n} w_{ij} \right) \sum_{i=1}^{n} \left(x_i - \bar{x} \right)^2}$$

where there are n locations and w_{ij} are the weights in the weight matrix w (Rogerson, 2004).

Moran's I is interpreted similarly to Pearson's correlation coefficient as values of Moran's I are bounded by -1 and +1. When Moran's I is negative, this indicates negative spatial autocorrelation and occurs when neighbors have very different values. When Moran's I is positive, this indicates positive spatial autocorrelation and occur when neighbors have similar values. When Moran's I is near zero, this indicates spatial randomness.

Calculations for Moran's I will be illustrated using the location identification numbers as defined in Figure 5.2, measured values in Figure 5.5, and the Rook's case to define neighbors as illustrated in Figure 5.3.

First, the development of the weight matrix, W, that summarizes the spatial relationship between locations. The entries in Table 5.4 are w_{ij} where i = 1, 2, 3, 4, 5, 6, 7, 8, 9 are rows and j = 1, 2, 3, 4, 5, 6, 7, 8, 9 are columns. To get started, we

---

PROBLEM 5.3

A quarter section is 160 acres. For this quarter section, there are 32,000 observations from the yield monitor. What is number of observations per acre?

ANSWER:
The number of observations per acre $= \dfrac{32,000}{160} = 200$.

For this same quarter section, there are measurements for 12 soil samples. What is the number of observations per acre?

ANSWER:
The number of observations per acre $= \dfrac{12}{160} = 0.075$

fix rows at Location 1. Since a location cannot be neighbors with itself, $w_{11} = 0$. Next, we see in Figure 5.5 that Location 1 and Location 2 share a boundary, so they are neighbors and $w_{12} = 1$. Location 1 is also neighbors with Location 4 so $w_{14} = 1$. However, Location 1 is not a neighbor with Locations 3, 5, 6, 7, 8, 9 so $w_{13} = w_{14} = w_{15} = w_{16} = w_{17} = w_{18} = w_{19} = 0$. Entries for the remaining rows are determined in a similar manner.

Next, we calculate the mean of the measurements:

$$\bar{x} = \frac{1}{n}\sum_{i=1}^{n} x_i = \frac{1}{9}(83+90+81+72+65+39+28+41+32)$$
$$= \frac{531}{9} = 59.0$$

Now, we calculate Moran's I for this spatial region:

$$I = \frac{n\sum_{i=1}^{n}\sum_{j=1}^{n} w_{ij}(x_i - \bar{x})(x_j - \bar{x})}{\left(\sum_{i=1}^{n}\sum_{j=1}^{n} w_{ij}\right)\sum_{i=1}^{n}(x_i - \bar{x})^2}$$

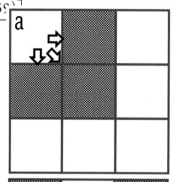

$$= \frac{9\cdot\left[0\cdot(83-59)(83-59)+1\cdot(83-59)(90-59)+\cdots+0\cdot(32-59)(32-59)\right]}{(0+1+\cdots+1+0)\left[(83-59)^2+\cdots+(32-59)^2\right]}$$
$$= \frac{9\cdot(5030)}{(24)(4640)}$$

$$= 0.4065 \cong 0.41$$

Since this value is positive, there is some indication of positive spatial autocorrelation.

The concept of neighbors defined by adjacency can also be applied to locations with irregular shapes that have lines for boundaries. The zones in Figure 5.6 are used to illustrate the construction of the weight matrix and calculations for Moran's I.

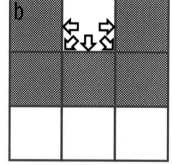

Calculations for Moran's I will be illustrated using the zones and measurement values in Figure 5.6 and the Rook's case to define neighbors.

First, the development of the weight matrix, w, that summarizes the relationship between zones. The entries in Table 5.5 are w_{ij} where $i = 1, 2, 3, 4, 5, 6$ are rows and $j = 1, 2, 3, 4, 5, 6$ are columns. To get started, we fix rows at Zone 1 and see in Figure 5.6 that Zone 1 and Zone 2 share a boundary, so they are neighbors and $w_{12} = 1$. Zone 1 is also neighbors with Zones 3, 4, and 5, so $w_{13} = w_{14} = w_{15} = 1$ and is not a neighbor with Zone 6 so $w_{16} = 0$. The remaining entries are determined in a similar manner.

Table 5.4. Spatial weight matrix, w, using Rook's Case, for Problem 5.4.

		Location							
		1	2	3	4	5	6	7	8	9
Location i	1	0	1	0	1	0	0	0	0	0
	2	1	0	1	0	1	0	0	0	0
	3	0	1	0	0	0	1	0	0	0
	4	1	0	0	0	1	0	1	0	0
	5	0	1	0	1	0	1	0	1	0
	6	0	0	1	0	1	0	0	0	1
	7	0	0	0	1	0	0	0	1	0
	8	0	0	0	0	1	0	1	0	1
	9	0	0	0	0	0	1	0	1	0

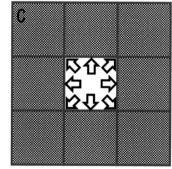

Figure 5.4. Neighbors defined using Queen's Case. (a) Corner location (b) Edge location (c) Interior location.

Next, we calculate the mean of the measurements:

$$\bar{x}=\frac{1}{n}\sum_{i=1}^{n}x_i=\frac{1}{6}(32+26+19+18+17+14)=21.0$$

Now we calculate Moran's I for this spatial region:

$$I=\frac{n\sum_{i=1}^{n}\sum_{j=1}^{n}w_{ij}\left(x_i-\bar{x}\right)\left(x_j-\bar{x}\right)}{\left(\sum_{i=1}^{n}\sum_{j=1}^{n}w_{ij}\right)\sum_{i=1}^{n}\left(x_i-\bar{x}\right)^2}$$

$$=\frac{6\cdot\left[0\cdot(32-21)(32-21)+1\cdot(32-21)(26-21)+\cdots+0\cdot(14-21)(14-21)\right]}{(0+1+\cdots+1+0)\left[(32-21)^2+\cdots+(14-21)^2\right]}$$

$$=\frac{6\cdot100}{18\cdot224}$$

$$=0.1488\cong0.15$$

83	90	81
72	65	39
28	41	32

Figure 5.5. Map of an attribute measured over a field.

Because this value is positive, there is some indication of positive spatial autocorrelation.

If spatial data does not exhibit spatial autocorrelation, then statistical methods that require independent observations, such as Analysis of Variance (ANOVA) or regression analysis, may be used for analyses. An example where there is lack of significant spatial autocorrelation is given by Sudduth et al. (2003), who apply Pearson correlation coefficients and regression analysis of profile soil electrical conductivity (E_Ca), as measured by two types of instruments, to soil physical and chemical properties. Chun and Griffith (2013) use ANOVA to analyze farm density of five agricultural administrative regions in Puerto Rico and verify that all assumptions are met.

The Semivariogram

The semivariogram is a way to characterize the spatial structure of data that varies continuously across a landscape (Plant, 2012). Semivariograms can be used

PROBLEM 5.4

Calculate Moran's I for the data in Figure 5.5.

Location ID	Response
1	83
2	90
3	81
4	72
5	65
6	39
7	28
8	41
9	32

ANSWER:
(See calculations below and R code in Appendix B.) $I = 0.41$

to assess the autocorrelation between the data points as well as provide information needed to adequately sample the system. It is based on spatial lag, which subdivides the distances between all locations with measurements into distinct groups. If the spatial correlation structure is the same in all directions, this is referred to as isotropic.

Two examples are given to illustrate the concepts and calculations for semi-variograms. The first example has four sample points along a straight-line transect (Figure 5.7) and the second example has four points arranged in a regular grid (Figure 5.9).

This information is also presented as a table in Problem 5.6 to illustrate mathematical notation used in various formulas.

The following equation is used for the experimental semivariogram, using the hat "^" symbol to indicate this is a value estimated from data:

$$\hat{\gamma}(h_j)=\frac{1}{2N_{h_j}}\sum_{i=1}^{N_{h_j}}\left[Z(s_i)-Z(s_{i+j})\right]^2$$

where
Z is the measured quantity,
s_i is the position along the transect,
h_j are distances between points, and
N_{h_j} is the number of pairs of points for a given spatial distance hjThis calculation is based on distance groups, each of which contains all of the pairs of points separated by a specified distance, and carrying out the summation within these distance groups. For this example, the distances are $h_1 = 1$, $h_2 = 2$, and $h_3 = 3$. The distance between two spatial points is defined as:

$$\text{Distance}=d_{ij}=\sqrt{\left(x_i-x_j\right)^2}\quad \text{for}\ i\neq j$$

Results for distance calculations for each pair of points and distance groups are presented in Table 5.6. Calculations for the first spatial distance group are:

$$\hat{\gamma}(h_1)=\frac{1}{2(3)}\sum_{i=1}^{3}\left[Z(s_i)-Z(s_{i+1})\right]^2=\frac{1}{6}\left[(123-124)^2+(124-113)^2+(113-108)^2\right]$$

$$=\frac{1}{6}\left(1^2+11^2+5^2\right)=\frac{1}{6}\left(1+121+25\right)$$

$$= 24.5$$

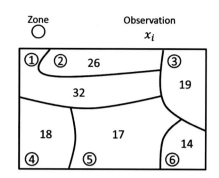

Zone ○   Observation x_i

Figure 5.6. Map of an attribute measured over a field.

Table 5.5. Spatial Weight Matrix, *w*, using Rook's Case, for Problem 5.5.

		Zone					
		1	2	3	4	5	6
	1	0	1	1	1	1	0
	2	1	0	1	0	0	0
Zone	3	1	1	0	0	1	1
	4	1	0	0	0	1	0
	5	1	0	1	1	0	1
	6	0	0	1	0	1	0

PROBLEM 5.5

Calculate Moran's I for the data in Figure 5.6.

Zone	Response
1	32
2	26
3	19
4	18
5	17
6	24

ANSWER:
(See calculations below and R code in Appendix B.) $I = 0.1488$

Calculations for the second spatial distance group $h_2 = 2$ are:

$$\hat{\gamma}(h_2) = \frac{1}{2(2)} \sum_{i=1}^{2} \left[Z(s_i) - Z(s_{i+2}) \right]^2 = \frac{1}{4} \left[(123-113)^2 + (124-108)^2 \right]$$

$$= \frac{1}{4}(10^2 + 16^2) = \frac{1}{4}(100+256) = 89.0$$

Calculations for the third spatial distance group $h_3 = 3$ are:

$$\hat{\gamma}(h_2) = \frac{1}{2(1)} \sum_{i=1}^{1} \left[Z(s_i) - Z(s_{i+3}) \right]^2 = \frac{1}{2} \left[(123-108)^2 \right]$$

$$= \frac{1}{2}(15^2) = \frac{1}{2}(225) = 112.5$$

The empirical semivariogram for the spatial data in Figure 5.7 is shown in Figure 5.8. Even though there are only four spatial points along a straight-line transect, it is seen that locations further apart exhibit more variation than observations closer together.

Huang et al. (2001) use a 400-m long transect to collect 40 soil samples on a 10 meter spacing to characterize the spatial variation of soil samples. Soil properties investigated included pH, available phosphorus, and soil total carbon.

This second example examines the mathematics used to create semivariograms using the spatial region in Figure 5.9. There are four locations with measured values that will be used to illustrate calculations. The coordinate axes are labeled and with integer increments to simplify calculations. Locations are represented by open circles and values of the measured variable are given for each location.

The following equation is used for the experimental semivariogram estimated from data based on distance lag intervals:

$$\hat{\gamma}(h_j) = \frac{1}{2N_{h_j}} \sum_{i=1}^{N_{h_j}} \left[Z(s_i) - Z(s_i + h_j) \right]^2$$

where

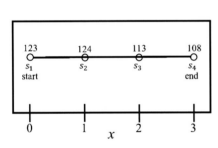

Z is the measured quantity,
s_i is the position vector with coordinates, (x_i, y_i)
h_j are spatial lag intervals, and
N_{h_j} is the number of pairs of points for a given distance interval.

This calculation is based on lag groups, each of which contains all of the lag vectors whose magnitude lies within a specified interval, and carrying out the summation of semi-variance within these lag groups. For this example, the lag groups are $h_1 = 1 \le h < 1.25$ and $h_2 = 1.25 \le h < 1.5$. Distance between spatial locations is defined as:

Figure 5.7. Data for Semivariogram calculations, points along a transect.

$$\text{Distance} = \sqrt{(x_i - x_j)^2 + (y_i - y_j)^2} \quad \text{for} \quad i \neq j$$

Results for distance calculations for each pair of points and spatial lag intervals are presented in Table 5.7. Calculations for the first spatial lag interval h1 = 1 £ h < 1.25 are:

$$\hat{\gamma}(h_1) = \frac{1}{2(8)} \sum_{i=1}^{8} \left[Z(s_i) - Z(s_i + h) \right]^2 = \frac{1}{16} \left(11^2 + 1^2 + 11^2 + 5^2 + 1^2 + 15^2 + 15^2 + 5^2 \right)$$

$$= \frac{1}{16}(121 + 1 + 121 + 25 + 1 + 225 + 225 + 25)$$

$$= 46.5$$

Calculations for the second spatial lag interval $h_2 = 1.25 \leq h < 1.5$ are:

$$\hat{\gamma}(h_2) = \frac{1}{2(4)} \sum_{i=1}^{4} \left[Z(s_i) - Z(s_i + h) \right]^2 = \frac{1}{8} \left(16^2 + 10^2 + 10^2 + 16^2 \right)$$

$$= \frac{1}{8}(256 + 100 + 100 + 256)$$

$$= 89.0$$

The semivariogram for the spatial data in Figure 5.9 is shown in Figure 5.10. Even though there are only four spatial locations in a grid with two spatial lag intervals, it is seen that observations closer together exhibit less variation than observations further apart.

Figure 5.8. Empirical semivariogram for points along a transect.

Table 5.6. Calculations for Spatial Distance Groups

Distance Group	Points	Distance	$Z(s_i)$	$Z(s_i + j)$	Difference	(Difference)2
			$h_1 = 1, N_{h_1} = 3$			
h_1	s_1, s_2	1	123	124	1	1
h_1	s_2, s_3	1	124	113	11	121
h_1	s_3, s_4	1	113	108	5	25
			$h_2 = 2, N_{h_2} = 2$			
h_2	s_1, s_2	2	123	113	10	100
h_2	s_2, s_4	2	124	108	16	256
			$h_3 = 3, N_{h_3} = 1$			
h_3	s_1, s_4	3	123	108	15	225

Table 5.7. Calculations for Spatial Lag Intervals.

Lag Group	Points	Distance	$Z(s_i)$	$Z(s_i + h)$	Difference	(Difference)^2
		spatial lag interval $h_1 = 1 \leq$ Distance < 1.25				
h_1	(1,1), (2,1)	1	124	113	11	121
h_1	(1,1), (1,2)	1	124	123	1	1
h_1	(2,1), (1,1)	1	113	124	11	121
h_1	(2,1), (2,2)	1	113	108	5	25
h_1	(1,2), (1,1)	1	123	124	1	1
h_1	(1,2), (2,2)	1	123	108	15	225
h_1	(2,2), (1,2)	1	108	123	15	225
h_1	(2,2), (2,1)	1	108	113	5	25
		spatial lag interval $h_2 = 1.25 \leq$ Distance < 1.5				
h_2	(1,1) (2,2)	1.414	124	108	16	256
h_2	(2,1), (1,2)	1.414	113	123	10	100
h_2	(1,2), (2,1)	1.414	123	113	10	100
h_2	(2,2), (1,1)	1.414	108	124	16	256

PROBLEM 5.7

Calculate the semivariogram for the data in Figure 5.9.

x	y	Measured value
1	1	124
2	1	113
1	2	123
2	2	108

ANSWER: (See calculations below.)

Spatial lag interval	Semivariance
$h_1 = 1.00 \leq h < 1.25$	46.5
$h_2 = 1.25 \leq h < 1.50$	89.0

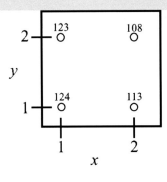

Figure 5.9. Data for semivariogram calculations for spatial lag intervals, points in a grid.

Interpreting Semivariogram Information

Semivariogram results can be presented graphically with spatial lag, on the horizontal axis and the corresponding experimental semi-variance, $\hat{\gamma}(h_j)$, on the vertical axis. The horizontal axis has the same units as x and y and the vertical axis has units equal to the units squared of the measured variable. Figure 5.11 is an example of a theoretical semivariogram with important properties labeled. The nugget represents the combination of sampling error and short-range variability at a scale finer than the sample spacing, the range represents the spatial lag at which spatial variability reaches a constant value (pairs of points that are at least this distance apart are not spatially correlated), and the sill represents long range spatial variability (Plant, 2012). A spherical model is often used to fit the empirical semivariogram as it rises at first and then levels off for distances beyond the range. Note that the semivariogram is based on the separation between points, not on the specific locations.

Sampling Strategies for Gathering Data

Information obtained from the semivariogram can be used to obtain meaningful information about the parameter being measured. Spatial points that are closer than the range exhibit spatial autocorrelation and therefore are not totally independent. Thus, one strategy would have all points farther apart than the range. Using this strategy the points will be spatially independent. The second strategy is to have points closer together than the range, and then use the spatial autocorrelation between the points to estimate values at unmeasured locations. In this approach, the additional information about the spatial autocorrelation between points is used to reduce the errors in predicted values at locations without measured values.

Identifying Patterns in the Data

Trends can be detected by using a directional semivariogram. If the spatial structure exhibits patterns in different directions, this is referred to as anisotropic. To check for trends, or directional dependence, in an empirical semivariogram requires simultaneously calculating semivariance values for data pairs contained within directional bands and prescribed lag intervals. This will result in fewer data pairs used to estimate semivariance for each direction and lag interval combination. An anisotropic empirical semivariogram will reach the sill more rapidly in some directions than others.

Software that will calculate directional semivariograms include ArcGIS Geostatistical Analyst and R packages gstat and geoR.

Interpolation Methods

Suppose that we have obtained seven soil samples from various locations within a field, such as those mapped in Figure 5.12(a). However, for precision farming purposes, we need to estimate the value at the point labeled with a question mark. We can do this using spatial interpolation methods that are used to predict values that are within the region where data has been collected, as shown in Figure 5.12(b). Predictions are improved by incorporating knowledge about spatial autocorrelation, since knowing the value at a location provides information about nearby locations. Three interpolation methods will be examined for creating maps. These are *Inverse Distance Weighted (IDW)*, *Kriging*, and *Co-Kriging*. The data in Figure 5.12 will be used to illustrate the calculations for the Inverse Distance Weighted and Ordinary Kriging methods. The purpose of the following examples is to examine the mathematics used for the different interpolation methods. The spatial region in Figure 5.12 consists of measured values for seven locations that will be used to illustrate calculations for predicting the value for the point z_o, indicated with a question mark, where data was not collected. The coordinate axes are labeled and with integer increments to simplify calculations. We will also assume the field is small enough to use Euclidean distance. Locations are represented by open circles and values of the measured variable are given for each location.

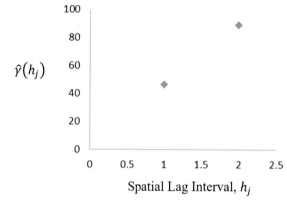

Figure 5.10. Empirical semivariogram for points in a grid.

Inverse Distance Weighted

The Inverse Distance Weighted (IDW) interpolation method is based on the assumption that the value at the prediction location is a distance-weighted average of the values from surrounding spatial data points that have measured values. The spatial points closest to the prediction location have greater influence on the predicted value than those farther away. This is accomplished by weighting each spatial point by the inverse of its distance from the prediction location. The IDW method is referred to as deterministic since there is not an underlying statistical model. Variations of this approach include the squared inverse distance or the cubed inverse distance. The formula for the IDW interpolated value is

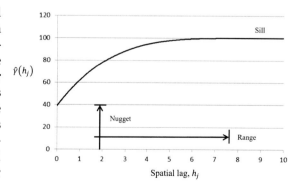

Figure 5.11. Theoretical semivariogram.

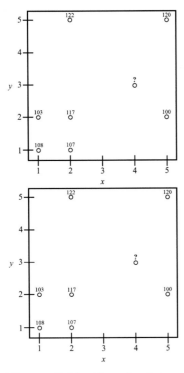

Figure 5.12. Data for interpolation calculations. (a) Measured values (b) Interpolation versus extrapolation. This data is also presented in Table 5.8.

$$z_0 = \frac{\sum_{i=1}^{s}\left(\dfrac{z_i}{d_i^k}\right)}{\sum_{i=1}^{s}\left(\dfrac{1}{d_i^k}\right)}$$

where

z_i is the measured valued at known point i,

d_i is the distance between point i and 0,

s is the number of points with measured values used in prediction, and

k is the specified power for distance, $k = 2$ is the value usually used (Atkinson and Lloyd, 2009).

The critical calculation for IDW is the distance from each spatial point with measured values to the point of prediction. Distance calculations are given in Table 5.9.

where $d_{0,i} = \sqrt{(x_0 - x_i)^2 + (y_0 - y_i)^2}$ is the distance between spatial points z_0 and z_i. The IDW interpolated value for z_0 is

$$z_0 = \frac{\dfrac{108}{13} + \dfrac{107}{8} + \dfrac{103}{10} + \dfrac{117}{5} + \dfrac{122}{8} + \dfrac{120}{5} + \dfrac{100}{2}}{\dfrac{1}{13} + \dfrac{1}{8} + \dfrac{1}{10} + \dfrac{1}{5} + \dfrac{1}{8} + \dfrac{1}{5} + \dfrac{1}{2}} = \frac{144.6327}{1.3269} = 109.0$$

That is, the predicted value for z_0 is 109.0 using IDW.

Kriging

Kriging is a geostatistical method that estimates both the predicted value and its standard error. Three common methods of kriging are simple, ordinary, and co-kriging. The method of simple kriging is used if the mean of the measured values is constant across the region of interest. This method is often unsuitable for many precision agriculture applications. The method of ordinary kriging allows the mean to vary spatially and will be used to predict the value of an unsampled location that lies within the spatial region of points with measured values. This is accomplished by limiting the stationarity of the mean to the local neighborhood of the unsampled location (Goovaerts, 1997). Among all linear interpolation methods, kriging is the one without systematic prediction error and with the smallest prediction variance. Weights for sampled locations are based on the distance to the unsampled location and the spatial relationships among the measured values around the unsampled location using the semivariogram. In order to maintain statistical properties, a model is fit to the empirical semivariogram. For illustrating mathematical calculations, a spherical model is used to determine semivariogram values for various distances and is the most widely used model (Atkinson and Lloyd, 2009). The spherical model is given by

$$\gamma(h) = \begin{cases} s\left[1.5\left(\dfrac{h}{r}\right) - 0.5\left(\dfrac{h}{r}\right)^3\right] & if\ h \le r \\ s & if\ h > r \end{cases}$$

where h is the spatial lag, s is the sill, and r is the range.

A spherical model that incorporates a nugget is given by:

$$\gamma(h) = \begin{cases} 0 & h = 0 \\ n + p\left[1.5\left(\dfrac{h}{r}\right) - 0.5\left(\dfrac{h}{r}\right)^3\right] & 0 < h \le r \\ p + n & h > r \end{cases}$$

where is the spatial lag, is the nugget, is the partial sill, the total sill is , and is the range. Parameters for both of these models can be estimated using any software that performs nonlinear estimation and modeling.

Other models used to fit the sample variogram include the exponential, Gaussian, and linear plateau models. Constraints for a semi-variogram model include monotonically increasing, the sill reaches an asymptotic max, the intercept is nonnegative which means the nugget is greater than or equal to zero, and isotropic which means the semi-variogram depends only on the lag distance and not on direction (Atkinson and Lloyd, 2009).

The statistical model for ordinary kriging is given by

$$z_0 = \sum_{i=1}^{k} \lambda_i z_i$$

where z_0 is the prediction point,

z_i is the measured valued at point i,

λ_i is the weight for point i, and

k is the number of points with measured values used in prediction.

To insure an unbiased prediction the weights are constrained to sum to one, that is

$$\sum_{i=1}^{k} \lambda_i = 1$$

The weights are determined using the coefficients of the model for the semivariogram and will minimize the kriging prediction error through the Method of Lagrange Multipliers. The Method of Lagrange Multipliers (Stewart, 2016) is a technique to solve optimization problems that have restrictions or constraints on the values that produce the optimal solution.

The kriging variance is an estimate of the prediction variance and is given by

$$\hat{\sigma}^2 = \sum_{i=1}^{k} \lambda_i \gamma\left[(x_i, y_i) - (x_0, y_0) \right] + \psi$$

where x_i, y_i are the x and y coordinates for point i, the points of measured values,
x_0, y_0 are the x and y coordinates for point 0, the prediction point,

$\gamma\left[(x_i, y_i) - (x_0, y_0)\right]$ is the semivariogram model evaluated at the distance between the two points,
λ_i is the weight for point i,
k is the number of points with measured values used in prediction, and
Y is the Lagrange multiplier (Atkinson and Lloyd, 2009).

Detailed instructions for creating an Excel spreadsheet to solve this problem are provided in Appendix A. The calculations for ordinary kriging and its estimated variance are achieved through the following steps:

1. Create a data table using the information in Table 5.8,
2. Fit a model to the empirical semivariogram or use a specified model (this example uses a spherical model with all parameter values specified),
3. Develop a $k \times k$ matrix of distances between all pairs of points with measured values,
4. Compute the matrix of semi-variances for each distance from Step 3 using the parameters for the semivariogram model,
5. Incorporate the Lagrange multiplier by appending a row of 1s to the bottom of the matrix of semivariances, a column of 1s to the right side of the matrix of semivariances, and a zero as the $(k + 1)$th diagonal element,
6. Compute the inverse of the $(k + 1) \times (k + 1)$ augmented matrix of semi-variances from Step 5,
7. Compute the vector of distances of each point with measured values, z_i, from the prediction point, z_0,
8. Compute a vector of semivariances using the vector of dis-

Table 5.8. Data for interpolation methods.

Spatial point i	x coordinate	y coordinate	Measured value (z_i)
1	1	1	108
2	2	1	107
3	1	2	103
4	2	2	117
5	2	5	122
6	5	5	120
7	5	2	100
0	4	3	?

tances from Step 7 and the semivariogram model. Then augment with a 1 as the $(k+1)$th element as the constraint for the Lagrange multiplier,

9. Compute the weights λ_i and the Lagrange multiplier by multiplying the inverse of the augmented matrix of semi-variances from Step 6 on the right by the augmented vector of semi-variances from Step 8 (note that this is a vector of length $k+1$),

10. Interpolate the value of by multiplying the transpose of the column vector of measured values by the column vector comprised of the first k elements of the vector from Step 9,

11. Compute the kriging variance by multiplying the augmented vector of semi-variances from Step 8 (as a $1 \times (k+1)$ row vector) by the augmented vector of weights in Step 9 (as a $(k+1) \times 1$ column vector).

The spreadsheet created will solve for the kriged value and its prediction variance for seven points with measured values, a spherical model with nugget as a model for the empirical semivariogram, and one prediction point without a measured value.

Required entries for the spreadsheet are shown in Figure 5.13 (a) and (b). Data for spatial location and measured values are entered in Columns B, C, and D and Rows 5 through 11. This information for the point of prediction is entered in Row 12. Parameter values for the semivariogram spherical model with nugget are entered in Column B and Rows 25 through 27. All calculations in the spreadsheet are carried out using formulas and functions. The interpolated value is in Column C and Row 68. The kriging variance is in Column C and Row 70.

Co-kriging

Co-kriging is applicable when additional observations of a covariable are incorporated into the prediction procedure. It is most useful when the covariable, which is highly spatially correlated with the measured variable, is sampled at a higher density than the measured variable (Stein and Corsten, 1991; Goovaerts, 1997). A more general formulation was developed based on the pseudo-cross-variogram that incorporates all spatial data even though there may not be any sample locations where data is available for both variables (Myers, 1991). For co-kriging to be beneficial, the covariable should be cheaper to obtain or more readily available (Atkinson and Lloyd, 2009; Webster and Oliver, 2007). The co-kriging estimator always has error variance that is less than or equal to kriging error variance since it incorporates additional information (Goovaerts, 1997).

Vaughan et al. (1995) provide a detailed analysis for apparent electrical conductivity measured at the surface with electrical conductivity of soil paste extract of soil samples to predict soil salinity at unsampled points. Other information that can be used in co-kriging includes soil reflectance, elevation, and yield measurements.

Other Kriging Methods

Mitas and Mitasova (1999) mention other kriging methods such as disjunctive kriging, zonal kriging, and

Table 5.9. Distance calculations from prediction point to points with measured values.

Spatial Points	Distance = $d_{0,i}$	$(d_{0,i})^2$
0,1	$d_{0,1}=\sqrt{(4-1)^2+(3-1)^2}=3.6056$	13
0,2	$d_{0,2}=\sqrt{(4-2)^2+(3-1)^2}=2.8284$	8
0,3	$d_{0,3}=\sqrt{(4-1)^2+(3-2)^2}=3.1623$	10
0,4	$d_{0,4}=\sqrt{(4-2)^2+(3-2)^2}=2.2361$	5
0,5	$d_{0,5}=\sqrt{(4-2)^2+(3-5)^2}=2.8284$	8
0,6	$d_{0,6}=\sqrt{(4-5)^2+(3-5)^2}=2.2361$	5
0,7	$d_{0,7}=\sqrt{(4-5)^2+(3-2)^2}=1.4142$	2

	A	B	C	D
		x	y	Measured
4	Spatial Point	coordinate	coordinate	Value
5	1	1	1	108
6	2	2	1	107
7	3	1	2	103
8	4	2	2	117
9	5	2	5	122
10	6	5	5	120
11	7	5	2	100
12	0	4	3	?

	A	B
25	range-	6
26	nugget=	10
27	partial sill=	90

	A	B	C
68	Interpolated Value		2
69	1	-0.023	0.0105
70	Kriging Variance		-0.024

Figure 5.13. Required Spreadsheet Entries for Kriging Calculations (a) Spatial location and measured values (b) Parameter values for semivariogram model.

spatio-temporal kriging. Moral et al. (2010) list methods that can be used when secondary data is exhaustive which include kriging with external drift, factor kriging analysis, and regression kriging. Hengl (2009) devotes a chapter to regression kriging and uses examples from various disciplines.

Other Interpolation Methods

Other deterministic methods include natural neighbor interpolation and interpolation based on a triangulated irregular network (TIN). There is also a variational approach that uses splines and often has a physical interpretation (Mitas and Mitasova, 1999).

Strengths and Weaknesses of Inverse Distance Weighting, Kriging, and Co-kriging

The major strength of inverse distance weighting is its ability to predict using only measured values at sampled locations. It can be used when it is not possible to perform statistical analyses to determine the underlying spatial variability. Weaknesses include its inability to reproduce the local shape implied by data and the introduction of local extrema at some data points (Mitas and Mitasova, 1999).

The major strength of kriging and co-kriging is their foundation on statistical models that minimize the errors at unmeasured locations. Kriging and co-kriging will be similar when variables are weakly correlated (Goovaerts, 1997). Kriging will not work well if the mean of the measured variable changes markedly over small distances. The most widely used approach to prediction where the mean is non-stationary is kriging with a trend model (Atkinson and Lloyd, 2009). A study using kriging and co-kriging to characterize soil properties across a field found that co-kriging did not consistently and substantially improve the characterization accuracy of soil variability. The improvement was greater for soil variables more highly correlated with the covariable soil electrical conductivity (Tarr et al., 2005). The main strengths of kriging are in the statistical quality of its unbiased predictions and its ability to predict the spatial distribution of uncertainty (Mitas and Mitasova, 1999).

Determining Errors

A common method to test kriging estimates is cross-validation (Olea, 1999). The most common method is leave-one-out validation. In the kriging literature this method is often called bootstrapping or the jackknife approach. We can apply this concept to Problem 5.9 by leaving out a location for a measured value and using the remaining six locations to predict z_0', utilizing the same semi-variogram model. The differences between each predicted value using all the data and the predicted value using subsets are squared and summed to determine the root mean square error. If the average squared difference is small, then the kriging model is doing a good job at predicting the unsampled location. For semivariogram modeling, cross-validation consists of removing one datum at a time and re-estimating this value from the remaining data using the different semivariogram models. Interpolated and actual values are compared and the semivariogram model is retained that results in the most accurate predictions (Goovaerts, 1997).

PROBLEM 5.10

Calculate the mean, median, variance, skewness, and kurtosis for the following data obtained by measuring percent moisture on random samples from a shipment of grain:

4.64, 3.59, 5.44, 2.29, 6.57, 9.92, 4.39, 10.90.

PROBLEM 5.11

A quarter section is 160 acres. For this quarter section, there are 8,000 observations on soil electrical conductivity in milliSiemens per meter (mS m^{-1}). What is the number of observations per acre?

PROBLEM 5.12

Calculate Moran's I using the Queen's case for the data in the following figure:

83	90	81
72	65	39
28	41	32

PROBLEM 5.13

Calculate Moran's I using the Rook's case for the data in the following figure:

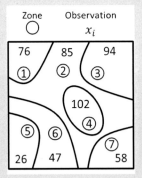

PROBLEM 5.14

Calculate the empirical semivariogram for the following data:

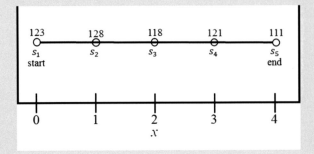

PROBLEM 5.15

Calculate the empirical semivariogram for the following data:

x	y	Measured value
1	1	92
2	1	83
3	1	92
1	2	75
2	2	105
3	2	99

PROBLEM 5.16

Map the following data and calculate the Inverse Distance Weighted interpolated value for z_0 in the following table:

Spatial point i	x coordinate	y coordinate	Measured values (z_i)
1	1	1	33
2	2	5	66
3	1	7	41
4	5	2	48
5	7	3	86
6	6	6	80
7	4	8	77
8	8	1	101
9	8	8	88
0	3	3	?

PROBLEM 5.17

Test the kriging prediction in Problem 5.6 by using cross-validation and the same semi-variogram model. For practical purposes, only do this for one measured location, Spatial Point 6. What is the difference between the predicted value of z_0 using all seven points and the predicted value of   using the remaining six points?

PROBLEM 5.18

Map the data in the following table and calculate the kriging interpolated value for z_0 and the kriging variance. Use a spherical semivariogram model with a nugget and use parameter values of the range $r = 10$, the nugget is $n = 50$, the partial sill is $p = 100$, and the total sill is: $s = p + n = 100 + 50 = 150$.

Spatial point i	x coordinate	y coordinate	Measured values (z_i)
1	2	3	103
2	1	4	131
3	6	3	120
4	3	5	125
5	6	7	101
6	6	8	110
7	9	6	120
0	6	5	?

APPENDIX A

Using Excel for Problem 5.9: Kriging Calculations

Detailed Instructions for Creating Spreadsheet

Using ordinary kriging and a semivariogram spherical model with nugget, calculate the interpolated value for z_0 in Figure 5.12 and the kriging variance. Use the following parameters: the range is $r = 6$, the nugget is $n = 10$, the partial sill is $p = 10$, and the total sill is $s = p + n = 90 + 10 = 100$.

1. Create a data table. The data have been entered into Excel as shown below, where the gray shaded cells represent rows and columns. Note that $k = 7$ since there are seven points with measured values.

	A	B	C	D
4	Spatial Point	x coordinate	y coordinate	Measured Value
5	1	1	1	108
6	2	2	1	107
7	3	1	2	103
8	4	2	2	117
9	5	2	5	122
10	6	5	5	120
11	7	5	2	100
12	0	4	3	?

2. Fit a model to the empirical semivariogram or use a specified model. For this example, a spherical model that incorporates a nugget is used with the following coefficients: the range is $r = 6$, the nugget is $n = 10$, the partial sill is $p = 90$, and the total sill is $s = p + n = 90 + 10 = 100$. This information is in the spreadsheet as Parameters for semivariogram spherical model:

	A	B
25	range=	6
26	nugget=	10
27	partial sill=	90

3. Develop a matrix of distances between all pairs of points with measured values. In general, the formula for computing distance between two spatial points (x_i, y_i) and (x_j, y_j) is

`=SQRT((xi-xj)^2+(yi-yj)^2).`

Note that all values on the main diagonal are zero. As an example, the following formula is in cell B19 and computes the distance between spatial point 1 and spatial point 4:

`=SQRT(($B$5-$B8)^2+($C$5-$C8)^2).`

	A	B	C	D	E	F	G	H
15	Spatial Point	1	2	3	4	5	6	7
16	1	0	1	1	1.414	4.123	5.657	4.123
17	2	1	0	1.414	1	4	5	3.162
18	3	1	1.414	0	1	3.162	5	4
19	4	1.414	1	1	0	3	4.243	3
20	5	4.123	4	3.162	3	0	3	4.243
21	6	5.657	5	5	4.243	3	0	3
22	7	4.123	3.162	4	3	4.243	3	0

4. Compute the matrix of semi-variances for each distance from Step 3 using the coefficients for the semi-variogram model from Step 2. As an example, the formula in cell B36 is =$B$26+$B$27*(1.5*($B19/$B$25)-0.5*($B19/$B$25)^3)

	A	B	C	D	E	F	G	H
32	Spatial Point	1	2	3	4	5	6	7
33	1	0	32.2917	32.2917	41.2305	88.1672	99.5669	88.1672
34	2	32.2917	0	41.2305	32.2917	86.6667	96.4583	74.5632
35	3	32.2917	41.2305	0	32.2917	74.5632	96.4583	86.6667
36	4	41.2305	32.2917	32.2917	0	71.875	89.5495	71.875
37	5	88.1672	86.6667	74.5632	71.875	0	71.875	89.5495
38	6	99.5669	96.4583	96.4583	89.5495	71.875	0	71.875
39	7	88.1672	74.5632	86.6667	71.875	89.5495	71.875	0

5. Incorporate the Lagrange multiplier by appending a row of 1s to the bottom of the matrix of semi-variances from Step 4, a column of 1s to the right side of this matrix, and a zero as the th diagonal element:

	A	B	C	D	E	F	G	H	I
32	Spatial Point	1	2	3	4	5	6	7	Lagrange
33	1	0	32.2917	32.2917	41.2305	88.1672	99.5669	88.1672	1
34	2	32.2917	0	41.2305	32.2917	86.6667	96.4583	74.5632	1
35	3	32.2917	41.2305	0	32.2917	74.5632	96.4583	86.6667	1
36	4	41.2305	32.2917	32.2917	0	71.875	89.5495	71.875	1
37	5	88.1672	86.6667	74.5632	71.875	0	71.875	89.5495	1
38	6	99.5669	96.4583	96.4583	89.5495	71.875	0	71.875	1
39	7	88.1672	74.5632	86.6667	71.875	89.5495	71.875	0	1
40	Lagrange	1	1	1	1	1	1	1	0

6. Compute the inverse of the augmented matrix of semi-variances from Step 5:

The function in Excel for matrix inversion is MINVERSE.

(i) Highlight the area where the matrix inverse will be located (must be same size as original matrix).

(ii) In the formula bar enter =MINVERSE(

(iii) Select the matrix to be inverted. This will populate the formula with the range of cells.

(iv) End the formula with a closing parenthesis =MINVERSE(B33:I40).

fx {=MINVERSE(b33:I40)}

(v) Press Ctrl+Shift+Enter on the keyboard to indicate we are working with a matrix.

(vi) The formula will be enclosed in curly braces

(vii) The inverse of this matrix is below.

	A	B	C	D	E	F	G	H	I
32	Spatial Point	1	2	3	4	5	6	7	Lagrange
33	1	-0.023	0.0105	0.0105	0.0011	-0.0001	0.0012	-0.0001	0.184
34	2	0.0105	-0.024	0.0013	0.0094	-0.0003	0	0.0031	0.1013
35	3	0.0105	0.0013	-0.024	0.0094	0.0031	0	-0.0003	0.1013
36	4	0.0011	0.0094	0.0094	-0.0253	0.0026	0.0001	0.0026	-0.0398
37	5	-0.0001	-0.0003	0.0031	0.0026	-0.0107	0.0045	0.0009	0.2139
38	6	0.0012	0	0	0.0001	0.0045	-0.0103	0.0045	0.2254
39	7	-0.0001	0.0031	-0.0003	0.0026	0.0009	0.0045	-0.0107	0.2139
40	Lagrange	0.184	0.1013	0.1013	-0.0398	0.2139	0.2254	0.2139	-65.0534

7. Compute the vector of distances of each point with measured values from the prediction point and augment with a 1 as the constraint for the Lagrange multiplier. As an example, the formula in Cell B62 is =SQRT(($B$12-$B8)^2+($C$12-$C8)^2).

8. Compute the vector of semi-variances using the vector of distances from Step 7 and the semi-variogram model. Then augment with a 1 as the constraint for the Lagrange multiplier. As an example, the formula in Cell C62 is =$B$26+$B$27*(1.5*($B62/$B$25)-0.5*($B62/$B$25)^3).

The results of Step 7 and Step 8 are shown below.

	A	B	C
	Spatial Point	Distance	Semivariances
58			
59	1	3.605	81.360
60	2	2.828	68.926
61	3	3.162	74.563
62	4	2.236	57.982
63	5	2.828	68.923
64	6	2.236	57.982
65	7	1.414	41.231
66	Constraint	1	1

9. Compute the weights λ_i and the Lagrange multiplier by multiplying the inverse of the augmented matrix of semi-variances from Step 6 on the right by the augmented vector of semi-variances from Step 8.

The function in Excel for matrix multiplication is MMULT.

(i) Highlight the area where the product of the two vectors will be located.

(ii) In the formula bar enter =MMULT(

(iii) Select the matrix from Step 6. This will populate the formula with the range of cells for the inverse matrix.

(iv) Select the vector created in Step 8.

(v) End the formula with a closing parenthesis =MMULT(B46:I53,C59:C66).

(vi) Press Ctrl+Shift+Enter on the keyboard to indicate we are working with a matrix.

(vii) The formula will be enclosed in curly brackets fx | {=MMULT(B46:I53,C59:C66)}

(viii) The result is shown at right:

	H
	Lambda Weights
58	
59	-0.0635
60	0.0507
61	0.0029
62	0.2296
63	0.1255
64	0.2266
65	0.4283
66	-1.2222

10. Interpolate the value for by multiplying the transpose of the column vector of measured values by the column vector of weights λ_i comprised of the first $k = 7$ elements of the vector in Step 9. This is accomplished by transposing the measured values to a row vector and multiplying on the right by the column vector of weights. The function MMULT is used to multiply the two vectors.

The function in Excel for matrix multiplication is MMULT.

(i) In the formula bar enter =MMULT(TRANSPOSE(

(ii) Select the vector of measured values from Step 1. This will populate the formula with the range of cells for the measured values.

(iii) Select the vector of lambda weights (only the first k elements) from Step 9.

(iv) End the formula with a closing parenthesis =MMULT(TRANSPOSE(D5:D11),H59:H65)

(v) Press Ctrl+Shift+Enter on the keyboard to indicate we are working with a matrix.

(vi) The formula will be enclosed in curly brackets

fx | {=MMULT(TRANSPOSE(D5:D11),H59:H65)}

(vii) The result is in Cell C68.

Interpolated Value: 111.1.

11. Compute the kriging variance by multiplying the transpose of the augmented vector of semi-variances from Step 8 by the column vector of augmented weights from Step 9.

The function in Excel for matrix multiplication is MMULT.

(i) In the formula bar enter =MMULT(TRANSPOSE(

(ii) Select the vector of measured values from Step 8. This will populate the formula with the range of cells for the prediction point semi-variances and the constraint.

(iii) Select the vector of weights and the Lagrange multiplier constraint from Step 9.

(iv) End the formula with a closing parenthesis =MMULT(TRANSPOSE(C59:C66),H59:H66)

(v) Press Ctrl+Shift+Enter on the keyboard to indicate we are working with a matrix.

(vi) The formula will be enclosed in curly brackets fx | {=MMULT(TRANSPOSE(C59:C66),H59:H66)}

(vii) The result is in Cell C70.

Kriging Variance: 50.1.

APPENDIX B

R Code for Selected Problems

```
####################################################################
#
#
#
####################################################################
#
#
# Problem 5.2
#
# R Code for calculating sample statistics
#
#
x = c(2.47, 1.34, 0.44, 1.45, 0.14, 1.51, 0.90, 2.47, 1.66)
x
n = length(x)
n
Mean = mean(x)
cat("The mean is", round(Mean, 3),"\n")
Median = median(x)
cat("The median is", round(Median, 3),"\n")
Variance = var(x)
```

```
cat("The variance is", round(Variance, 3),"\n")
Skewness1 = skewness(x, na.rm=TRUE, type=1)
cat("Type 1 skewness is", round(Skewness1, 3),"\n")
Skewness2 = skewness(x, na.rm=TRUE, type=2)
cat("Type 2 skewness is", round(Skewness2, 3),"\n")
Skewness3 = skewness(x, na.rm=TRUE, type=3)
cat("Type 3 skewness is", round(Skewness3, 3),"\n")
#
# The R Packacge e1071 is used to calculate kurtosis
#
library(e1071)
Kurtosis1 = kurtosis(x, na.rm=TRUE, type=1)
cat("Type 1 kurtosis is", round(Kurtosis1, 3),"\n")
Kurtosis2 = kurtosis(x, na.rm=TRUE, type=2)
cat("Type 2 kurtosis is", round(Kurtosis2, 3),"\n")
Kurtosis3 = kurtosis(x, na.rm=TRUE, type=3)
cat("Type 3 kurtosis is", round(Kurtosis3, 3),"\n")
#
#
#
###################################################################
#
#
#
###################################################################
#
# Problem 5.4
#
# R Code for analyzing a 3x3 regular lattice using Rook's case
#
#
# Locations IDs
# -------------------
# | 1 | 2 | 3 |
# -------------------
# | 4 | 5 | 6 |
# -------------------
# | 7 | 8 | 9 |
# -------------------
#
#
# Weight matrix W
#
# Location
# j
# | 1 2 3 4 5 6 7 8 9
# --------------------
# 1 | 0 1 0 1 0 0 0 0 0
# 2 | 1 0 1 0 1 0 0 0 0
# 3 | 0 1 0 0 0 1 0 0 0
# 4 | 1 0 0 0 1 0 1 0 0
# i 5 | 0 1 0 1 0 1 0 1 0
# 6 | 0 0 1 0 1 0 0 0 1
# 7 | 0 0 0 1 0 0 0 1 0
```

```
# 8 | 0 0 0 0 1 0 1 0 1
# 9 | 0 0 0 0 0 1 0 1 0
#
#
# Data for Problem 5.4
# ---------------
# | 84 | 91 | 92 |
# ---------------
# | 73 | 65 | 39 |
# ---------------
# | 28 | 41 | 32 |
# ---------------
#
# Enter data as Location 1, 2, 3, 4, 5, 6, 7, 8, 9
# Location | 1 | 2 | 3 | 4 | 5 | 6 | 7 | 8 | 9 |
# |-----|-----|-----|-----|-----|-----|-----|-----|-----|
# Data |  |  |  |  |  |  |  |  |  |
#
#
#
x = c(83,90,81,72,65,39,28,41,32)
#
xbar = mean(x)
xbar
#
W = matrix(rbind(c(0, 1, 0, 1, 0, 0, 0, 0, 0),
c(1, 0, 1, 0, 1, 0, 0, 0, 0),
c(0, 1, 0, 0, 0, 1, 0, 0, 0),
c(1, 0, 0, 0, 1, 0, 1, 0, 0),
c(0, 1, 0, 1, 0, 1, 0, 1, 0),
c(0, 0, 1, 0, 1, 0, 0, 0, 1),
c(0, 0, 0, 1, 0, 0, 0, 1, 0),
c(0, 0, 0, 0, 1, 0, 1, 0, 1),
c(0, 0, 0, 0, 0, 1, 0, 1, 0)), ncol=9)
W
#
num1 = length(x)
den1 = sum(W)
num2=0
for(i in 1:num1){
for(j in 1:num1){
num2 = num2 + W[i,j]*(x[i]-xbar)*(x[j]-xbar)
}
}
den2 = sum((x-xbar)^2)
Morans.I = (num1/den1)*(num2/den2)
cat("Moran's I is", round(Morans.I, 2),"\n")
#
#
################################################################
#
#
#
################################################################
```

```
#
# Problem 5.5
#
# R Code for analyzing 6 zones with irregular shapes
#
#
# Zone Response
# 1 32
# 2 26
# 3 19
# 4 18
# 5 17
# 6 14
#
# Weight matrix W
#
# Zone
# j
# | 1 2 3 4 5 6
# ----------------
# 1 | 0 1 1 1 1 0
# 2 | 1 0 1 0 0 0
# i 3 | 1 1 0 0 1 1
# 4 | 1 0 0 0 1 0
# 5 | 1 0 1 1 0 1
# 6 | 0 0 1 0 1 0
#
#
# Enter data as Zone 1, 2, 3, 4, 5, 6
#
# Location | 1 | 2 | 3 | 4 | 5 | 6 |
# |-----|-----|-----|-----|-----|-----|
# Data | | | | | | |
#
#
x = c(32,26,19,18,17,14)
#
xbar = mean(x)
xbar
#
W = matrix(rbind(c(0, 1, 1, 1, 1, 0),
c(1, 0, 1, 0, 0, 0),
c(1, 1, 0, 0, 1, 1),
c(1, 0, 0, 0, 1, 0),
c(1, 0, 1, 1, 0, 1),
c(0, 0, 1, 0, 1, 0)), ncol=6)
W
#
num1 = length(x)
den1 = sum(W)
num2=0
for(i in 1:num1){
for(j in 1:num1){
num2 = num2 + W[i,j]*(x[i]-xbar)*(x[j]-xbar)
```

```
}
}
den2 = sum((x-xbar)^2)
Morans.I = (num1/den1)*(num2/den2)
cat("Moran's I is", round(Morans.I, 2),"\n")
#
#
######################################################################
#
#
#
######################################################################
#
#
# Data for IDW and Kriging
#
# Spatial x y Measured
# Point coordinate coordinate Value
# 1 1 1 108
# 2 2 1 107
# 3 1 2 103
# 4 2 2 117
# 5 2 5 122
# 6 5 5 120
# 7 5 2 100
# 0 4 3 ?
#
#
# Problem 5.8
#
# R Code for IDW
#
#
library(gstat)
library(sp)
x=c(1,2,1,2,2,5,5)
y=c(1,1,2,2,5,5,2)
z=c(108,107,103,117,122,120,100)
xyz=data.frame(cbind(x,y,z))
coordinates(xyz) = ~x+y
# new obervation
x=4;y=3
xy=data.frame(cbind(x,y))
coordinates(xy) = ~x+y
# IDW
krige(z~1,xyz,xy)
#
#
######################################################################
#
# Problem 5.9
#
# R Code for kriging
#
```

```
#
# Method 1 Using functions
x=c(1,2,1,2,2,5,5)
y=c(1,1,2,2,5,5,2)
z=c(108,107,103,117,122,120,100)
# look at scatterplot of coordinates
plot(x,y)
points(4,3,pch="?")
# spherical variogram
sph.vario=function(h,p,r,n)
{
ifelse(h==0,0,ifelse(h>r,p+n,
n+p*(1.5*h/r-0.5*(h/r)^3)))
}
n=length(x)
# distance
(D=as.matrix(dist(cbind(x,y))))
# variogram
A=sph.vario(D,90,6,10)
(A=cbind(rbind(A,rep(1,n)),c(rep(1,n),0)))
# new observation
(d=as.matrix(dist(rbind(c(4,3),cbind(x,y))))[-1,1])
b=sph.vario(d,90,6,10)
(b=c(b,1))
# weight
w=solve(A)%*%b
# prediction
w[1:n]%*%z
# variance
b%*%w
#
#
#
# Method 2 Using R Packages
#
library(gstat)
library(sp)
x=c(1,2,1,2,2,5,5)
y=c(1,1,2,2,5,5,2)
z=c(108,107,103,117,122,120,100)
xyz=data.frame(cbind(x,y,z))
coordinates(xyz) = ~x+y
# new obervation
x=4;y=3
xy=data.frame(cbind(x,y))
coordinates(xy) = ~x+y
# kriging
m <- vgm(psill=90, "Sph", range=6, nugget=10)
krige(z~1,xyz,xy,model=m)
#
#
#
###############################################################
```

ACKNOWLEDGMENTS

Support for this document was provided by South Dakota State University, Precision Farming Systems community in the American Society of Agronomy, International Society of Precision Agriculture, and the USDA-AFRI Higher Education program (2014-04572). Selected R code was provided by SDSU Computational Sciences and Statistics doctoral students Toby Flint and Dongmin Jung.

REFERENCES

Atkinson, P.M., and C.D. Lloyd. 2009. Geostatistics and spatial interpolation. In: A.S. Fotheringham and P.A. Rogerson, editors, The SAGE handbook of spatial analysis. SAGE Publications Ltd., London. p. 159–181. doi:10.4135/9780857020130.n9

Bivand, R.S., E.J. Pebesma, and V. Gomez-Rubio. 2008. Applied spatial data analysis with R. Springer: New York.

Bolker, B.M. 2008. Ecological models and data in R. Princeton University Press: Princeton, New Jersey.

Chun, Y., and D.A. Griffith. 2013. Spatial statistics & geostatistics. SAGE: Los Angeles, CA.

Fotheringham, A.S., and P.A. Rogerson. 2009. Introduction. In: A.S. Fotheringham and P.A. Rogerson, editors, The SAGE handbook of spatial analysis. SAGE Publications Ltd., London, UK. p. 1–4. doi:10.4135/9780857020130.n1

Goovaerts, P. 1997. Geostatistics for natural resources evaluation. Oxford University Press: Oxford, UK.

Griffith, D.A. 2003. Spatial autocorrelation and spatial filtering: gaining understanding through theory and scientific visualization. Springer-Verlag: Berlin, Germany. doi:10.1007/978-3-540-24806-4

Hengl, T. 2009. A practical guide to geostatistical mapping. University of Amsterdam, Amsterdam, the Netherlands.

Huang, X., E.L. Skidmore, and G. Tibke. 2001. Spatial Variability of Soil Properties along a Transect of CRP and Continuously Cropped Land. In D. E. Stott, R. H. Mohtar, and G. D. Steinhardt, editors, Sustaining the global farm – Selected papers from the 10th International Soil Conservation Organization Meeting, May 24-29, 1999, West Lafayette, IN. International Soil Conservation Organization in cooperation with the USDA and Purdue University. West Lafayette, IN. p. 641-647.

Kolaczyk, E.D., J. Ju, and S. Gopal. 2005. Multiscale, Multigranular statistical image segmentation. J. Am. Stat. Assoc. 100:1358–1369. doi:10.1198/016214505000000385

Meyer, D., E. Dimitriadou, K. Hornik, A. Weingessel, and F. Leisch. 2015. e1071: Misc Functions of the department of statistics, Probability Theory Group (Formerly: E1071), TU Wien. R package version 1.6-7. https://CRAN.R-project.org/package=e1071(verified 5 June 2017).

Mitas, L., and H. Mitasova. 1999. Spatial Interpolation. In: P.L. Longley, M.F. Goodchild, D.J. Maguire, and D.W. Rhind, editors, Geographical information systems: Principles and technical issues. John Wiley & Sons, Inc., New York.

Moral, F.J., J.M. Terron, and J.R. Marques da Silva. 2010. Delineation of management zones using mobile measurements of soil apparent electrical conductivity and multivariate geostatistical techniques. Soil Tillage Res. 106:335–343.

Myers, D.E. 1991. Pseudo-cross variograms, positive-definiteness, and cokriging. Math. Geol. 23:805–816. doi:10.1007/BF02068776

Olea, R.A. 1999. Geostatistics for engineers and Earth scientists. Springer: New York. doi:10.1007/978-1-4615-5001-3

Plant, R.E. 2012. Spatial data analysis in ecology and agriculture using R. CRC Press: Boca Raton, FL. doi:10.1201/b11769

R Core Team. 2016. R: A language and environment for statistical computing. R Foundation for Statistical Computing, Vienna, Austria. https://www.R-project.org/ (verified 5 June 2017).

Rogerson, P.A. 2004. Statistical methods for geography. Sage Publications Ltd.: London, UK.

Sadler, E.J., W.J. Busscher, P.J. Bauer, and D.L. Karlen. 1998. Spatial scale requirements for precision farming: A case study in the Southeastern USA. Agron. J. 90:191–197. doi:10.2134/agronj1998.00021962009000020012x

Shekhar, S., V. Gandhi, P. Zhang, and R.R. Vatsavai. 2009. Availability of spatial data mining techniques. In: A.S. Fotheringham and P.A. Rogerson, editors, The SAGE handbook of spatial analysis. SAGE Publications Ltd., London. p. 159–181. doi:10.4135/9780857020130.n5

Stein, A., and I.C.A. Corsten. 1991. Universal kriging and cokriging as a regression procedure. Biometrics 47:575–587. doi:10.2307/2532147

Stewart, J. 2016. Calculus: Early transcendentals. 8th Edition. Cengage Learning: Boston, MA.

Sudduth, K.A., N.R. Kitchen, G.A. Bollero, D.G. Bullock, and W.J. Wiebold. 2003. Comparison of electromagnetic induction and direct sensing of soil electrical conductivity. Agron. J. 95:472–482. doi:10.2134/agronj2003.0472

Tarr, A.B., K.J. Moore, C.L. Burras, D.G. Bullock, and P.M. Dixon. 2005. Improving map accuracy of soil variables using soil electrical conductivity as a covariate. Precis. Agric. 6:255–270.

Vaughan, P.J., S.M. Lesch, D.L. Corwin, and D.G. Cone. 1995. Water content effect on soil salinity prediction: A Geostatistical study using cokriging. Soil Sci. Soc. Am. J. 59:1146–1156. doi:10.2136/sssaj1995.03615995005900040029x

Webster, R., and M.A. Oliver. 2007. Geostatistics for environmental scientists. 2nd Edition. John Wiley & Sons, Ltd: West Sussex, England. doi:10.1002/9780470517277

Westfall, P.H. 2014. Kurtosis as peakedness, 1905 – 2014. R.I.P. Am. Stat. 68:191–195. doi:10.1080/00031305.2014.917055

Soil Sampling and Understanding Soil Test Results for Precision Farming

David E. Clay,* Clay Robinson, and Thomas M. DeSutter

Chapter Purpose

Fertilizer recommendations often start with collecting soil samples and submitting those samples for laboratory analysis. The recommendations are only as good as the samples and associated chemical analysis. This chapter discusses: (i) how to interpret soil test results reports; (ii) the importance of a good sampling protocol, (iii) soil texture and the cation exchange capacity (CEC), (iv) soil nutrient extraction technique, (v) sources of acidity and the impact of acidity on nutrient and herbicide solubility, and (vi) electrical conductivity. In addition, a number of sample problems associated with interpreting and converting soil test results into recommendations are provided.

Key Terms

Soil test recommendation, calibration, available nutrients, cation exchange capacity (CEC), soil texture, part per million (ppm), base saturation, lime requirements, salinity, exchangeable sodium percent, active acidity, cation, anion, mean, median, soil sampling, electrical conductivity (EC).

Mathematical Skills

Understanding soil test results, converting units from one to another, converting concentrations to amount per land unit, understanding the difference between the mean and median, and calculating fertilizer recommendations.

Soil Testing Goal

The goal of soil nutrient testing is to provide information that can be used to improve fertilizer recommendations. Soil nutrient recommendations are built on extensive research that involves developing efficient soil sampling protocols, calibration, model development, and model validation. Calibration is required to determine the relationship between the amount of extractable nutrient and crop yield response. Based on these calibrations, empirical models are developed and validated. However, due to differences in soils, crops, and climatic conditions, the empirical models should not be used outside of the boundaries of where the experiments were conducted.

Soil Sampling Impact on Soil Test Results

Many fertilizer recommendations start with collecting soil samples that, when analyzed, provide the basis for accurate recommendations. In precision farming, this involves collecting multiple samples that may follow protocols for grids, grid cells, or management zones (Fig. 6.1). In grid sampling, a composite sample from the area surrounding each of the points on a grid map are collected, whereas in grid cell and management zone sampling, random soil samples are collected from within the

D. Clay, South Dakota State University, Brookings, SD 57006; C. Robinson, Illinois State University, Normal IL, 61790; T.M. DeSutter, North Dakota State University, Fargo, ND 58102. *Corresponding author (dclay@sdstate.edu)

doi: 10.2134/practicalmath2017.0022

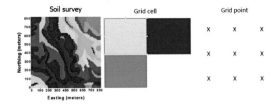

Fig. 6.1. A comparison of soil sampling techniques. In the soil survey map, a composite sample is collected from each soil type, where in the grid cell approach, a composite sample is collected from each grid cell. In the grid point approach a composite sample is collected from the area surrounding each point.

borders of the grid cell or management zone (Chang et al., 2003). The approach used to collect the information influences the interpretation of the data (Table 6.1).

If a soil type is considered a management zone, then a composite soil sample can be collected from each soil (i.e., samples from the soil types shown in red would be taken and combined into a single sample for analysis, then samples from the brown areas would be composited etc.). In the grid cell approach, multiple samples are taken from the grid cell and composited into a single soil sample for analysis, note that within a grid cell there may be multiple soil types. For the grid point sampling approach, a composite soil sample is collected from the area surrounding each point. Soil samples from each grid point would then be analyzed separately. Grid point sampling is the most expensive protocol and may generate the most samples and highest analysis costs, depending on the number of points in a field. Consultation with agronomists is suggested to determine the number and location of grid points and the specific suite of analyses to obtain the most useful data. Both grid cell and grid point analysis should consider the type of equipment that will be used in the field.

The values returned from soil testing laboratories represent the mean value for each soil parameter (e.g., pH or EC) or nutrient tested. The value reported for the sample is only as good as the sample submitted. Many fields contain irregularities or variations that can result in laboratory test results that do not accurately represent the area sampled. In the Midwest and Great Plains regions of the United States, many fields contain old homesteads (or other irregularities such as animal feeding pens) that may have been incorporated into the cropping area decades ago, but

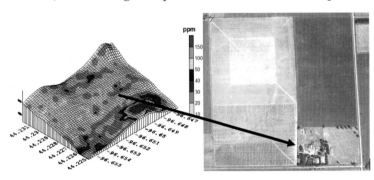

Fig. 6.2. A soil test P contour map and historic aerial image of that field. In this field the old homestead is located in an area containing very high P concentrations. The high concentrations are attributed to livestock associated with the homestead.

these areas can still impact the soil nutrient concentrations today (Fig. 6.2). For example, a field was sampled that contained an old homestead that had been incorporated into the cropping area prior to 1950. Samples from the homestead area were either combined with the "regular" field samples or kept separate. For nitrate N concentrations, sampling the old-homestead separately from the rest of the field had a minimal impact on the mean, median, and variance (Table 6.1) because N does not build up in the soil. However, P concentration was almost twice has high when soil samples were combined with soil from the homestead (26 vs. 14 ppm). In addition, there can be a large difference between the mean (the value reported in the soil test report) and the median (value where 50% of the samples are greater than or less than).

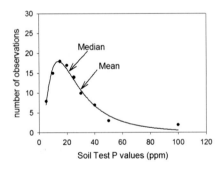

Fig. 6.3. A comparison of the median and mean value in a hypothetical data set. The median is the point where 50% of the values are greater than or less than, whereas the mean is the average value of the samples.

These data illustrate the importance of understanding field histories before sampling and how inappropriate mxing of soil samples compositing can change field recommendations. If there is a large difference between the mean and the median, the sample is not representative of the sampled area (Fig. 6.3). Increasing the size of the sampled area usually increases this difference. In addition, nitrate levels are less effected by old homesteads or animal pastures than P levels.

Soil Texture

Soil texture is the percentage of sand, silt, and clay in the soil, and it does not consider the amount of organic matter or soil mineralogy (Table 6.2). Sand, silt, and clay percentages are found by measuring the particle size distribution of a sample. Particles less than 0.002 mm are considered the clay fraction. The silt fraction is made up of particles ranging from 0.002 and 0.05 mm in diameter. The sand fraction are particles from 0.05 to 2 mm in size and can further be divided into very fine (0.05–0.1 mm), fine (0.1–0.25 mm), medium (0.25–0.50 mm), coarse (0.50–1.00 mm), and very coarse (1.00–2.00 mm) categories. Soil texture can be determined by a sedimentation procedure or by feel (Whiting et al., 2015). If the percentage of sand, silt, and clay contained in the soil is known, the texture can be determined using Fig. 6.4.

Soil texture influences many factors including water flow, the ability of the soil to retain soil nutrients, the amount of plant available water (see Clay and Trooien, 2017), the specific surface area of the soil particles, and the risk of compaction. For example, sandy soils (coarse) have faster water movement, less plant available water, and lower compaction risk than soils containing high amounts of clays and silts (finer-textured soils). Most of the nutrient-holding capacity and chemical activity in a soil is related to the amount of soil surface area and the charges contained within each particle. Finer-textured soils have more surface area and generally hold more nutrients.

Soil particles can contain positive, negative, or no charge. The amount of negative charges are called the cation exchange capacity (CEC) and the amount of positive charges is called the anion exchange capacity (AEC). On the soil test results the CEC may be reported as the sum of the bases or as CEC (Table 6.3). The CEC is related to the ability of the soil to retain nutrients.

Table 6.1. The influence of the old homestead on the mean, median and variance of grid soil samples collected from a South Dakota field (modified from Chang et al., 2003).

	With Homestead		Without Homestead	
	Olsen P	Nitrate–N	Olsen P	Nitrate–N
	mg kg⁻¹	mg kg⁻¹	mg kg⁻¹	mg kg⁻¹
Mean concentration	25.4	10.2	13.9	9.4
Median concentration	12	8	11	7.7
Mean–median	13.4	2.2	2.9	1.7
Variance	1611	54	86.9	18.1

Table 6.2. Relative characteristics of sand, silt, and clay particles.

Soil separate	Size	Surface area	Soil charge (CEC)
	mm		
Sand	2 to 0.05	low	low
Silt	0.05 to 0.002	mod	low to mod
Clay	< 0.002	high	high

Soil Test Reports

Many soil test reports provide information on soil texture, soil test levels, soil pH, electrical conductivity, and some estimate of the soil's CEC (Table 6.3). These measurements may be reported differently by different laboratories. Understanding the units and associated values are critical for understanding the soil test report (Clay et al., 2011b). Most nutrient values are reported in parts per million (ppm), although some are reported in the equivalent units as milligrams per kilogram (mg kg⁻¹). The ppm values can be converted to lb acre⁻¹ by multiplying the value times the amount of soil contained in the depth interval. In many soil test reports it is assumed that each 6 inches of soil contains 2 million pounds of soil and 12 inches of soil contains 4 million pounds of soil (Chang et al., 2011). In Table 6.3, the person interpreting the soils report needs to understand how the samples were collected and their associated sampling depths.

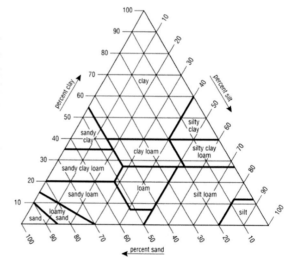

Figure 6.4. The USDA soil textural triangle. The soil texture is determined based on the amount of sand, silt, and/or clay. For example, a soil containing 30% clay and 15% silt is classified as a sandy clay loam.

PROBLEM 6.1.

A sample contains 10 g sand, 50 g silt, 30 g clay, and 5 g organic matter. What is the percentage of sand, silt, and clay in the sample? Using the texture chart in Fig. 6.4, determine the texture.

ANSWER:

Soil texture only considers the mineral portions of the soil, and therefore in these calculations soil organic matter is not considered.

$$\text{Total mineral components} = 10 + 50 + 30 = 90 \text{ grams}$$

$$\text{Sand (0.05-2.0 mm)} = 100 \times \left(\frac{10}{90} \right) = 11\%$$

$$\text{Silt (0.002-0.05 mm)} = 100 \times \left(\frac{50}{90} \right) = 56\%$$

$$\text{Clay (<0.002 mm)} = 100 \times \left(\frac{30}{90} \right) = 33\%$$

Silty clay loam

PROBLEM 6.2A.

Based on a sand, silt, and clay content of 20, 35, and 45%, respectively, what is the soil texture?

ANSWER:

Clay

PROBLEM 6.2B.

Based on a sand, silt, and clay content of 50, 30, and 20%, respectively, what is the soil texture?

ANSWER:

Loam

In a loam soil, the sand, silt, and clay properties all have similar impacts on the physical properties of the soil?

Table 6.3. Hypothetical soil test laboratory report from a submitted sample. For this analysis soil samples were collected from two depths (0–6 and 6–24 inches). Each sample consisted of a composite sample collected from a whole field, point, grid point, grid cell, or management zone. In this case, each composite sample contained 10 individual cores that were combined to produce 1 sample. The findings may be reported many different ways including ppm, lb acre^{-1}, and percent.

Sam. Id	Depth	SOM	EC or soluble salts (1:1)	Nitrate–N	P	K	Ammonium acetate			CEC by Sum cations	% Bases			
		%	mmho cm^{-1}	lb acre^{-1}	ppm	ppm	Ca ppm	Mg ppm	Na ppm	meq 100 g^{-1}	K	Ca	Mg	Na
2275	0–6	4.2	0.57	45	22	1037	2273	236	20	16.2	17	70	12	1
2275	6–24	1.8	0.88	22	15	557								

Soil Nutrients

Extraction Technique

A goal when testing soil for nutrients is to use reagents that only extract the plant available nutrients, not the total amount of nutrient in the soil sample. Research has shown that there is not a universal solution that mimics plant availability for all nutrients and soils. Therefore, the analysis approach and sometimes soil type influences the chemicals used to extract the nutrients. For example, reagents used when nitrate is analyzed with an electrode differs from the reagent used for extraction when using a colorimetric test. In addition, if the laboratory is analyzing both NO_3^- and NH_4^+, then a solution containing between 1 and 2 M KCl might be used. The relatively high concentration of KCl is needed to extract NH_4^+ from the soil.

For soil test P, many different extractions are used. For example, in the North Central Region of the United States soil-testing laboratories commonly use the Bray-1 (B-P1) reagent if the soil pH < 7.4, and the Olsen (O-P) method if the soil pH > 7.4 (Sawyer and Mallarino, 1999). Different extractants are used because their ability to extract only plant-available P changes with pH and mineralogy. Midwest Laboratories, Inc. (2007; Omaha, NE) reports values for P1 and P2. The P1 (weak Bray) represents plant-available P, whereas P2 (strong Bray) measures readily available + a portion of the active reserve P. For the P2 value, it is suggested that a range between 40 and 60 ppm (40 to 60 mg kg⁻¹) is desirable (Midwest Laboratories, 2007). In the Southern Region of the United States, soil P extractants include Mehlich-1, Bray P2, Morans, Landcaster, and Mehlich-3 (Sikora et al., 2005). Fertilizer recommendation guides from the appropriate region should be consulted when preparing a diagnosis for a field.

A range of extractants are used for soil K including ammonium acetate, Mehlich-1, and Mehlich-3. In addition, some laboratories may offer the analysis of moist samples (Mallarino, 2012). Moist soil extraction provides numbers that are generally lower than the same dried soil sample. In summary, due to the wide range of extractants possible, it is important to select a laboratory and a fertilizer guideline that uses methods and interpretations that are meaningful for your location.

PROBLEM 6.3A.

Soil nitrate–N is reported as ppm and the values for the 0 to 6 and 6 to 24 in soil depths are 10 and 15 ppm. Convert to lb acre⁻¹. Assume each 6 in depth contains 2 million lb of soil.

ANSWER:

$$\frac{10 \text{ lb } NO_3 - N}{1{,}000{,}000 \text{ lb soil}} \times \frac{2{,}000{,}000 \text{ lb}}{acre} = \frac{20 \text{ lb } NO_3^-}{acre}$$

$$\frac{15 \text{ lb } NO_3 - N}{1{,}000{,}000 \text{ lb soil}} \times \frac{6{,}000{,}000 \text{ lb}}{acre} = \frac{90 \text{ lb } NO_3^-}{acre}$$

Total nitrate-N is 110 lb acre⁻¹.

PROBLEM 6.3B.

How is nitrate-N different from nitrate?

ANSWER:

Nitrate considers only the N in the nitrate molecule, while nitrate considers both the N and oxygen. To convert nitrate-N to nitrate, multiply nitrate-N by the molecular weight of nitrate (62 g) divided by the weight of N (14 g) or 4.428.

PROBLEM 6.4A.

Calculate how much soil is contained in the surface 6 inches of soil per acre if the bulk density is 1.3 g cm⁻³

ANSWER:

Bulk density is used to measure the amount of soil per unit volume and it is defined as the dry weight of soil per unit volume. For this problem, we will solve using metric units followed by a conversion to English units. (1 in contains 2.54 cm, 1 m² contains 10,000 cm², 1 ha contains 10,000 m², 1 ha contains 2.471 acres, and 1 kg contains 2.205 lb.

$$15.24 \text{ cm soil sample} \times \frac{1.3 \text{ g}}{cm^3} \times \frac{10{,}000 \text{ cm}^2}{m^2} \times \frac{10{,}000 \text{ m}^2}{ha} \times \frac{1 \text{ ha}}{2.471 \text{ acres}} \times \frac{kg}{1{,}000 \text{ g}} \times \frac{2.205 \text{ lb}}{1 \text{ kg}} = \frac{1.77 \text{ million soil lb}}{acre}$$

PROBLEM 6.4B.

If a 6-in soil sample contains 10 ppm soil test P, how much P is there in lb acre⁻¹?

ANSWER:

$$\frac{10 \text{ lbs}}{1{,}000{,}000 \text{ lbs soil}} \times \frac{1{,}770{,}000 \text{ lbs soil}}{acre} = \frac{17.7 \text{ lbs}}{acre}$$

PROBLEM 6.5.

Convert 0.01 ppm to ppb.

ANSWER:

Rewrite 0.01 ppm as g per g.

$$\frac{0.01 \text{ g}}{1,000,000 \text{ g}} \times \frac{1,000}{1,000} = 10 \text{ ppb}$$

Note, in this calculation the denominator has increased from 1 million to 1 billion by multiplying by 1000.

PROBLEM 6.6.

Convert 5 ppm to lb 1,000,000 lb^{-1} of soil

ANSWER:

$$\frac{5 \text{ lb}}{1,000,000 \text{ lb}}$$

PROBLEM 6.7.

Convert 200 kg ha^{-1} to lb acre$^{-1.}$

ANSWER:

$$\frac{200 \text{ kg}}{\text{ha}} \times \frac{\text{ha}}{2.471 \text{ acre}} \times \frac{2.205 \text{ lb}}{\text{kg}} = \frac{178.5 \text{ lb}}{\text{acre}}$$

PROBLEM 6.8.

Convert 10 ppm to lb acre^{-1} for the surface 6 in of soil if each 6 in contains 2 million lb of soil

ANSWER:

$$\frac{10 \text{ lb}}{1,000,000 \text{ lb}} \times \frac{2,000,000 \text{ lb}}{\text{acre}} = \frac{20 \text{ lb}}{\text{acre}}$$

Soil Test Phosphorus

In soil test reports, soil P and several other nutrients generally are reported as ppm; however, nutrient concentrations also may be reported as lb acre^{-1}. If the numbers are reported in lb acre^{-1} unit, they are calculated from the ppm values by multiplying the ppm value by 2. This conversion assumes that a 6 in soil layer over 1 acre weighs 2 million lb. Therefore, when multiplying the ppm value by 2, the calculated value represents the amount of nutrient in lb acre^{-1} expected to be present in the top 6 in of soil. Sample problems for this conversion are below. The ppm value is a dimensionless number that can be used to calculate the total amount of a nutrient in the soil. Parts per 100, or percent is designated by the % sign, while parts per 1000 is designated by the ‰ symbol. Part per million (parts per 1,000,000) and parts per billion (parts per 1,000,000,000) are designated by ppm and ppb, respectively.

Phosphorus generally exists in the soil as the $H_2PO_4^-$ and HPO_4^{2-} ions. Even though phosphate is an anion (negatively charged) it can be strongly sorbed to the soil particles. Phosphorus loss through leaching is less of a problem because it is often strongly sorbed to the soil particles. However, at very high concentrations, both leaching and erosion losses can occur. Phosphorus has maximum availability to the plant at soil pH values from 6.5 to 7. Phosphorus fertilizer efficiency can be improved by liming acid soils, and by banding P in many soils.

Inorganic Nitrogen

Soil nitrogen is separated into two pools, organic and inorganic. The dominant inorganic N species in soil are ammonium (NH_4^+) and nitrate (NO_3^-). Unless requested, most laboratories only report soil nitrate N (NO_3^--N). Soil nitrate N (NO_3^--N) has a negative charge and it often reported as lb N acre^{-1}. NO_3^--N and NO_3^- have different meaning. When written as NO_3^--N, only the N is considered, whereas when written as NO_3^- the entire molecule is considered. NO_3^- is converted to NO_3^--N by multiplying NO_3^- times 14/62, where 14 is the molecular weight of N and 64 is the molecular weight of NO_3^-. The reported N value does not include soil ammonium (NH_4^+).

Due to its negative charge and solubility, nitrate moves through negatively charged soils with water and can be leached below the root zone to the groundwater. Nitrate also can be biologically denitrified to N_2O or N_2 if oxygen becomes limiting, such as following a rainfall when most of the soil pores are filled with water. Following the application of ammonium based fertilizers (urea, anhydrous ammonia), the ammonia form of N (NH_3) is converted to the nitrate form. This conversion can be slowed by applying a nitrification inhibitor.

Estimates of soil organic matter (SOM) provide an indirect estimate of the amount of organic N in the soil. Most soil samples are not analyzed for total organic N, however, it can be estimated by converting SOM to soil organic carbon (SOC) [SOC = SOM × 0.58] (it is assumed that SOM contains 58% carbon) which is then used to estimate

PROBLEM 6.9A.

Convert $\dfrac{150 \text{ kg}}{\text{ha}}$ to $\dfrac{\text{lb}}{\text{acre}}$.

ANSWER:

$$\frac{150 \text{ kg}}{\text{ha}} \times \frac{\text{ha}}{2.471 \text{ acre}} \times \frac{2.205 \text{ lb}}{1 \text{ kg}} = \frac{133.8 \text{ lb}}{\text{acre}}$$

PROBLEM 6.9B.

Convert $\dfrac{150 \text{ lb}}{\text{acre}}$ to $\dfrac{\text{kg}}{\text{ha}}$.

ANSWER:

$$\frac{150 \text{ lb}}{\text{acre}} \times \frac{2.471 \text{ acre}}{\text{ha}} \times \frac{\text{kg}}{2.205 \text{ lb}} = \frac{168 \text{ k}}{\text{ha}}$$

PROBLEM 6.10.

A soil sample report indicates the 0-6 and 6-24 inch soil depths contain 10 and 15 ppm NO_3-N, respectively. How many lb of nitrate-N/acre (NO_3-N) are contained in this soil? Assume that each 6 inches of soil contain 2 million pounds and that 18 inches contain 6 million pounds of soil

ANSWER:

$$\frac{10 \text{ lb}}{1{,}000{,}000 \text{ lb soil}} \times \frac{2{,}000{,}000 \text{ lb}}{6 \text{ in}} = \frac{20 \text{ lb}}{6 \text{ in}}$$

$$\frac{15 \text{ lb}}{1{,}000{,}000 \text{ lb soil}} \times \frac{6{,}000{,}000 \text{ lb}}{18 \text{ in}} = \frac{90 \text{ lb}}{18 \text{ in}}$$

$$20 + 90 = 110 \text{ lb total } NO_3\text{–N}$$

Total amount of NO_3–N is 110 lb acre^{-1}. This calculation converted 10 ppm to 20 lb of nitrate–N that is contained in 2 million lb of soil.

organic N. A common approach for estimating total N is to divide total C by 10 (Malo et al., 2005; Kirkby et al., 2011) (for every 10 pounds of C the soil contains 1 pound of N). In many soils, approximately 1 to 3% of the organic N is annually converted to inorganic N through mineralization. If it is assumed that 2% of the organic N from the organic matter is mineralized annually, then each % of organic matter will provide approximately 20 lb of N.

PROBLEM 6.12.

Surface (0-6 inches) soil contains 4% organic matter. Estimate the amount of N that will be available during the growing season if 2% of the N in the organic matter is mineralized. Assume the soil weighs 2,000,000 lb.

ANSWER:

$$\frac{0.04 \text{ lb organic matter}}{\text{lb soil}} \times \frac{0.58 \text{ lb C}}{1 \text{ lb organic matter}} \times \frac{2,000,000 \text{ lb soil}}{\text{acre}} = \frac{46,400 \text{ lb C}}{\text{acre}}$$

If the C/N ratio is 10, then the soil contains 4640 lb N. If 2% of the N is mineralized (0.02 × 4640) in one year then 92.8 lb of N is mineralized in one year. Note that the estimated N from 4% organic matter is 80 lb (20 lb N from each 1% OM, = 20×4 = 80 lb). Based on these calculations, many agronomists assume that for each 1% of organic matter, about 20 lb of N per acre, will be mineralized, although no-tilled systems may have lower mineralization amounts during the first five years while the residue (high C/N) is being stabilized.

Potassium

Potassium (K) is a positively charged nutrient in soil that generally is extracted using a salt solution such as ammonium acetate. With a single positive charge, it is not held in soil as tightly as Ca^{+2}. Potassium is a macronutrient and should be applied when the soil K levels cannot meet the needs of the crop. In soil, K may be unavailable, slowly available or fixed, readily available, or exchangeable. Due to the many fates of K, it is important to understand the specific reactions in your area's soils.

Secondary and Micronutrients

Deficiencies of these secondary and micronutrients (Ca, Mg, S, Zn, Fe, Mn, and Cu) can limit growth in many soils. Deficiencies for these nutrients have been observed in soils that are highly weathered, have high pH, sandy, low organic matter, and are eroded. In many situations, it may be economical to target applications to problem areas.

Cation Exchange Capacity (CEC)

In many soil test reports, the sum of the bases is reported as the cation exchange capacity (CEC). The CEC is a sum of the soil's negative charges, however, because adding up the negative charges directly is very difficult, CEC is oftentimes estimated by extracting the total cations in the soil, which includes cations on the exchange sites soil solutions and soluble salts. The CEC value provides an estimate of the soil's ability to hold onto essential nutrients (e.g., a nutrient reserve) and helps buffer the soil against acidification. Within the United States Great Plains, soils with high clay and organic matter contents generally have high CEC values. For example, CEC for sandy soils may be less than 10 $cmol_c$ kg^{-1} while CEC values for high clay soils may be 40 to 50 $cmol_c$ kg^{-1}. Cation Exchange Capacity values in old soils can be very low. Also of importance to CEC is the contribution of organic matter to the total number of a soil's exchange sites. In some soils, the organic matter can provide the exchange sites for about 30% of the soil's total CEC.

The soil charges are attributed to the types of chemical bonds contained within the soil organic matter, the amount of clay, and the type of clay minerals present. Positive charges attract anions and the total amount of positive charges are called the anion exchange capacity (AEC). The negative charges attract positively charged ions such as H^+, Ca^{2+}, Mg^{2+}, K^+, Na^+, and NH_4^+. The CEC provides an index for the ability of the soil to retain cations (positively charged ions).

In Table 6.3, the apparent CEC value is reported as milliequivalents per 100 g but could be reported as cmol of charge per kg of soil ($cmol_c$ kg^{-1}); both units have identical values (1 meq 100 g^{-1} = 1 $cmol_c$ kg^{-1}). The CEC is often estimated as the sum of the H, K, Ca, Mg, and Na cations. First, the concentration (ppm) of K, Ca, Mg, and Na is converted to units of charge (meq 100 g^{-1}). Sample calculations for a soil containing 2000 ppm Ca and 20 ppm Na are provided below. Note that Ca has a +2 charge, so that the value is multiplied by 2. Na with a +1 charge is multiplied by 1.

$$\frac{2000 \text{ mg Ca}}{kg} \times \frac{\text{mmol Ca}}{40 \text{ mg Ca}} \times \frac{2 \text{ meq}}{1 \text{ mmol}} \times \frac{1 \text{ kg}}{10 \times 100 \text{ g}} = \frac{10 \text{ meq}}{100 \text{ g}}$$

$$\frac{20\,\text{mg Na}}{\text{kg}} \times \frac{\text{mmol Na}}{23\,\text{mg Na}} \times \frac{1\,\text{meq}}{1\,\text{mmol}} \times \frac{1\,\text{kg}}{10 \times 100\,\text{g}} = \frac{0.09\,\text{meq}}{100\,\text{g}}$$

or as

$$\frac{2000\,\text{mg Ca}}{\text{kg}} \times \frac{\text{g}}{1000\,\text{mg}} \times \frac{\text{mol}}{40\,\text{g}} \times \frac{2\,\text{eq}}{\text{mol}} \times \frac{100\,\text{cmol}_c}{\text{eq}} = \frac{10\,\text{cmol}_c}{\text{kg}}$$

$$\frac{20\,\text{mg Na}}{\text{kg}} \times \frac{\text{g}}{1000\,\text{mg}} \times \frac{\text{mol}}{23\,\text{g}} \times \frac{1\,\text{eq}}{\text{mol}} \times \frac{100\,\text{cmol}_c}{\text{eq}} = \frac{0.09\,\text{cmol}_c}{\text{kg}}$$

This calculation shows that 2000 ppm Ca is converted to 10 meq 100 g⁻¹. The molecular weights used in this calculation are provide in Table 6.4. In short, to convert Ca, Mg, K, and Na ppm to meq 100 g^{-1} or cmol$_c$ kg^{-1} divide the ppm by 200, 120, 390, or 230, respectively. The CEC can be estimated from the %Clay and organic matter using the following:

1. Soil organic matter typically contains 200 meq 100 g^{-1} or 200 cmol$_c$ kg^{-1} soil
2. Smectite clay (2:1 layer silicate, expanding) contains 100 meq 100 g^{-1} or 100 cmol$_c$ kg^{-1}
3. Kaolinite clay (1:1 layer silicate) contains 10 meq 100 g^{-1} or 10 cmol$_c$ kg^{-1}
4. Illite (2:1 layer silicate, non-expanding) and chlorite (2:1 layer silicate, nonexpanding) contain 30 meq 100 g^{-1} or 30 cmol$_c$ kg^{-1}.

Percent Base Saturation

The CEC is used to calculate the Percent Base Saturation $\%\,BS = 100 \times \dfrac{\text{Sum of basic cations}}{\text{CEC}}$, and the exchangeable sodium percentage $\%\,ESP = 100 \times \dfrac{\text{Na}}{\text{CEC}}$. The %BS is used to assess the percentage of exchange sites occupied by basic cations (Ca, Mg, K, Na). However, when the sum of the extractable cations is used, the %BS is the sum of the cations (positively charged nutrient) excluding H and Al. Soils with high soil pH, greater than 7.5, tend to have %BS greater than 80% while %BS in soils with low pH (<5.5) tend to be less than 50%. At low pH, H$^+$ and Al^{3+} concentrations can become more prominent. As such, weathering of soil minerals, the decomposition of plant residues, and the application of ammonia-based fertilizer tends to decrease %BS. For example, the biological conversion of NH$_4^+$ to NO$_3^-$ releases two H+, thus lowering the soil pH.

Apparent Exchangeable Sodium Percentage (ESP$_a$)

In Table 6.3, the apparent ESP (ESPa) is approximated with the %Na extracted with ammonium acetate and it is the ratio between Na and the total number of basic cations extracted from the soil. For the data provided in Table 6.3,

ESP$_a$ is 6.2% $\left(ESPa = \dfrac{\left(1.0\,\dfrac{\text{meq Na}}{100\text{g}}\right)}{16.2\,\dfrac{\text{meq}}{100\text{g}}} \times 100 \right)$ or $\left(ESPa = \dfrac{1.0\,\dfrac{\text{cmol}_c\,\text{Na}}{\text{kg}}}{16.2\,\dfrac{\text{cmol}_c}{\text{kg}}} \times 100 \right)$.

Table 6.4. The symbols and atomic weights of elements routinely found in soil. These are used to find the meq per 100 grams or 100 cmolc kg⁻¹.

Element	Symbol	Atomic Weight (g)	Element	Symbol	Atomic Weight (g)
Calcium	Ca^{+2}	40.1	Nitrogen	N	14.0
Carbon	C	12.0	Oxygen	O	16.0
Chloride	Cl$^-$	35.5	Phosphorus	P	31.0
Copper	Cu^{+2}	64.5	Manganese	Mn	54.9
Hydrogen	H$^+$	1.0	Potassium	K$^+$	39.1
Iron	Fe^{+3}	55.8	Sodium	Na$^+$	23.0
Lead	Pb^{+2}	207.2	Sulfur	S	32.1
Magnesium	Mg^{+2}	24.3	Zinc	Zn^{+2}	65.4

Table 6.5. Estimated cation exchange capacity (CEC) for different soils. The term "superactive" indicates a cation exchange capacity (by NH$_4$OAc [NH$_4$–acetate] at pH 7) to clay (percent by weight) ratio (i.e., %clay/CEC) of 0.60 or more. The terms kaolinitic and smectitic refer to the type of clay in the soil. Additional information on soil classification is available in Soil Survey Staff (1999).

Soil series	State	Family classification	cmol$_c$ kg^{-1} or meq per 100 g
Cecil	North Carolina	fine, kaolinitic, thermic Typic Kanhapludults	10
Drummer	Illinois	fine-silty, mixed, superactive, mesic Typic Endoaquoll	100
Houdek	South Dakota	fine-loamy, mixed, superactive, mesic Typic Argiustolls	100
Houston Black	Texas	fine, smectitic, thermic Udic Haplusterts	100

PROBLEM 6.13.

If a soil has 2% soil organic matter (SOM; 200 cmol$_c$ kg^{-1}) and 30% clay, of which 70% was smectite (100 cmol$_c$ kg^{-1}) and 30% was illite (30 cmol$_c$ kg^{-1}), what is the CEC?

ANSWER:

$$\frac{0.02 \text{ kg SOM}}{\text{kg soil}} \times \frac{200 \text{ cmol}_c}{1 \text{ kg SOM}} + \frac{0.3 \times 0.7 \text{ kg smectite}}{\text{kg soil}} \times \frac{100 \text{ cmol}_c}{1 \text{kg smectite}} + \frac{0.3 \times 0.3 \text{ kg illite}}{\text{kg soil}} \times \frac{30 \text{ cmol}_c}{\text{kg illite}} = \frac{27.7 \text{ cmol}_c}{\text{kg}} = \frac{27.7 \text{ meq}}{100 \text{ g}}$$

In this method, the sum of extractable cations is used to estimate the CEC.

Sodic soils are soils with high concentrations of Na, whereas saline soils have high concentration of cations and anions in the soil solution. The ESP value is used to assess the risk of soil dispersal and the degradation of soil aggregates. Classically sodic soils were defined as having ESP greater than 13. However, recent research has shown that the critical value depends on management, soil, and climatic conditions. Research conducted in the northern Great Plains suggests that the soil can disperse when the ESP increases above 4 or 5 (He et al., 2013, 2015a, 2015b) and the EC (1:1 soil to water dilution) is less than 1 mmho cm^{-1} (0.1 S m^{-1}). The dispersion risk is greatly controlled by soil EC.

Gypsum (CaSO$_4$·2H$_2$O) is the most widely-used amendment for improving the productivity potential of sodic soils. However, if the soil contains gypsum it may be of limited value. Gypsum adds Ca^{2+} to counteract the dispersive effects of Na$^+$ and Mg^{2+}, and also increases the EC, which facilitates water movement into and within the soil. Techniques for calculating gypsum requirements are provide by Oster and Frenkel (1980) and Carlson et al. (2016).

Soil pH and Lime

The soil relative acidity and alkalinity is characterized by the soil pH. Water dissociates weakly into H$^+$ and OH$^-$ ions, and pH is a measure of the concentration of the hydrides in the solution. The pH is defined as the negative logarithm of the hydrogen ion concentration (pH = -log[H$^+$] or 1/log[H$^+$]. Likewise, the pOH is defined as the negative logarithm of the hydroxyl concentration (pOH = -log[OH$^-$] or 1/log[OH$^-$]). The sum of the pH and pOH is 14. Pure water has a pH close to 7, which means that it is relatively neutral. Soil pH values < 7 means that the soil is acidic (the concentration of H$^+$ ions are greater in solution), whereas soil pH values > 7 indicate that the soil is alkaline (greater concentrations of OH$^-$ ions in solution).

Soil pH is generally measured by mixing air-dried soil with deionized water (DI) in ratios that are defined for one's region (e.g., 1:1, 1:2, or 1:5 soil to DI water). For example, 10 g air-dried soils is mixed with with 20 mL of pure water (1:2 w/v) or by mixing 10 g air-dried soils with 20 mL of 0.01 M CaCl$_2$. The weak salt solution is often used because soil water (and rainwater) is not pure, but contains small amounts of ions. Thus the 0.01 M solution is used to more closely mimic soil water. Soil acidity has been described using three different pools: active, exchangeable, and residual. Active acidity is the H$^+$ concentration in the soil solution, whereas exchangeable acidity is the H$^+$ on the soil cation exchange sites that is replaced with an unbuffered salt. Residual activity is the acidity that is not active or exchangeable. When water is used it measures active acidity, whereas a 0.01 M CaCl$_2$ measures the active and exchangeable pools. Different laboratories measure pH differently.

The soil pH influences nutrient availability to plants. Most plant nutrients are most available at pH values between 6.6 and 7.3 (Table 6.6). Selected soil nutrient rules of thumb as impacted by pH include:

1. The conversion of organic N to ammonium can be reduced at very low and very high soil pH values.

2. Reactions between phosphate and calcium, phosphate and iron, and phosphate and aluminum can reduce P concentrations. At low pH values PO_4 is reduced by Fe and Al, whereas at high pH values PO_4 is reduced by Ca.

3. Potassium activity may be reduced at very low pH values.

4. Sulfur: the microbial transformation of organic S to sulfate may be reduced at very low and very high pH values.

5. The availability of magnesium and calcium can be reduced at very low soil pH values.

6. Micronutrient availability can be impacted by soil pH.

7. While most plants grow best in neutral soils, some prefer acid soils (e.g., blueberries, pine trees), and others (honeysuckle, clematis, poppy) prefer alkaline soils. The pH in low pH soils can be increased by the addition of lime and high pH soils can be lowered by adding ammonium based fertilizers.

Adding Lime

In many soils, pH gradually decreases from the application of ammonium-based fertilizers (Table 6.7). For example, over a 6-yr period, annual applications of ammonium sulfate lowered soil pH from 6.5 to less than 4.0 (Clay et al., 1993). At low soil pH values, nitrification (i.e., the oxidation of ammonia to nitrate) is slowed. Different N and P sources produce different amounts of acidity. Calcium carbonate (lime; $CaCO_3$) can be added to counteract this effect.

The size of exchangeable H^+ pool determines the ability of the soil to resist a change in pH. The ability to resist a pH change is related to the soil's buffering capacity. The size of exchangeable H^+ pool is determined by the CEC and the concentration of H^+ and Al^{+3} on the exchange sites. A high buffering capacity means that the soil has the ability to resist a pH change when an acid or base is added and is related to the soils CEC. Soils with greater CEC typically have greater buffering capacity (high clay soils), and require more lime addition than soils with lower buffering capacity (low clay soils). For lime recommendations, the buffering capacity of the soil should be measured. A common technique for determining the lime requirement involves testing the pH of the soil in water, then testing the pH of a soil that has been mixed with a standard buffer solution. Comparing the pH in the two solutions provides an estimate of the soil buffering capacity. Various laboratories use different buffering reagents (Mallarino et al., 2013; Bly and Gelderman, 2016), and one of the most common is the SMP Buffer Solution. Lime requirements for different SMP readings are presented in Table 6.8.

All lime is not created equal, the chemical composition varies with the source (Table 6.9). Some limestone formations are relatively pure calcium carbonate, while others contain dolomite, a calcium–magnesium carbonate mineral. Some industrial byproducts may be used as lime products, such as calcium oxide or magnesium oxide or waste lime from the processing of sugar beets. Additionally, most lime products have inert materials, other minerals, or even soil particles that dilute the lime effectiveness. All agricultural lime is compared to a standard of pure calcium carbonate, which has a defined effectiveness of 100%.

Table 6.6. Suggested pH for selected crops.

Crop	pH
Corn, soybean	6.5
Alfalfa	6.9
Forage grasses, pastures	6.0

Table 6.7. Acidity produced by different fertilizer products. The calculations are based on pounds of calcium carbonate needed to neutralize the acidity per pound of fertilizer and per pound of N.

	%N	lb $CaCO_3$ lb fert^{-1}	lb $CaCO_3$ lb N^{-1}
Anhydrous ammonia	82	1.48	1.80
Urea	46	0.84	1.83
Ammonium nitrate	33	0.63	1.91
Ammonium sulfate	21	1.12	5.33
Diammonium phosphate (DAP)	18	0.74	4.11
Monoammounium phosphate (MAP)	11	0.65	5.91
Ammonium polyphosphate	10	0.53	5.30
KCl	0	0	0.00

Problem 6.15.

The pH of a soil is 6.0, and the SMP Buffer pH is 6.6. Using Table 6.8, how many pounds of $CaCO_3$ must be applied to raise the pH of a 6-in plow layer to 6.5?

Answer:

2,100 $CaCO_3$ lb acre^{-1}.

Problem 6.16.

A producer applies 175 lb P_2O_5 acre^{-1} using MAP, 11-52-0. How many pounds of lime will be required to neutralize the acidifying capacity of the fertilizer?

Answer:

Determine amount of MAP needed to apply 175 lb of phosphate.

$$\text{amount} = \frac{175 \, \dfrac{\text{lb } P_2O_5}{\text{acre}}}{0.52 \, \dfrac{\text{lb } P_2O_5}{\text{lb MAP}}} = 337 \, \frac{\text{lbs MAP}}{\text{acre}}$$

Determine amount of lime required to neutralize 340 $\dfrac{\text{lbs MAP}}{\text{acre}}$. (Rounded to nearest 10 lb.) From Table 6.7, 0.65 lb $CaCO_3$ lb MAP^{-1} is required.

$$\text{amount} = 340 \, \frac{\text{lbs MAP}}{\text{acre}} \times 0.65 \, \frac{\text{lbs } CaCO_3}{\text{lbs MAP}} = 221 \, \frac{\text{lbs } CaCO_3}{\text{acre}}$$

(Rounded up to nearest 10 lb.)

Table 6.8. Hypothetical lime recommendations for a specified depth using an SMP Buffer to raise the pH to 6.5.

SMP Buffer pH	CaCO3 to apply, lb acre^{-1}	
	6-in layer	8-in layer
6.9	0	0
6.8	600	800
6.7	1300	1700
6.6	2100	2800
6.5	2800	3700

Table 6.9. Characteristics of agricultural lime stone from three quarries. Information on Calcium carbonate equivalent (CCE) and fineness is provided.

Quarry	CCE (%)	Passing through mesh (sieve)		
		100	60	8
Product 1	99	94	30	26
Product 2	93	84	42	38
Product 3	85	99	29	20

The composition and purity of the agricultural lime determines the calcium carbonate equivalent (CCE) relative to pure calcium carbonate.

Particle fineness is another factor that contributes to the effectiveness of agricultural lime. To be effective, the lime must dissolve. Although the solubility of lime is the same irrespective of the particle size, the smaller particles dissolve more quickly compared to larger particles due to a greater surface area to volume ratio in smaller particles. All agricultural lime materials are analyzed for these two components: CCE and fineness (calculated below). Each

PROBLEM 6.17.

What is the mass of $CaCO_3$ in one mole of $CaCO_3$?

ANSWER:

$Ca = 40$ g mol^{-1}, $C = 12$ g mol^{-1}, $O = 16$ g mol^{-1}.
Mass of $CaCO_3$ in one mole $= 40 + 12 + 16 \times 3 = 40 + 12 + 48 = 100$ g

PROBLEM 6.18.

Determine ECCE of the lime from product 1. Use these fineness factors (efficiency) of the lime materials: materials passing 8 mesh sieve = 0.3; passing 60 mesh = 0.6; passing 100 mesh = 1.0. The calcium carbonate equivalent (CCE) is 99%.

ANSWER:

So, the ECCE = CCE × Total fineness efficiency = 99% ×72% = 71%

PROBLEM 6.19.

If the SMP Buffer method indicated a lime requirement of 2,100 $CaCO_3$ lb acre^{-1}, and the ECCE is 71%, what is the adjusted ECCE value if the moisture content is 4%?

ANSWER:

ECCE must be adjusted for moisture content

$$ECCE_{adj} = 71\% \times \frac{100-4}{100} = 68\%$$

PROBLEM 6.19B.

What is the Ag lime recommendation?

$$Ag\ lime\ recommendation = \frac{\dfrac{2,100\ lbs\ CaCO_3}{acre}}{\dfrac{68}{100}} = \frac{3,100\ lbs}{acre}$$

laboratory may have its own standards for rating lime effectiveness. The lime recommendation is made using the lime requirement and the Effective Calcium Carbonate Equivalent (ECCE), the product of the CCE and the fineness factor. The fineness factor is determined by screening the product through sieves with appropriate openings. Screen meshes are rated as the number of openings per square inch. If the lime contains moisture, the ECCE must be adjusted for the moisture content ($ECCE_{adj}$). The agricultural lime recommendation is determined by dividing the lime requirement by the adjusted ECCE. Equations for the Effective Calcium Carbonate Equivalent (ECCE), a ECCE that is adjusted for moisture (ECCEadj), and how to determine the lime recommendation are below.

ECCE = %CCE/100 × (total fineness efficiency/100)
$ECCE_{adj}$ = ECCE * ([100− moisture]/100)

$$Ag\ lime\ recommendation = \frac{lime\ requirement\ \dfrac{lbs\ CaCC}{acre}}{\dfrac{ECCE_{adj}}{100}}$$

Soil pH and Agrichemicals

The soil pH influences the ability of the soil to retain nutrients and herbicides (Table 6.10). Because many agricultural chemicals are weak acids or bases, pH influences their concentration in solution. The equation for a weak acid is, HA → H$^+$ + A^{-1}, where H is hydrogen and A$^-$ is the salt of a weak acid. Weak acids do not completely dissociate in soil

and they can be described by the acid dissociation equation, $Ka = \dfrac{[H][A^-]}{[HA]}$. As shown in Table 6.10, the Ka (acid dissociation constant) is often reported a pKa, a constant which is the negative logarithm of the Ka, and which varies by acid type. At a pH value [-log (H)] less than the pKa, the chemical species on the left side of the equation is the

Table 6.10. Equations and pKa values of phosphorus, ammonia, atrazine, and 2,4-D in soil. At pH values less than the pKa, the material on the left is the dominant form. At pH values greater than the pKa, the dominant form is on the right. For example, at pH 6 H_2PO_4 is dominant; whereas at pH 8, HPO_4^{2-} is dominant.

Equation	pKa
$H_3PO_4 \leftrightarrow H^+ + H_2PO_4^-$	2.23
$H_2PO_4^- \leftrightarrow H^+ + HPO_4^{2-}$	7.2
$NH_4^+ \leftrightarrow H^+ + NH_3$	9.26
Atrazine-$NH_3^+ \leftrightarrow$ Atrazine-$NH_2 + H^+$	1.6
2,4-D-COOH $\leftrightarrow$ 2,4-D-COO⁻ + H^+	2.8

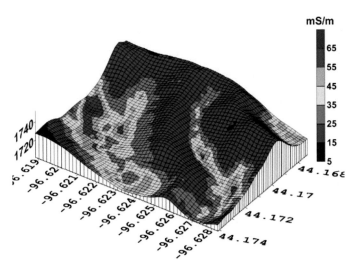

Fig. 6.5. Electrical conductivity superimposed on the topography map from the Moody field.

dominant species, whereas at pH values greater than the pKa, the species on the right side of the equation is the dominant species. Because many soils have a negative charge, negatively-charged compounds typically are not sorbed and move rapidly through the soil. For example, 2,4-D has a pKa of 2.8, meaning that it will have a negative charge if the pH is greater than 2.8 (Table 6.10). As a negatively charged ion, it will readily move with water through the soil profile.

Electrical Conductivity

The EC provides an index of all the anions and cations contained in the soil solution. At high values (i.e., high salt concentrations), seed germination and plant growth can be slowed or stopped. Solutions having high EC values also have high concentrations of soluble salts. Electrical conductivity may have identical values when reported as dS m⁻¹ (deciSiemens per meter), mS m⁻¹ (milliSiemens per m), or mmho cm⁻¹ (millimho per cm) (Clay et al., 2001; Carlson et al., 2016).

In Table 6.2, soil EC is provided under the headings, EC or soluble salts (1:1). Classically, saline soils were characterized as soils having EC values > 4 dS m⁻¹ as measured using a saturated paste method (ECe). However, most soil testing laboratories do not use the saturated paste method and therefore to assess salinity risks, the reported values may need to be converted to the saturation paste value. (Lee et. al, 2017). This conversion is soil dependent. In South Dakota medium-textured soils, the EC value of the 1:1 solution is converted to the EC of the saturated paste by multiplying the 1:1 value by 2.14 (Carlson et al., 2016).

In the field, EC can also be measured using an electromagnetic sensor (Fig. 6.5; Clay et al., 2011a). Special care is required to interpret the results of field EC measurements; the electromagnetic sensors measure the bulk EC of the soil, which is a function of water content, clay content, bulk density, salt concentration, organic matter content, and even some soil horizons. Based on the relative amounts of salts in the soil, appropriate remediation strategies can be developed (Oster et al., 1999; DeSutter et al., 2015)

PROBLEM 6.21.

The EC value of a soil measured in a 1:1 solution is 1 dS m^{-1}. Would the EC of a saturation paste extract (ECe) be higher or lower than 1 dS m^{-1}?

ANSWER:
Higher.
The estimated ECe value is 2.14 dS m^{-1} (1×2.14). At this EC value, yields and germination can be reduced.

Conversion of Soil Test Results to Fertilizer Recommendations

The conversion of the nutrient soil test value to fertilizer recommendations requires the field size and equation relating the soil test value to fertilizer recommendation. For some nutrients, recommendations are based on ppm, whereas for other nutrients recommendations are based on total amounts. Different agronomists use different models for these calculations (Kim et al., 2013). Different mathematical models have been developed for different soils and climatic conditions. To minimize errors, check with local agronomists about the appropriate calculations.

Summary

In summary, historically soil test reports are the starting point for assessing the chemical and nutrient status of the site. Information included in the report can be used for a wide range of purposes including assessing the risk of agrichemical leaching, determining fertilizer requirements, and estimating saline and sodic risks. Different laboratories use different methods for analyzing and reporting the values. All numbers must be standardized for comparison. Additional information for understanding the soil test results is available at Espinozo et al. and Horneck et al. (2011).

ACKNOWLEDGMENTS

Support for this document was provided by South Dakota State University, South Dakota Soybean Research and Promotion Council; Precision Farming Systems community in the American Society of Agronomy, International Society of Precision Agriculture, and the USDA-AFRI Higher Education Grant (2014-04572).

Additional Problems

6.22. The EC of a 1:1 solution is 0.8 dS m^{-1}. If a saturation paste would have been used, what is the likely EC of the sample?

6.23. A sample contains 60% silt and 30% clay, what is its soil texture?

6.24. The surface 6 inches contains 10 ppm (10 mg kg^{-1}) NO$_3$–N (nitrate–nitrogen) and the 6- to 18-in depth contains 5 ppm NO$_3$–N. How much NO$_3$–N is contained in the surface 18 in? Assume each 6 in contains 2 million pounds.

6.25. Convert 10 ppm to ppb.

6.26. Convert 50 ppm NO$_3$ (nitrate) to NO$_3$–N (nitrate–nitrogen).

6.27. The cation exchange capacity is 25 meq 100 g^{-1} (25 cmol$_c$ kg^{-1}), and the H + Al is 3 meq 100 g^{-1} (25 cmol$_c$ kg^{-1}), and Na is 4 meq 100 g^{-1} (4 cmol$_c$ kg^{-1}). What is the base saturation and exchangeable sodium percentage?

REFERENCES AND ADDITIONAL READINGS

Bly, A., and R. Gelderman. 2016. Chapter 25: Liming South Dakota soils. In: Clay, D.E., S.A., Clay, and E. Byamukama, editors, iGROW Corn: Best management practices, South Dakota State University Extension, Brookings, SD.

Carlson, C.G., D.E. Clay, D. Malo, J. Chang, C. Reese, R. Kerns, T. Kharel, G. Birru, and T. DeSutter. 2016. Chapter 32: Saline (salts) and sodium problems and their management in dryland corn production. In: D.E. Clay, C.G. Carlson, S.A. Clay, and E. Byamukama, editors, iGROW Corn: Best management practices. South Dakota State University, Brookings, SD.

Chang, J., D.E. Clay, C.G. Carson, S.A. Clay, D.D. Malo, R. Berg, J. Kleinjan, and W. Wiebold. 2003. Different techniques to identify management zone impact nitrogen and phosphorus sampling variability. Agron. J. 95:1550–1559. doi:10.2134/agronj2003.1550

Chang, J., D.E. Clay, B. Arnall, and G. Reicks. 2017. Essential plant nutrients, fertilizer sources, and application rate calculations. In: D.E. Clay, S.A. Clay, and S. Bruggeman, Practical Mathematics and Agronomy for Precision Farming. American Society of Agronomy, Madison, WI.

Clay, D.E., J. Chang, D.D. Malo, C.G. Carlson, C. Reese, S.A. Clay, M. Ellsbury, and B. Berg. 2011a. Factors influencing spatial variability of apparent electrical conductivity. Comm. Plant and Soil Analysis 32:2993–3008. doi:10.1081/CSS-120001102

Clay, D.E., S.A. Clay, C.G. Carlson, and S. Murrell. 2011b. Mathematics and science for agronomists and soil scientists. International Plant Nutrition Institute, Norcross, GA.

Clay, D.E., C.E. Clapp, R.H. Dowdy, and J.A.E. Molina. 1993. Mineralization of nitrogen in fertilizer-acidified lime-amended soil. Biol. Fertil. Soils 15:249–252. doi:10.1007/BF00337208

Clay, D.E., and T.P. Trooien. 2017. Understanding soil water and yield variability in precision farming. In: D.E. Clay, S.A. Clay, and S. Bruggeman, Practical Mathematics and Agronomy for Precision Farming. American Society of Agronomy, Madison, WI.

DeSutter, T., D. Franzen, Y. He, A. Wick, J. Lee, B. Deutsch, and D. Clay. 2015. Relating sodium percentage to sodium adsorption ratio and its utility in the Northern Great Plains. Soil Sci. Soc. Am. J. 79:1261–1264. doi:10.2136/sssaj2015.01.0010n

Lee, J., D.E. Clay, C. Reese, D.D. Malo, and T.M. DeSutter. 2017. Predicting electrical conductivity of the saturation extract from a 1:1 solution to water ration. Commun. Soil Sci. Plant Anal. Taylor and Francis Group, Batavia, IL.

Espinozo, L., N. Slaton, and M. Mozaffari. Understanding the numbers on your soil test report. FSS2118, University of Arkansas. Fayetteville, AR.

He, Y., T.M. DeSutter, and D.E. Clay. 2013. Dispersion of pure clay minerals as influenced by calcium/magnesium ratios, sodium adsorption ratios, and electrical conductivity. Soil Sci. Soc. Am. J. 77:2014–2019. doi:10.2136/sssaj2013.05.0206n

He, Y., T.M. DeSutter, F. Casey, D.E. Clay, D. Franzen, and D. Steele. 2015a. Field capacity water as influenced by Na and EC: Implications for subsurface drainage. Geoderma 245-246:83–88. doi:10.1016/j.geoderma.2015.01.020

He, Y., T.M. DeSutter, D.G. Hopkins, and D.E. Clay. 2015b. The relationship between SAR1:5 and SARe of three extraction methods. Soil Sci. Soc. Am. J. 79:681–687. doi:10.2136/sssaj2014.09.0384

Horneck, D.A., D.M. Sullivan, J.S. Owens, and J.M. Hart. 2011. Soil test interpretation guide. EC 1478. Oregon State University, Corvallis, OR.

Kim, K., D. Clay, S. Clay, G.C. Carlson, and T. Trooien. 2013. Testing corn (Zea Mays L.) preseason regional nitrogen recommendation models in South Dakota. Agron. J. 105:1619–1625. doi:10.2134/agronj2013.0166

Kirkby, C.A., J.A. Kirkegaard, A.E. Richardson, L.J. Wade, C. Blanchard, and G. Batton. 2011. Stable organic matter: A comparison of C:N:P:S ratios in Australia and other world soils. Geoderma 163:197–208. doi:10.1016/j.geoderma.2011.04.010

Mallarino, A.P. 2012. The moist soil test for potassium and other nutrients: What it all about. Iowa Integrated Crop Management, Ames, IA.

Mallarino, A.P., J.E. Sawyer, and S.K. Barnhrt. 2013. A general guide for crop nutrient and limestone recommendations in Iowa. PM 1688. Iowa State University Extension and Outreach, Winterset, IA.

Malo, D.D., T.E. Schumacher, and J.J. Doolittle. 2005. Long-term cultivation impacts on selected soil properties in the northern Great Plains. Soil Tillage Res. 81:277–291. doi:10.1016/j.still.2004.09.015

Midwest Laboratories. 2007. Interpreting soil analysis. Midwest Laboratories, Inc. Omaha, NE.

Oster, J.D., I. Shainberg, and I.P. Abrol. 1999. Reclamation of salt-affected soils. Agricultural Drainage. Agron. Monogr. 38. ASA, CSSA, SSSA, Madison, WI.

Oster, J.D., and H. Frenkel. 1980. The chemistry of the reclamation of sodic soils with gypsum and lime. Soil Sci. Soc. Am. J. 44:41–45. doi:10.2136/sssaj1980.03615995004400010010x

Sawyer, J.E., and A. Mallarino. 1999. Differentiating and understanding the Mehlich 3, Bray and Olsen soil phosphorus tests. In: Proceedings of the University of Minnesota Crop Pest Management Short Course Program St. Paul, MN. p. 22–23.

Sikora, F.S., R.S. Mylavarapu, D.H. Hardy, M.R. Tucker, and R.E. Franklin. 2005. Conversion equations for soil test extractants. Southern Regional Fact Sheet, SERA IEG-6*5. University of Kentucky, Lexington, KY.

Soil Survey Staff. 1999. Soil taxonomy: A basic system of soil classification for making and interpreting soil surveys. 2nd ed. U.S. Department of Agriculture Handbook 436. USDA-NRCS. Washington, DC.

Spargo, J., and T. Allen. 2013. Interpreting your soil test report. Univ. of Massachusetts Amherst, Amherst, MS.

Whiting, D., A. Card, C.W. Wilson, and J. Reeder. 2015. Estimating soil texture. Master Gardener # 214. Colorado State University. Fort Collins, CO.

Calculations Supporting Management Zones

7

David E. Clay,* Newell R. Kitchen, Emmanuel Byamukama, and Stephanie A. Bruggeman

Chapter Purpose

In precision farming, a management zone is a subfield area that expresses a relatively homogeneous combination of yield limiting factors for which a practice or input can be tailored to that zone (Doerge, 1999). Management zones are often defined based on set of soil or plant properties that are used to identify areas with similar characteristics. The management zone concept be used to vary seeding rates, cultivars, fertilizers, and pesticides. Different problems (e.g., fertility, weeds, diseases or insect infestations) may require different management zones. This chapter reviews the management zone concept and the calculations associated with delineating the zone boundaries. Included are examples of how different types of crops and soil data may be used for management zone delineation.

Key Terms

Grid soil sampling, management zones, remote sensing, yield maps, directed sampling, NDVI, GNDVI, variance.

Mathematical Skills

Calculations associated with defining management zones.

Collecting Spatial Information

Since the early 1990s the tools of precision farming (GPS, yield monitors, soil sensors, etc.) have documented how spatial and temporal yield variability are important factors for efficient and profitable crop management systems. Once variability is measured, it can be used to separate the field into smaller subfield areas where tailored recommendations for each zone can be created and implemented. These subfield areas are commonly referred to as "management zones". Obtaining and transforming measurements of within-field variability into management zones has historically been accomplished using two different strategies:

1) Direct sampling at grid points, and
2) Ancillary measurements or maps as a proxy of crop and soil variability.

Direct Sampling

Spatial information for creating management zones can be obtained by many approaches including grid point sampling. In grid point sampling, sampling

D. Clay, E. Byamukama, and S. Bruggeman, Agronomy, Horticulture, and Plant Science Department, Box 2207A, 57007; N. Kitchen, Division of Plant Sciences, University of Missouri, Columbia, MO. *Corresponding author (david.clay@sdstate.edu).
doi:10.2134/practicalmath2017.0024

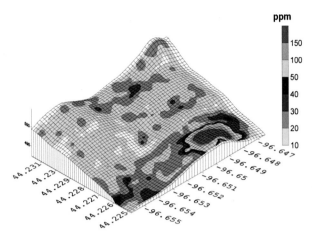

Fig. 7.1. A contour map of soil test P values in a 160 acre field located in South Dakota.. This map was created based on collecting point samples from a 100 ft. grid (30 m) (1/4 acre grid). Spatial statistics was then used to estimate values at unknown locations.

points are identified and information is collected from the point and surrounding area. The density of the sampling protocol depends on costs as well as the problem and location, and can range from two- to four-acre grids for fertility maps to large regions for insect control (Franzen and Kitchen, 1999; Clay et al., 2002, 2004; Chang et al., 2003, 2004; Fridgen et al., 2004; Schepers et al., 2004). For grid point sampling, information collected at each sampling point along with the location of the sampling point are processed within a software program to generate a map that represents the field variation (Fig. 7.1). Examples of this method include the use of soil test results collected from a two-acre grid to create soil nutrient availability and fertilizer application maps. Other examples include the generation of weed, insect, or disease intensity maps that are used to create chemical or cultural management treatment maps.

The disadvantages of grid point sampling include the expense of collecting and analyzing the data (Vaughan. 1999), and the need for highly trained people to develop the application maps. Whereas the advantages are that highly trained experts are not required to collect the data and this approach can locate areas that have historically received heavy manure applications (Fig. 7.1). To keep the costs manageable, farmers or their consultants often increase the spacing between the grid points. However if samples are too sparse, one loses the ability to accurately represent differences within a field.

Ancillary Measurements or Maps

The second strategy for creating management zones is to use other spatial information that may be available for a field. This information may include electrical conductivity data, yield maps, elevation, drainage class, or remote sensed data. This information may be less expensive to obtain than physically walking the field. Within a management zone, it is expected that the crop will respond to management similarly. Examples of management zones include landscape position classes (e.g., summit, sideslope, toeslope), yield maps classed into productivity zones, remote sensed image maps, or publicly-available soil survey maps. The boundaries of these zones are then used as the basis for additional sampling and variable management within the field. However, the reliability of the zone depends on known reliable relationship between the information used to create the zones, and the desired management practice. For example, apparent soil electrical conductivity (EC_a) sensing could be used as a proxy for nutrient variability (Franzen and Kitchen, 1999; Chang et al., 2004; Sudduth et al., 2005; Bobryk et al., 2016). To do so, one would need confidence that for the fields where this is available, there are relationships between soil EC_a and variation in the soil nutrient to be managed. Similarly, if a known relationship between landscape position and incidence of a particular fungal disease exists; landscape classes could be a valuable method of creating management zones for variably applying fungicide.

Regardless of which strategy is used for gaining information about in-field variability, management zone boundaries may vary with each problem addressed and their location can move annually from one year to the next. For example, Park and Tollefson (2005) reported that based on correlation analysis, a good predictor for identifying corn rootworm problems was the number of adults in the ear during the current year. Once the boundaries are identified, the appropriate information can be collected, and site-specific recommendations developed.

The different methods for measuring and mapping within-field variation have unique strengths and weaknesses. As noted, grid sampling can be expensive yet the method can be reliable and accomplished with little statistical or technical training. However, identifying the right information and procedures for creating management zones according to the second strategy requires a highly skilled practitioner who has a fundamental understanding of soil and agronomic sciences.

Useful Information for Delineating Management Zones

In management delineation, questions that must be answered include: (i) what information should be collected (ii) how the information should be processed, and (iii) how many meaningful unique management zones can be identified (Chang et al., 2004; Fridgen et al., 2004; Bobryk et al., 2016; Fig. 7.2). For management zone delineation, it is critical to match the information to the problem. Common information layers used to create zone boundaries are yield maps, remote sensing, electrical conductivity, and soil surveys. Each of these data layers and how they might be used for creating management zones are discussed below.

Yield Monitor Data

Yield monitor data is a dense data layer consisting of points that have the dimensions of the combine head by the distance traveled, during a specified time period. Prior to conducting an analytical assessment, errors in the data layer should be corrected or removed. Information on the potential errors is available in Colvin and Arslan (1999), Lems et al. (2001), Reese et al. (2001), Sudduth and Drummond (2007), and Franzen et al. (2008).

Yield data is an excellent source for management zone identification, because for many fields, yield represents the dominant expression of the interaction between soil, climate, and management. Yield data can be evaluated using multiple techniques. The first approach is to analyze a single yield map. In many situations, superimposing the yield map on an elevation map simplifies the ability to identify the factors responsible for the yield variation. The second approach is to reduce the prediction error by combining multiple data sets into a common analysis (Chang et al., 2003). As discussed in Franzen et al. (2008), using multiple years of yield monitor data is a preferred technique in the development of more meaningful management zone maps. For example, in Illinois, meaningful relationships between soil P, K, and pH was only apparent after combining multiple years of data. For fields with

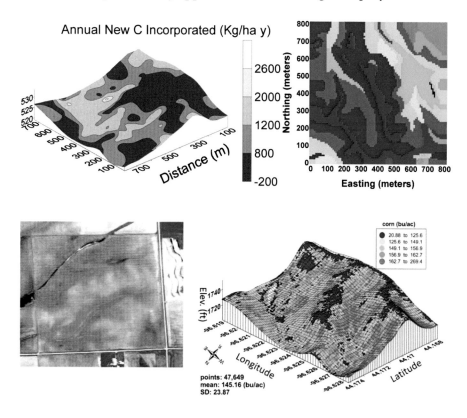

Fig. 7.2. Potential information that can be used to create a management zone map. All of these maps were collected from the same field. This figure provides information on the amount of new soil organic carbon added to the soil annually (top left), the order 1 soil survey map (top right), color image of the bare soil (bottom left), and yield map (bottom right) of a South Dakota production field. Additional information on this field is available in Fig. 7.3, Chang et al. (2003), and Clay et al. (2005).

PROBLEM 7.1.

What is the size of the area harvested by a combine with a 40-ft head in 1 s if it is traveling at 3 miles per hour (mph) and 4 mph?

ANSWER:

At 3 mph: $40\ \text{ft} \times \dfrac{3\ \text{miles}}{\text{hr}} \times \dfrac{5280\ \text{ft}}{\text{mile}} \times \dfrac{1\ \text{hr}}{60\ \text{min}} \times \dfrac{1\ \text{min}}{60\ \text{s}} = 176\ \text{ft}^2\ \text{s}^{-1}$

At 4 mph: $40\ \text{ft} \times \dfrac{4\ \text{miles}}{\text{hr}} \times \dfrac{5280\ \text{ft}}{\text{mile}} \times \dfrac{1\ \text{hr}}{60\ \text{min}} \times \dfrac{1\ \text{min}}{60\ \text{s}} = 235\ \text{ft}^2\ \text{s}^{-1}$

PROBLEM 7.2.

The combine slows down from 4 mph to 2 mph as it approaches the edge of the field. During this time period, the apparent yield monitor yield increases from 200 to 400 bu acre^{-1}. Why?

ANSWER:

As the combine slows down the area harvested in a given time period decreases. However, the amount of grain impacting the sensor remains relatively constant because there is a lag time between the header and the yield monitor sensor. The apparent yield increase is because yield is defined as amount of grain contained in a given area.

PROBLEM 7.3.

In years 1, 2, and 3, the corn, soybean, and wheat yields from zone 6 are 180, 55, and 70 bu acre^{-1}, respectively. If the highest corn, soybean, and wheat yields are 200, 70, and 90, what are the normalized yields from these years?

ANSWER:

When determining yields and variances, the combination of actual yield data from different crops, results in very large variance values which in turn reduces the ability to characterize management zones. This is solved by putting all of the values on a common scale ranging from 0 to 1.

$$\text{corn}: \frac{180\ \dfrac{\text{bu}}{\text{acre}}}{200\ \dfrac{\text{bu}}{\text{acre}}} = 0.9 \quad \text{soybean}: \frac{55\ \dfrac{\text{bu}}{\text{acre}}}{70\ \dfrac{\text{bu}}{\text{acre}}} = 0.79 \quad \text{wheat}: \frac{70\ \dfrac{\text{bu}}{\text{acre}}}{90\ \dfrac{\text{bu}}{\text{acre}}} = 0.78 \quad ,$$

multiple crops, this can be accomplished by expressing yield on a relative basis. For example, if the highest corn yield in a yield map is 220 bu acre^{-1}, divide all points in the yield map by 220. This will produce a map with values ranging from 0 and 1. A value of 1 represents the highest yield possible, whereas a zero is the lowest yield posible.

One approach to define management zones is to use the variance to identify zones that should be similar or dis-similar. For example, if a field has yields at four points along a transect that are 100, 105, 150, and 147 bu acre^{-1}, based on a visual interpretation, a manager would think that first two should be treated as one zone and the second two as a second zone. Numerically, this same solution can be developed by calculating the variance of the four points (711) and the pooled variance of the two zone (8.5). By treating this transect as 2 management zones, the yield variance is reduced 99%.

Determining the management zones containing two variables follows the approach discussed in problem 7.4, except in this case, use two variables (Problem 7.5). The goal is to identify boundaries that minimizes the pooled variance for both data layers. In Layer 1, the variance for Zone 1 and Zone 2 are 0.57 ($n = 6$) and 0.33 ($n = 3$). The pooled variance for Layer 1 is 0.5. In Layer 2, the same split is used and the resulting variances are 1.37 ($n = 6$) and 1.00 ($n = 3$). The resulting pooled variance is 1.26. The two data sets are combined by adding the variance. For both layers the combined variance is 5.22 (1.61+3.61). The combined pooled variance is 1.76 (0.5+1.26). The percent variance reduction is,

$$100 \times \left(1 - \frac{1.76}{5.22}\right) = 66\%$$

PROBLEM 7.4.

In this problem, use the data provided below and our goal is to identify two management zones. In this field, the yields from nine points are identified. First, visually identify similar and dissimilar areas. Second, calculate the variance of the entire data set. In Excel, this is conducted using the command = var(start, end). The variance is 361. Third calculate the pooled variances where Zone 1 contains all values except 130 and 140, and Zone 2 contains the 130 and 140 values.

North		
100	110	130
90	100	140
80	110	120

ANSWER:

Zone 1: 100, 90, 80, 110, 100, 110, and 120

$s^2 = 181$

$n = 7$

Zone 2: 130, 140

$s^2 = 50$

$n = 2$

$$s^2_{pooled} = \frac{\left[(n_1 - 1)s_1^2 + (n_2 - 1)s_2^2\right]}{n_1 + n_2 - 2}$$

In this equation, the n_1 and n_2 values are associated with the number of points in Zone 1 and Zone 2. The s^2 values are the variances in the two management zones. Variance can be calculated in Microsoft Excel using the = var(start,end) command. For the entire data set the variance is 361.1. When yields (130 and 140 bu acre^{-1}) are treated as a separate zone, the pooled variance decreases to 162.2, and the percent variance reduction was:

$$100 \times \left(1 - \frac{s^2_p}{s^2_{field}}\right) = 55\%$$

It is important to separating the field into two zones if the yield is used to make a recommendation. In this example, zone 1 had an average yield = average(start, end) of 101 bu acre^{-1} and zone 2 had a yield of 135 bu acre^{-1}. All that the variance did was provide a numeric approach to identify these two areas.

An alternative approach to complete this analysis is to convert both data sets to a relative scale ranging from 0 to 1 and then complete the above analysis. Again once the zones are identified, the next question is to identify the limiting factor.

Remote Sensing

Remote sensing is a method of collecting information without physical contact (Chang et al., 2013). In remote sensing, spectral reflectance can be measured from wave bands that may be visible or not visible to the human eye. Visible wave bands generally are those between 390 and 770 nm. Widely used visible band are blue (450–510 nm), green (530–590 nm), and red (640–670 nm). These primary colors are used to create the secondary colors and many of the other colors. When blue, green, and red are combined, white is the product. The secondary colors have various combinations of the primary colors. For example yellow is produced by combining green and red (Fig. 7.3).

Different crop stresses that can limit yield may have different reflectance patterns or signatures. Remote sensing can be used to identify zone boundaries because different problems have spectral signatures which may or may not be unique (Fig. 7.2, 7.4). Remote-sensing can be obtained from handheld devices, manned aircraft, UAV (Unmanned Aerial Vehicle), and space-based (satellite) platforms (Fig. 7.5), each with unique strengths and weaknesses (Table 7.1 and 7.2). Understanding the benefits and limitations with each platform is critical for selecting the appropriate data source. Cost for remote sensing information is highly variable, and can range from free to very expensive (Table 7.2; Chang et al., 2004, 2013; Reese et al., 2003; Reese, 2004; Clay et al., 2012).

An excellent source of archived information is the Landsat program that has been continuously acquiring information since 1972. The Landsat program collects and archives data from multiple wavebands, at resolutions ranging from 15 m for panchromatic (black and white) to 30 m for multispectral (many different wave bands). The thermal infrared band has a resolution of 60 m. The data are archived in scenes that are approximately 180 by 190 km and the

PROBLEM 7.5.

A data set contains two different information layers, which are shown below. Determine two management zones for this field. In this example the color indicates how to split the field.

Layer 1			Layer 2		
3	4	5	10	9	7
2	3	6	11	8	5
3	4	5	9	8	6

Answer:

Layer 1	Layer 2
Zone 1: 3, 2, 3, 4, 3, 4	Zone 1: 10,11,9, 9, 8, 8
$n = 6$	$n = 6$
mean = 3.2	mean = 9.17
$s^2 = 0.567$ $s^2 = 0.57$	$s^2 = 1.37$

Zone 2: 5, 6, 5	Zone 2: 7, 5, 6
$n = 3$	$n = 3$
mean = 5.3	mean = 6
$s^2 = 0.33$	$s^2 = 1$
Whole field variance	Whole field variance
$n = 9$	$n = 9$
$s^2 = 1.61$	$s^2 = 3.61$

Determining the management zones follows the approach discussed in problem 7.4 except in this case, use two variables. Step 1: Add the variance for both data sets together (1.61+3.61 = 5.22). In this case the variance for the top data set is 1.61 and the variance for the bottom data set is 3.61. The goal in this management zone approach is to minimize the pooled variance. In Layer 1, the variance for the gray and white areas are 0.57 (n = 6) and 0.33 (n = 3). The pooled variance for Layer 1 is 0.5. In Layer 2, the same split is used and the resulting variances are 1.37 (n = 6) and 1.00 (n = 3). The resulting pooled variance is 1.26. The two data sets are combined by adding the variance. For both layers the combined variance is 5.22 (1.61+3.61). The combined pooled variance is 1.76 (0.5+1.26). The percent variance reduction is,

An alternative approach to complete this analysis is to convert both data sets to a relative scale ranging from 0 to 1 and then complete the above analysis. Again once the zones are identified, the next question is to identify the limiting factor.

images are collected when the satellite is directly overhead. However, the latest satellite, Landsat 8, can also collect off-nadir (not directly overhead) information. The satellite follows a sun-synchronous orbit with a 16 d revisit time. For comparison, data is also collected from the moderate resolution imaging spectroradiometer (MODIS) every two to three days. This system has a viewing swath of 2330 m and collects 36 spectral bands ranging from 405 to 14385 nm. The spatial resolution ranges from 250 to 1000 m.

Bands that may be of interest are near-infrared (750–1400 nm), short wavelength infrared (1400–3000 nm), mid-wavelength infrared (3000–8000 nm), long wavelength infrared (8000–15,000 nm), and far infrared (15,000–1,000,000 nm). The thermal imaging region is associated with the long-wavelength infrared, and the mid wavelength infrared or thermal infrared, is the region used by heat seeking missiles.

Digital images consist of pixels located in columns and rows. For each waveband, a pixel will have an associated solar reflectance or intensity value. The radiometric resolution varies and depends on the bits allocated for each number. For example, an 8-bit number can have numbers that range from 0 to 255 (2^8-1), whereas an 11-bit number has values that range from 0 to 2047. The pixel intensity values can be converted to reflectance by dividing the reflected value by the value representing solar radiation. Reflectance is the ratio between the reflected light and the non-reflected light (incoming). Reflectance values from the different bands can be combined to produce multispectral images.

The advantages of using remote sensing to identify management zones include that: (i) archived free informationis being collected from LandSat, and (ii) data collected from a whole field can be used to target scouting. Disadvantages with archived images are that the signatures may not be unique and it (cont. on p. 130)

PROBLEM 7.6.

Determine management zones based on yields and variability. Yields are measured over four years and it is separated into eight zones. The standard deviation is the square root of the variance and is calculated in Microsoft Excel with the command = stdev(start,end).

Sampling points	1996	1998	Year 2000	2002	Ave	Standard deviation
			Bu acre^{-1}			
1	111	177	151	111	137.5	32.4
2	113	176	156	95	135	37.4
3	108	218	155	90	142.9	57.2
4	114	213	153	107	146.7	48.6
5	115	205	165	113	149.5	44.1
6	109	216	186	110	155.2	54.2
7	140	208	183	105	159.0	45.7
8	97	226	184	103	152.5	63.0

ANSWER:

Yields (great or less than average)

        Low: < 147.3 identified as a 1

        High > 147.3 identified as a 2

Variability (greater or less than the standard deviation)

        Low < 43.7 identified as a 1

        High > 43.7 identified as a 3

Areas with a yield + variability sum of 2 are areas with low yields and low variability

Areas with a yield + variability sum of 3 are areas with high yield and low variability

Areas with a yield + variability sum of 4 are areas with low yields and high variability

Areas with a yield + variability sum of 5 are areas with high yields and high variability

Approach simplified

Step 1: Determine average yield and standard deviation of the entire data set over the four years.

Command = average(start,end), Average = 147

Command = stdev(start,end), standard deviation = 43.7

Step 2: Determine the mean and standard deviation for each point

Step 3: Based on the mean and standard deviation determine the management zone score. Areas with high yields and low variability should be managed as high yield areas. Areas with low yield and low variability should be managed as low variability. High variability areas management depends on expectations.

Zone	Ave	Standard deviation	Yield zone	Variability zone	Sum Yield + var	Zone
1	137	32.39	1	1	2	low yield low variability
2	135	37.44	1	1	2	low yield low variability
3	143	57.16	1	3	4	low yield high variability
4	147	48.58	2	3	5	high yield high variability
5	149	44.13	2	3	5	high yield high variability
6	155	54.23	2	3	5	high yield high variability
7	159	45.66	2	3	5	high yield high variability
8	152	63.05	2	3	5	high yield high variability

PROBLEM 7.7.
What is the amount of reflected light if the incoming red (640–670 nm) light has a value of 240 and reflected light reflected from bare soil has a value of 40?

ANSWER:
Reflected light = 40/240 = 0.167. The percentage of reflected light is 16.7%.

A

Color	Blue	Green	Red	NIR
Band width (nm)	450-510	530-590	640-670	750-1400

Fig. 7.3. Band width for near infrared (NIR) and the primary colors of blue, green, and red.

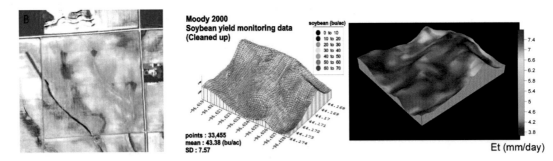

Fig. 7.4. Landsat image of a growing soybean crop (19 June), yield map, and an evaporation transpiration map of the identical field. All maps show common features in summit/shoulder and footslope areas (Mishra et al., 2008).

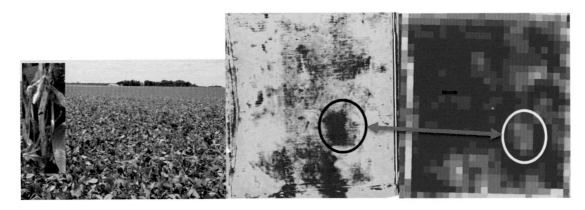

Fig. 7.5. White mold in a South Dakota production field (left). The center image was collected with a UAV and the right image was collected with the Landsat satellite. These images show that the ability to detect differences depends on spectral resolution.

is impossible to conduct ground truthing after the fact. In addition, multiple factors may impact the reflectance characteristics (Kim et al., 2008; Mishra et al., 2008). Ground truthing involves confirming that the spectral signature is accurately identified. For example, the plants appear yellow because they are N stressed. However, other limiting factors such as S and Fe deficiencies or infected plants may also appear yellow.

Depending on the platform, remote sensing sensors collect information from multiple wavebands and resolutions (Table 7.2). Wave bands generally used in agriculture include the visible (blue, green, and red) and the near infrared (NIR) regions of the electromagnetic spectrum. The electromagnetic spectrum of an (cont. on p. 132)

Table 7.1 Advantages and disadvantages of various platforms for remote-sensing data collection.

Platform	Advantages	Disadvantages
Hand or ground	· Can be used to identify the reflectance characteristics of a leaf, plant, or small (< 2 m²) area. · Flexible · Cost can range from low to high · Useful for real-time treatments.	· Collect the reflectance characteristic from a single point, · Point measurement are not designed for creating image that show landscape features. · Collecting data from a whole field requires a mobile vehicle.
UAV	· Flexible. · Relatively low cost. · Very high spatial resolution. · Changeable sensors.	· Relatively unstable platform can create blurred images. · Geographic distortion. · May require certification to operate. · Limited to 400 ft. elevation · Processing the data into field images may contain "stitching" errors.
Aircraft	· Relatively high spatial resolution. · Changeable sensors.	· Geographic distortion. · Cost may be very high. · Availability depends on climatic condition.
Satellite	· Some free images. · Clear and stable images. · Large area within each image. · Good historical data.	· May have high costs. · Clouds may hide ground features. · Fixed schedule. · Data may not be collected at critical times. · Requires sorting to identify useful data.

Table 7.2. Spaced-based sensors that can be used for agriculture. The spectral band supported by a sensor may include: B, Blue (470 nm); G, Green (550 nm); R, Red (650 nm); N, NIR (860 nm); R-edge, Red-edge (730 nm); and IR, Infrared (1550–2350 nm).

	Spatial Resolution (m)		Multi-Spectral Bands	Temporal	Relative Cost
	Pan	Multi		Revisit days	
High Spatial Resolution Images					
GeoEye-1	0.46	1.84	B, G, R, N	2.1 to 8.3	High
WorldView-1	0.55			1.7 to 5.9	High
WorldView-2	0.52	2.4	B, G, R, N, R-edge, 3 others	1.1 to 3.7	High
WorldView-3	0.34	1.38	B, G, R, N, R-edge, 23 others	1 to 4.5	High
Pleiades-1A	0.5	2	B, G, R, N	Daily	High
Pleiades-1B	0.5	2	B, G, R, N	Daily	High
QuickBird	0.73	2.9	B, G, R, N	1 to 3.5	High
IKONOS	1	4	B, G, R, N	3	High
SPOT-6	1.5	6	B, G, R, N	1 to 5	High
SPOT-7	1.5	6	B, G, R, N	1 to 5	High
Plant Labs (previously RapidEye)		5	B, G, R, N, R-edge	1 to 6	Med to High
Moderate Spatial Res. Images					
Sentinel-2		10	B, G, R, N, R-edge, 5 others	5 to 10	Free
SPOT-5	5	10	G, R, N, Shortwave IR	2 to 3	High
LANDSAT 7 ETM+	15	30	B, G, R, N, 3 others	16	Free
LANDSAT 8 OLI	15	30	B, G, R, N, 6 others	16	Free

PROBLEM 7.8.

Determine the NDVI value of plants located in an upland and bottom slope areas. In the upland area, plants have reflectance values in the red and NIR bands of 0.1 and 0.3, whereas in the bottom land area, plants have reflectance values of 0.08 and 0.60 in the red and NIR bands, respectively.

ANSWER:

Upland: $NDVI = \dfrac{(NIR - RED)}{(NIR + RED)} = \dfrac{(0.3 - 0.1)}{(0.3 + 0.1)} = 0.5$

Bottom land: $NDVI = \dfrac{(NIR - RED)}{(NIR + RED)} = \dfrac{(0.6 - 0.08)}{(0.6 + 0.08)} = 0.76$

The lower value in the upland area indicates that at this location, the plant is not as healthy as the bottom land area. The relative lower value could result from a multitude of reasons, including low plant survival, diseases, water stress, and nutrient stress.

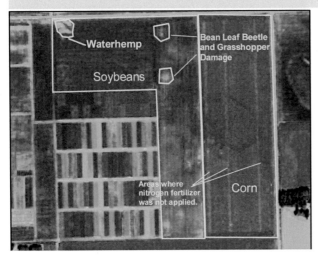

Fig. 7. 6. False color IKONOS image (green, red, and near infrared) of corn and soybeans on 17 July 2002 (Source: Upper Midwest Aerospace Consortium). In this image all problems were confirmed by ground scouting.

object refers to electromagnetic radiation that is emitted or adsorbed by that object. Different objects have different characteristic reflectance signatures.

Remote-sensing data can be visualized and processed in a variety of ways. For example, true color images can be created by displaying the primary color blue (470 nm), green (550 nm), and red (650 nm) bands as the blue, green, and red colors, respectively. However, a false color composite image is produced when the green, red, and NIR bands are displayed as blue, green and red colors, respectively (Fig. 7.7). In a false color image, healthy plants may appear red, whereas in a true color image, healthy plants appear green. Based on the color intensity, management zones for nutrients, water stress, weeds, insects, and diseases can be identified. However, because many problems have similar reflectance characteristics additional information is needed to confirm problems. This may involve ground scouting, tissue sampling, or using the pest biological characteristics to confirm problems.

For each pixel, the different bands can be combined into indices (Scharf et al., 2002). Two widely used indices are the normalized difference vegetation indices (NDVI) and the green normalized difference vegetation indices (GNDVI). These indices are based on reflectance in three bands, red, green, and near infrared (NIR). The NDVI and GNDVI indices are calculated with the equations,

$$NDVI = \frac{(NIR - RED)}{(NIR + RED)}$$

$$GNDVI = \frac{(NIR - GREEN)}{(NIR + GREEN)}$$

The NDVI and GNDVI values can range from -1 to 1 and different targets have different values. Previous work suggests that NDVI and GNDVI are sensitive to multiple factors including water and N stress (Scharf et al., 2002; Clay et al., 2005).

When using remote sensing, it is important to understand several key terms. Radiometric correction is the process of correcting or calibrating the collected reflectance data to a standard value. Approaches used for radiometric correction include the use of multiple sensors that measure incoming and reflected light simultaneously, the use of reflectance standards for correcting point measurements, and the use of computer programs to correct for atmospheric distortions. Atmospheric distortion result from atmospheric gases changing the reflectance value. Georectification is the process of using points with known locations to reduce distortions resulting from: (i) sensor height and velocity variability, (ii) nonlinearity in the data set, and (iii) Earth curvature.

Apparent Soil Electrical Conductivity

The ability of soil to conduct an electrical current is related to many factors including bulk density, compaction, depth of the soil profile, soil pH, soil water content, soil clay content, and salt concentration (Heiniger et al., 2003; Sudduth et al., 2005). In production fields, apparent soil electrical conducivity (EC_a) is measured using a mobile sensor either using electromagnetic induction sensors, or using electrically-charged coulter sensors that are in contact with the soil (Sudduth et al., 2005). This information can be used alone or in combination with other mapped information to create management zones for seeding rates, fertilizer rates, and cultivar selection (Clay et al., 2001; Heiniger et al., 2003; Chang et al., 2004; Sudduth et al., 2005). The map shown in Fig. 7.6 is based on soil EC_a. In this

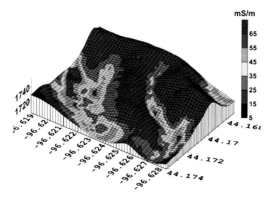

Fig. 7.7. Elevation impact on apparent soil electrical conductivity (EC_a) in a South Dakota field (Clay et al., 2001; Heiniger et al., 2003; Sudduth et al., 2005).

case, areas with high soil EC_a were associated with lower yields, which could have resulted from soil compaction or high salt concentrations.

Soil Surveys

Soil surveys do not require processing because the boundaries are already identified and these classed areas can serve as management zones. In the United States, soil surveys have been collected, digitized, and distributed in the SSURGO and STATSGO2 data layers. SSURGO scales can range from 1:12,000 to 1:63,360, whereas STATSGO2 scales can range from 1:250,000 to 1:1,000,000 in Alaska. The surveys provide information on available water capacity, soil pH, soil paste electrical conductivity (lab), flooding frequency, productivity, and land-use limitations (Reitsma et al., 2016). The different data layers can produce different results (Mednick, 2010) and for precision farming, SSURGO is the preferred data source. The soil survey attempts to separate the soil into mapping units with similar characteristics. Each mapping unit can be treated as a management zone.

Additional Problems

7.9. Do healthy plants have higher or lower NDVI values than nonhealthy plants?

7.10. What is radiometric correction?

7.11. Why is radiometric correction conducted?

7.12. What is georectification, and why is it important?

7.13. How would you separate the following corn yields from a field into management zone?

160	200	250	250
140	150	160	160
130	145	155	158
100	110	115	140

7.14. How would you assess the effectiveness of the management zones?

7.15. Which is more important, increasing the yield of low yielding area or increasing the yield of high yielding area?

By themselves, soil surveys have had a mixed ability to accurately identify management zones. However, the accuracy of the individual soil boundary can be improved by using EC or other information layers to adjust the boundaries (Mount, 2001; Bobryk et al., 2016).

Software Programs

There are many different software programs and statistical approaches that can be used in management zone calculations (Fridgen et al., 2004; Mount, 2001; Schepers et al., 2004). However, a discussion of these programs is beyond the scope of this chapter.

Summary

In summary, the purpose of management zones is to identify areas that can be managed similarly. The approach and information used to identify the zone boundaries are dependent on the problem. For example, the weed and disease management zones may be based on reflectance (Reese, 2004; Chang et al., 2004), corn root worm management zones based on modeled adult populations (Park and Tollefson, 2005), and fertilizer recommendations may be based on soil information (Franzen and Kitchen, 1999; Scharf et al., 2002; Chang et al., 2003; Clay et al., 2002, 2004). Within a zone, it is assumed that a single management practice is appropriate, and the goals of the management zone approach is to target the treatments to where and when they are needed (Chang et al., 2004; Fridgen et al., 2004).

ACKNOWLEDGMENTS

Support for this document was provided by South Dakota State University, South Dakota Soybean Research and Promotion Council, Precision Farming Systems community in the American Society of Agronomy, International Society of Precision Agriculture, and the USDA-AFRI Higher Education Grant (2014-04572).

REFERENCES

Bobryk, C.W., D.B. Myers, N.R. Kitchen, J.F. Shanahan, K.A. Sudduth, S.T. Drummond, B. Gunzenhauser, and N.N. Gomez Raboteaux. 2016. Validating a digital soil map with corn yield data for precision agriculture decision support. Agron. J. 108:957–965. doi:10.2134/agronj2015.0381

Chang, J., S.A. Clay, D.E. Clay, and K. Dalsted. 2004. Detecting weed-free and weed-infested areas of a soybean field using near-infrared spectral data. Weed Sci. 52:642–648. doi:10.1614/WS-03-074R1

Chang, J., D.E. Clay, S.A. Clay, and C.L. Reese. 2013. Using remote sensing technique to assess soybean yield limiting factors. In: D.E. Clay, C.G. Carlson, S.A. Clay, L. Wagner, D. Deneke, and C. Hay, editors, iGrow soybean: Best management practices for soybean production. South Dakota State University, SDSU Extension, Brookings, SD.

Chang, J., D.E. Clay, C.G. Carson, S.A. Clay, D.D. Malo, R. Berg, J. Kleinjan, and W. Wiebold. 2003. Different techniques to identify management zones impacts nitrogen and phosphorus sampling variability. Agron. J. 95:1550–1559. doi:10.2134/agronj2003.1550

Clay, D.E., J. Chang, D.D. Malo, C.G. Carlson, C. Reese, S.A. Clay, M. Ellsbury, and B. Berg. 2001. Factors influencing spatial variability of apparent electrical conductivity. Commun. Soil Sci. Plant Anal. 32:2993–3008. doi:10.1081/CSS-120001102

Clay, D.E., C.G. Carlson, S.A. Clay, J. Chang, and D.D. Malo. 2005. Soil organic C maintenance in a corn (Zea mays L.) and soybean (Glycine max L.) as influenced by elevation zone. J. Soil Water Conserv. 60:342–348.

Clay, D.E., T.P. Kharel, C. Reese, D. Beck, C.G. Carlson, S.A. Clay, and G. Reicks. 2012. Winter wheat crop reflectance and N sufficiency index values are influenced by N and water stress. Agron. J. 104:1612–1617. doi:10.2134/agronj2012.0216

Clay, D.E., N.R. Kitchen, C.G. Carlson, J.L. Kleinjan, and W.A. Tjentland. 2002. Collecting representative soil samples for N and P fertilizer recommendations. Crop Management doi:10.1094/cm-2002-1216-01-MA.

Clay, D.E., C.G. Carlson, and J. Chang. 2004. Determining the "best" approach to identify nutrient management zones: A South Dakota example. SSMG 41. In: D.E. Clay, editor, Site specific management guidelines. Potash and Phosphate Institute, Norcross, GA.

Colvin, T.S., and S. Arslan. 1999. Yield monitor accuracy. SSMG-9. In: D.E. Clay, editor, Site specific management guidelines. Potash and Phosphate Institute, Norcross, GA.

Doerge, T.A. 1999. Management zone concept. SSMG-2. In: D.E. Clay, editor, Site specific management guidelines. Potash and Phosphate Institute, Norcross, GA.

Franzen, D., F. Casey, and N. Derby. 2008. Yield mapping and use of yield map data. SF-1176-3. North Dakota State University Extension, Fargo, ND.

Franzen, D.W., and N.R. Kitchen. 1999. Developing management zones to target nitrogen applications. SSMG-5. In: D.E. Clay, editors, Site specific management guidelines. Potash and Phosphate Institute, Norcross, GA.

Fridgen, J.J., N.R. Kitchen, K.A. Sudduth, S.T. Drummond, W.J. Wiebold, and C.W. Fraisse. 2004. Management zone analysis (MZA): Software for subfield management zone delineation. Agron. J. 96:100-108. https://www.ars.usda.gov/research/software/download/?softwareid=24

Heiniger, R.W., R.G. McBride, and D.E. Clay. 2003. Using soil electrical conductivity to improve nutrient management. Agron. J. 95:508–519. doi:10.2134/agronj2003.0508

Kim, K., D. E. Clay, C. G. Carlson, S. A. Clay, and T. Trooien. 2008. Do synergistic relationships between nitrogen and water influence the ability of corn to use nitrogen derived from fertilizer and soil? Agron. J. 100:551–556. doi:10.2134/agronj2007.0064

Lems, J., D.E. Clay, D. Humburg, T.A. Doerge, S. Christopherson, and C.L. Reese. 2001. Yield monitors-basic steps to ensure system accuracy and performance. SSMG #31. In: D.E. Clay, editors, Site specific management guidelines. Potash and Phosphate Institute, Norcross, GA.

Mednick, A.C. 2010. Does soil data resolution matter: State soil geographic data base versus soil survey geographic data in rainfall-runoff modeling across Wisconsin. J. Soil Water Conserv. 65:190–199. doi:10.2489/jswc.65.3.190

Mishra, U., D.E. Clay, T. Trooien, K. Dalsted, D.D. Malo, and C.G. Carlson. 2008. Assessing the value of using a remote sensing-based evaportranspiration map in site-specific management. J. Plant Nutr. 31:1188–1202. doi:10.1080/01904160802134491

Mount, H.R. 2001. Obtaining soil information needed for site-specific management decisions. SSMG-35. In: D.E. Clay, editors, Site specific management guidelines. Potash and Phosphate Institute, Norcross, GA.

Park, Y.L., and J.J. Tollefson. 2005. Spatial prediction of corn rootworm (Coleoptera chrysomelidae) adult emergence in Iowa cornfields. J. Econ. Entomol. 98:121–128. doi:10.1093/jee/98.1.121

Reese, C.L., S. Christopherson, C. Fossey, J. Gray, A. Hager, R. Morman, G. Schmitt, B. Showalter, C.G. Carlson, and D.E. Clay. 2001. Trouble-shooting yield monitor systems. SSMG #32. In: D.E. Clay, editors, Site specific management guidelines. Potash and Phosphate Institute, Norcross, GA.

Reese, C. 2004. Using remote sensing to make insurance claim for crop damage due to hail event. UMAC Success Story. Upper Midwest Aerospace Consortium, University of North Dakota, Grand Forks, ND.

Reese, C.L., D.E. Clay, S.A. Clay, and G.C. Carlson. 2003. Using remote sensing to detect weeds and diseases in a soybean field. UMAC Success Story. Upper Midwest Aerospace Consortium, University of North Dakota, Grand Forks, ND.

Reitsma, K.D., D.E. Clay, S.A. Clay, B.H. Dunn, and C.L. Reese. 2016. Does the U.S. cropland data layer provide an accurate benchmark for land-use change estimates? Agron. J. 108:266–272. doi:10.2134/agronj2015.0288

Scharf, P.C., J.P. Schmidt, N.R. Kitchen, K.A. Sudduth, S.Y. Hong, J.A. Lory, and J.G. Davis. 2002. Remote sensing for nitrogen management. J. Soil Water Conserv. 57:518–524.

Schepers, A.R., J.F. Shanahan, M.A. Liebig, J.S. Schepers, S.H. Johnson, and A. Luchiari, Jr. 2004. Appropriateness of management zones for characterizing spatial variability of soil properties and irrigated corn yields across years. Agron. J. 96:195–203. doi:10.2134/agronj2004.0195

Sudduth, K.A., and S.T. Drummond. 2007. Yield editor: Software for removing errors from crop yield maps. Agron. J. 99:1471–1482.

Sudduth, K.A., N.R. Kitchen, W.J. Wiebold, W.D. Batchelor, G.A. Bollero, D.G. Bullock, D.E. Clay, H.L. Palm, F.J. Pierce, R.T. Schuler, and K.D. Thelen. 2005. Relating apparent electrical conductivity to soil properties across the north-central USA. Comput. Electron. Agric. 46:263–283. doi:10.1016/j.compag.2004.11.010

Vaughan, B. 1999. How to determine an accurate soil testing laboratory. SSMG No. 4. In: D.E. Clay, editor, Site specific management guidelines. Potash and Phosphate Institute, Norcross, GA.

Understanding Soil Water and Yield Variability in Precision Farming

David E. Clay* and T.P. Trooien

Chapter Purpose

Water can be managed using precision techniques that include variable irrigation treatments, controlled drainage, and seeding rates. For example, in many fields, yields in summit and/or shoulder areas are limited by too little water, whereas yields in footslope areas are limited by too much water. Precision techniques such as varying the population or fertilizer rates can be used to modify when and how much water is used. Understanding this variability is critical for optimizing production efficiency. This chapter discusses and demonstrates the mathematics associated with estimating plant available water, water use efficiency, and other water-related terms.

Key Terms

Plant available water, permanent wilting point, saturation point, drainage, soil moisture, precipitation use efficiency, irrigation.

Mathematical Skills

Using soil and plant information to calculate plant available water, calculating maximum allowable depletion (MAD).

An Introduction to Soil Water

After rainfall or irrigation, water can infiltrate into the soil or runoff. Water that infiltrates the soil can be characterized as freely draining (gravitational), plant available (after free drainage stops but before the roots can no longer take up remaining water), and unavailable water (water films around soil particles that is held so tightly by tension that it is not available for root uptake). As soil dries, the tension holding the water in the micropores increases. This tension (the water potential) is measured in bars with different cutoff values used to define the various types of soil water (Table 8.1).

In many fields, yield variability is driven by water variability (Fig. 8.1, Mishra et al., 2008). Yields in summit areas are limited by too little water, whereas yields in footslope areas are limited by too much water. In areas with too much water, plants wilt because the soil has little or no air present. Plant roots need air for growth and cellular respiration. In areas where plants are limited by too little water, plants close their stomata, stopping water uptake from roots, and begin to wilt. Closing the stomates also reduces CO_2 exchange with the atmosphere, which slows and then stops photosynthesis so that sugars for growth are no long produced. In addition, reduced transpiration decreases nutrient uptake and reduces the plants ability to mitigate stress responses to wounding and pests (Hansen

South Dakota State University, Department of Agronomy, Horticulture and Plant Science, Brookings, SD 57007-2201. *Corresponding author (david. clay@sdstate.edu).

doi:10.2134/practicalmath2016.0025

Table. 8.1. Table relating the soil characteristics at the saturation point, field capacity, and permanent wilting point. Additional information is available in Hay and Trooien (2016).

Water potential	Description	Characteristics
0 to -1/3 bars	Saturation	All pores are filled with water
	Water between saturation and -1/3 bars is the drainable porosity.	Free drainage of water
-1/3 to -15 bars	Plant available water is the difference between the amount of water at field capacity and the permanent wilting point.	The amount of water remaining after free drainage The water can be absorbed by plant roots
-15 bars	Permanent wilting point	Point where plants will not recover from water stress

Moody 1999
Corn yield monitoring data

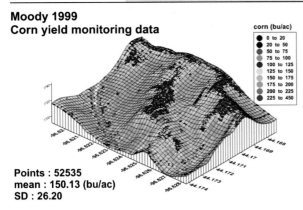

corn (bu/ac)
- 0 to 20
- 20 to 50
- 50 to 75
- 75 to 100
- 100 to 125
- 125 to 150
- 150 to 175
- 175 to 200
- 200 to 225
- 225 to 450

Points : 52535
mean : 150.13 (bu/ac)
SD : 26.20

Fig. 8.1. A corn yield map in bushels per acre from 1999 that is overlaid on an elevation map. This map shows that yields in summit and/or shoulder areas are lower than those in footslope areas. Additional work has shown that yields in summit areas are reduced by water stress (Mishra et al., 2008).

et al., 2013). One of the reasons for understanding soil water variability is to improve our ability to match fertilizers, cultivars, and seeding rates to the available resources.

Evapotranspiration

Water movement from the soil to the atmosphere can be classified as evaporation and transpiration. Evaporation is water lost to the atmosphere directly from soil and does not contribute to plant growth or yield. Transpiration is the water that moves from the roots to the aerial plant portions and moves to the atmosphere as water vapor through the stomata in the leaves. Transpiration is needed for the uptake of nutrients and water and contributes to plant growth and yield. The combination of both processes, evaporation and transpiration, without separation of which process is occurring, is called evapotranspiration (ET).

Gravitational Water

Soil water-holding properties are similar to a sponge: when a sponge is placed in a bucket of water, all the pores in the sponge are filled to the saturation point. When the saturated sponge is removed from the bucket, some of the water freely drains due to the force of gravity. The amount of water that freely drains from the soil is called gravitational water. A related concept is drainable porosity. The drainable porosity is the volume of unfilled voids after the gravitational water has drained. The water potential of gravitational water is between 0 and -1/3 bar. When tile drainage is installed, gravitational water is the water that is removed. In field soil, after a precipitation event that saturates the soil, the drainage of gravitational water occurs over several hours in a sandy soil or over several days in a clay soil. When this free-water drainage stops, the soil is said to be at field capacity (Table 8.1).

Plant-available Water

Plant available water is the amount of water remaining after free drainage but before the roots can no longer extract water from the soil. As the water content decreases, it becomes increasingly harder to extract. Plant available water is often defined as the difference (in inches or cm of water) between field capacity (-1/3 bar) and the permanent wilting point (-15 bar) of a plant. Different soils have different amounts of plant available water (Table 8.2) and the total amount is influenced by pore size, pore space, soil depth, soil texture, and organic matter content. Generally, increasing soil organic matter and clay content increases plant available water, with maximum plant available water of 2.6 inches per foot (21.7 cm m^{-1})of soil in a silt loam soil.

Plant-unavailable Water

At the permanent wilting point, there is a small film of water left around soil particles. However, this water is held so tightly to the soil particles that the water is unavailable to the plant. This point is called the permanent wilting point because even if additional water is added to the soil, the plant will not recover.

Calculating and Measuring Soil Water

Calculating plant available water from soil samples requires an understanding of soil bulk density, porosity, particle density, and the soil properties that influence water retention in soil. For information collected from soil samples to make sense, the samples should be collected following appropriate protocols (chapter 6, 7, and 8). The bulk density is measured by collecting a known volume of soil, which is oven dried at 105 C. Bulk density, particle density, and porosity (the amount of space in a soil that can be filled with water and/or air) are related and defined by the following equations:

$$\text{Bulk density} = \frac{\text{dry weight of solids}}{\text{volumes of solids + pores}}$$

$$\text{Particle density} = \frac{\text{dry weight of solids}}{\text{volume of the solids}}$$

$$\%\text{ Porosity} = 100 \times \frac{\text{Volume of the pores}}{\text{Volumes of solids + pores}}$$

$$\%\text{ Porosity} = \left(1 - \frac{\text{Bulk Density}}{\text{Particle Density}}\right) \times 100$$

Most mineral soils have bulk densities that range from 1 to 1.6 g cm^{-1} (Table 8.3), particle densities that range from 2.5 to 2.7 g cm^{-3}, and % porosities that range from 40 to 60%. Particle and bulk densities provide an index of soil compaction and the ability of plant roots to penetrate compacted zones (Table 8.3). The amount of space available for plant available water depends on soil depth and soil texture (Table 8.4) and is equal to or less than the % pore space. At field capacity, all pore space is filled with water and as the soil dries, water is replaced with air. Sandy soils have less plant available water than silt loam soils.

Table 8.2. The influence of soil texture on plant-available water. In this chart, cm per m are shown in parenthesis.

Soil Texture	Plant-available water
	in per ft soil
fine sands	0.7–1.0 (5.8-8.3)
Loamy sands	0.9–1.5 (7.5-12.5)
Sandy loams	1.3–1.8 (10.8-15)
Loam	1.8–2.5(10.8-20.8)
silt loam	1.8–2.6 (10.8-21.7)
clay loam	1.8–2.5 (10.8-21.7)

Table 8.3. Bulk densities where root growth is restricted in sandy, silty, and clayey soils.

Soil textures	Bulk density (g cm^{-3})
Silty	> 1.8
Sandy	> 1.65
Clayey	> 1.47

PROBLEM 8.1.

What is the % porosity if the bulk density is 1.30 g cm^{-3} and the particle density is 2.70 g cm^{-3}?

ANSWER:

$$\%Porosity = \left(1 - \frac{BulkDensity}{ParticleDensity}\right) \times 100$$

Insert the bulk density and particle density into the equation and solve for % pore space.

$$Porosity \% = \left(1 - \frac{1.30\ g/cm^3}{2.70\ g/cm^3}\right) \times 100 = 51.9\%$$

PROBLEM 8.2.

If the soil bulk density is 1.35 g cm^{-3}, how many pounds of soil are contained in an acre that is 6 inches deep?

1 acre = 2.47 ha, 2.21 lb = 1 kg, 2.54 cm = 1 inch, 10,000 cm^2 = 1 m^2, 10,000 m^2 = 1 ha

ANSWER:

$$\left(\frac{1.35g}{cm^1 cm^2}\right)\left(\frac{10,000\ cm^2}{1\ m^2}\right)\left(\frac{10,000\ m^2}{1\ ha}\right)\left(\frac{1ha}{2.47\ acre}\right)\left(\frac{kg}{1000g}\right)\left(\frac{2.21\ lbs}{1\ kg}\right)\left(\frac{2.54\ cm}{1\ in}\right)(6\ in)$$

=1.84 million pounds per 6-in acre (2,060,000 kg per 15.2-cm ha)
Information on unit conversions is available in Chapter 7.

PROBLEM 8.3.

If the soil bulk density is 1.2 g cm^{-3}, how many pounds of soil are contain in an acre that is 6 in deep?

ANSWER:

$$\left(\frac{1.20}{cm^3}\right)\left(\frac{10,000\ cm^2}{m^2}\right)\left(\frac{10,000\ m^2}{ha}\right)\left(\frac{1\ ha}{2.47\ acre}\right)\left(\frac{kg}{1000\ g}\right)\left(\frac{2.21\ lbs}{kg}\right)\ or\ 1,800,000\ kg\ soil\ per\ 15.2\text{-}cm\ ha$$

$$\left(\frac{2.54\ cm}{in}\right)(6\ in) = \frac{1,630,000\ lbs\ Soil}{6\ in\ acre}$$

Table 8.4. Range of drainable porosity values by soil texture (Sands, 2001).

Soil texture	Drainable porosity % by volume
Clays, clay loams, silty clay	3 to 11%
Well-structured loams	10 to 15%
Sandy	18 to 35%

Measuring Soil Water

Soil water content can be measured (i) using sensors that measure the water potential (the strength that water is held in soil in bars), (ii) using sensors that measure the water content of the soil, or (iii) by taking soil samples and determining the amount of water in the soil. Soil samples can be analyzed to determine the gravimetric or volumetric soil water contents. Gravimetric soil water contents are reported by weight, whereas volumetric soil water contents are reported by volume. Gravimetric values are converted to volumetric values by multiplying the gravimetric values by the bulk density. Gravimetric water contents are determined by drying a measured weight of a sample at 105 C° until it reaches a constant weight. The equation for calculating gravimetric and volumetric water contents are:

$$\%\ Gravimetric\ water\ content = 100 \times \frac{\left[weight\ of\ wet\ soil\ (g) - weight\ of\ dry\ soil\ (g)\right]}{g\ of\ dry\ soil}$$

$$\% \text{ Volumetric water content} = 100 \times \frac{(\text{weight of wet soil} - \text{weight of dry soil})}{\text{g of dry soil}} \times \frac{\text{g dry soil}}{\text{cm}^3}$$

In this equation, grams dry soil per cm^{-3} is the bulk density the weight difference between wet and dry soil represents the amount of water lost during drying.

Calculating Plant-available Water

Plant available water represent the amount of water that is available to the plant and it is defined as the water contained in the soil between field capacity (-1/3 bar) and the permanent wilting point (-15 bars). This value can be used to calculate the cm or inches of plant available water contained in a soil column. For example, the surface 3 ft soil contains 6 in of water. The amount of plant available water (PAW) is calculated with the equation,

$$\text{cm of plant available water} = \text{sampling depth (cm)} \times \frac{\text{g soil}}{\text{cm}^3} \times \frac{(\%GW_{fc} - \%GW_{pwp})}{100 \times \text{g soil}}$$

where the weight of the soil in g soil cm^3 is the bulk density, GW_{fc} is the gravimetric water content at field capacity (-1/3 bar), GW_{pwp} is the gravimetric water at the permanent wilting point (-15 bars), and 100 converts the percent to a decimal. The difference between GW_{fc} and GW_{pwp} represents the amount of plant available water in a gram of soil.

Precipitation Use Efficiency

The amount of water required to produce a crop is reported as the precipitation use efficiency (PUE), and it is calculated as the amount of biomass (or grain) produced per unit of rainwater. In field studies, PUE is often reported as water use efficiency (WUE). Precipitation use efficiency is dependent on both evaporative and transpiration losses, and it is influenced by tillage, plant genetics, and other factors. Plant breeding programs have attempted to increase the PUE, and thereby increase biomass production in water-limited systems. For example, Clay et al. (2014) reported that from 1960 to 2012, precipitation use efficiency of corn increased from less than 100 kg (ha× cm)$^{-1}$ to over 200 kg (ha ×cm)$^{-1}$. There is a limit to a plant's PUE because reducing transpiration decreases photosynthesis rates and, ultimately, reduces biomass production. However, PUE also can be increased by reducing evaporation, thus increasing the amount of water available for transpiration. This can be accomplished by reducing tillage intensity and leaving residue on the soil surface. For example, Hatfield et al. (2000) reported that evaporative water

PROBLEM 8.4.

What is the percent water content if the dry soil weight is 10 g and the wet soil weight is 14 g?

ANSWER:

$$\% \text{ moisture} = \frac{14 - 10 \text{ g}}{10 \text{ g}} \times 100\% = 40\%$$

PROBLEM 8.5.

What is the percent water content if the dry soil weight is 25 g and the wet soil weight is 30 g?

ANSWER:

$$\% \text{ moisture} = \frac{30 - 25 \text{ g}}{25 \text{ g}} \times 100\% = 20\%$$

PROBLEM 8.6.

What is the volumetric water content if the gravimetric water content is 20% and the bulk density is 1.25 g cm^{-3}? Volumetric water (VWC) is the gravimetric water multiplied times the bulk density,

ANSWER:

$$\frac{1.25 \text{ g Soil}}{\text{cm}^3} \times \frac{0.20 \text{ g Water}}{\text{g Soil}} = \frac{0.25 \text{ g Water}}{\text{cm}^3}$$

In this solution, the bulk density is 1.25 g cm^{-3} and 0.2 g water per g soil is the % gravimetric water content divided by 100.

PROBLEM 8.7.

If the permanent volumetric wilting point (-15 bars) is 10% and volumetric water content at field capacity of the surface 12 in is 0.25 g cm⁻³, how much plant-available water is in the surface 12 in if the bulk density is 1.25 g cm⁻³. Note the volumetric water content is equal to the gravimetric water content times the bulk density.

ANSWER:

Available water is:

$$12 \text{ in} \times \frac{(0.25 - 0.1 \text{ g water})}{\text{cm}^3} \times \frac{1 \text{ cm}^3}{1 \text{ g water}} = 1.8 \text{ in water}$$

If the soil water was provided as gravimetric water, then bulk density would be included in the calculation

$$12 \text{ in} \times \frac{(0.25 - 0.1 \text{ g water})}{\text{g}} \times \frac{1.25 \text{ g}}{\text{cm}^3} \times \frac{1 \text{ cm}^3}{1 \text{ g water}} = 2.25 \text{ in water}$$

PROBLEM 8.8.

Calculate the total amount of available water in the 0 to 36 in soil profile for the following data.

ANSWER:

In this problem it is important to know that the gravimetric water at -1/3 bar is field capacity and the gravimetric water at -15 bars is the permanent wilting point.

Step 1. Convert all percentages to decimal forms, for example, 20% = 0.20 g per g soil.

Step 2: Calculate available water for depth 1.

$$6 \text{ in} \times \frac{1.1 \text{ g Soil}}{\text{cm}^3} \times \frac{(0.2 - 0.10) \text{ g Water}}{\text{g Soil}} \times \frac{\text{cm}^3}{1 \text{ g water}} = 0.66 \text{ in water} \quad (1.67 \text{ cm water})$$

Step 3: Calculate available water for depth 2.

$$30 \text{ in} \times \frac{1.3 \text{ g Soil}}{\text{cm}^3} \times \frac{(0.25 - 0.08) \text{ g water}}{\text{g Soil}} \times \frac{\text{cm}^3}{1 \text{ g water}} = 6.63 \text{ in water} \ (18.4 \text{ cm water})$$

Step 4: Add water in depths 1 and 2

0.66 + 6.63 = 7.29 in water (18.52 cm water in the surface 76.2 cm of soil)

losses following cultivation in Iowa were four to five inches over a three day period, whereas evaporative water losses in no-tillage systems were less than one inch. The calculation of PUE requires accurate measurement of yield, precipitation, and changes in soil water. Yield estimates can be obtained from a yield monitor, whereas changes in soil water can be obtained by sensors or soil sampling.

In field studies, water is often reported as cm per hectare or inches per acre (Clay et al., 2014). Biomass production is calculated by harvesting and measuring the plant biomass or yield, whereas the water use is the sum of the growing season rainfall and difference between the amount of water contained in the soil in the spring and fall. Any water lost as runoff or drainage should also be subtracted. In these calculations, the entire root zone should be considered.

Management Techniques to Increase PUE

No-tillage, when compared to a moldboard plow system, increases snow catch and water infiltration, and reduces runoff and evaporation from the soil surface (Triplett and Dick, 2008). In addition, no-till reduces erosion, and reduces the risk of detached soil particles forming impermeable crusts or seals. Reduced evaporation is attributed to crop residues that remain on the soil surface (Klocke et al., 2009) that disrupt a direct link between the soil water and the atmosphere. In addition, it helps maintain cooler temperatures in the spring and warmer temperatures in the fall. The impacts to plant available water and yields can be significant. For example, one inch of stored (saved) water can increase corn yield 8 to 14 bushels (Clay et al., 2014).

Over the past 25 yr, decreases in tillage intensity and increases in crop yields have contributed to a 24% increase in the South Dakota soil's organic matter content (Clay et al., 2012). The bottom line is that building the soil organic matter content improves drought resilience. Precision tips for reducing evaporation and increasing the amount of water available for transpiration are provided below.

1. Use rotations and tillage systems that increase water storage and increase soil organic matter, which helps conserve water.
2. Do not harvest crop residues.
3. Apply sufficient rates of P and K fertilizer.
4. Control pests.
5. Select a planting date that minimizes early frost damage, but maximizes season length.
6. Use realistic yield goals that match seeding rates and planting depth to available water.
7. Select cultivars with the proper maturity for your area that have drought tolerance and pest resistance. If drought is a concern, consider seeding a cultivar that will mature earlier in the season.
8. Consider a seeding strategy that improves soil water management. An approach that was tested in Nebraska was the skip-row technique (Klein, 2017). In skip-row planting, rows are skipped, while the plant population per acre remains the same. Three configurations that were tested included plant two rows and skip two rows, plant a row and skip a row, and plant two rows and skip one row. Skip row planting is most effective in deep soils located in low rainfall environments.

Irrigation

To be most effective, the appropriate amount of irrigation water should be applied to the crop at the right time (Heeren et al., 2011; Irmak and Rudnick, 2014; Kranz and Specht, 2012, Ostrem et al., 2016). Different strategies should be used for different crops and locations. For example, corn, soybean, and wheat have different irrigation needs because of

PROBLEM 8.9.

A field contains 1 inch of plant available water. How many gallons are contained in 1 acre? For this calculation it is important to know that 7.481 gallons are contained in 1 ft³.

ANSWER:

$$\left(1.0 \text{ inch water} \times \frac{1 \text{ ft}}{12 \text{ in}} \times \frac{43560 \text{ ft}^2}{\text{acre}} \times \frac{7.481 \text{ gal}}{\text{ft}^3}\right) = 27,154 \text{ gallons } (102,789 \text{ L})$$

PROBLEM 8.10A.

A silt loam soil is 3 ft thick. Estimate how much plant available water could be in the soil? For this question use data in Table 8.2.

ANSWER:

Each foot of soil contains between 1.8 and 2.6 in of water, total water ranges between 5.4 to 7.8 in (Table 8.2).

PROBLEM 8.10B.

What is the range in amounts of plant available water if a sandy soil is three feet thick? For this problem use data in Table 8.2.

ANSWER:

The range of AW per foot of soil contains between 0.7 to 1.8 in (Table 8.2). Therefore the range is (0.7 × 3) to (1.8×3) = 2.1 to 5.4 in of AW per foot (17.5 – 45 cm m⁻¹)

PROBLEM 8.11.

Determine the PUE if the soil water content at planting is 7 inches and the water content at harvest is 3 in. During the growing season it rained 16 inches and the yield was 200 bu acre⁻¹. This calculation assumes that runoff was minimal.

ANSWER:

Determine water available water= 16+7-3= 20 inches
Calculate PUE = 200 bu per acre per 20 in = 10 bushels per in water (478 kg cm⁻¹)

Problem 8.12A.

Calculate the precipitation use efficiency if growing season rainfall is 65 cm, soil water at planting is 10 cm, soil water at harvest is 5 cm, and the amount of biomass produced is 5000 kg.

Answer:

Rainfall + soil water = total water available for the season

Water at end of season is still available.

So water used = 65 + 10 – 5 = 70 cm of water was used to produce 5000 kg.

Answer:

$$= \frac{5{,}000 \text{ kg}}{(65+10-5)} = \frac{71.4 \text{ kg}}{\text{cm}} \quad (400 \text{ lbs inch-1 or } 7.15 \text{ bu inch-1})$$

This calculation assumes that runoff and deep drainage of water does not occur.

Problem 8.12B.

During one rainfall event, 3 cm was measured as runoff. Calculate the PUE using this information.

Answer:

Water used = 65+10 -5-3 = 67 cm of water used to produce 5000kg (11,050 lbs)

PUE = 5000 kg/67 cm = 74.6 kg cm^{-1} (7.48 bu inch^{-1})

differences in rooting depths ET rates, and other factors. For corn, early in the growing season, the corn roots may be concentrated in the surface 12 in. As the season progresses, corn roots can extend down to five feet. Most of the roots, however, are found in the surface three feet. A general guideline for corn is to schedule irrigations based on the amount of plant available water (PAW) in the surface 2 feet prior to R1 (silk) and three feet thereafter.

For soybeans, irrigation or high rainfall that occurs during vegetative growth stages may not increase yields unless the soil water contents are extremely low. In many moderate- to fine-textured soils, early season irrigation can actually stimulate vegetative growth without increasing soybean yields (Kranz and Specht, 2012). For soybeans, drought stress during the reproduction stages (between R2 to R6) can greatly reduce yields. Irrigation during this period of time typically increases the number of seeds per plant and yield per acre. However, it may also increase the disease potential. A good approach for irrigating soybeans is to match irrigation watering with the most sensitive growth stages of the plant.

For winter wheat grown in Colorado, irrigation in fine-textured soils should be considered in the fall to fill the soil profile, in the spring to refill the soil profile, and after the boot stage (Al-Kaisi and Shanahan, 1999). If applied during the early vegetative stage, irrigation water can stimulate vegetative growth, lodging, and diseases.

Another way to think of soil water content is soil water depletion. The soil water depletion is the amount of water required to bring the root zone back to field capacity. When the soil is at field capacity, depletion is zero. In other words, if you apply an irrigation that is the same depth as the soil water depletion, the soil profile will be at field capacity when the irrigation is completed.

When irrigating, it is appropriate to keep the soil profile from getting too dry. When between 30 and 70% (depending on crop, growth stage, and other factors) of the water has been removed from the soil, the plant starts to experience water stress. To minimize this stress, many managers only allow a portion of the plant available water to be depleted by the plant. This value is called the maximum allowable depletion (MAD), which generally range from 30 to 70% of the plant available water (Heeren et al., 2011).

Irrigation is used to replenish water used by the crop and removed from the root zone. An appropriate method to schedule irrigations is the check book method. This method uses rainfall, evapotranspiration (ET) estimates, and a soil water worksheet to estimate remaining plant-available water. There are many spreadsheets available to implement the check book approach. Precision farming offers another method to schedule irrigation using sensors that continuously monitor soil water.

Drainage

In many fields, yields are limited by too much water. Lower landscape positions (foot and toe slopes) are frequently areas containing excess water. Subsurface tile drainage can be used to remove this excess water. Tile drainage systems are designed to remove gravitational water from the soil. Drainage can provide many advantages to a field including timely planting, harvesting, and higher crop yields. However, it also has several disadvantages including the potential loss of wetlands and nitrate—nitrogen loss to off-site locations. Research is being conducted to assess the ability of various conservation drainage practices including bioreactors, drainage water management, and other practices to reduce the amount of nitrate-nitrogen transported from production fields to streams and rivers (Hay, 2016). When planning to install tile drainage systems, it is important to consider water disposal.

The rate that water is removed from the soil profile is reported as the drainage coefficient. This value is based on the soil and system design. Additional information on tile drainage systems is available in Skaggs (2007) and Hay and Trooien (2016).

The drainable porosity is the pore space from which water (the gravitational water) is removed by a drainage system. It is numerically equal to the difference between water at saturation (0 bars) and field capacity. Drainable porosity varies with soil texture (Table 8.4), with finer-textured (e.g., clay soils) soils having smaller drainable porosity values than coarse-textured soils.

PROBLEM 8.13.

An irrigation system pumps at a rate of 800 gal/minute. How long will it take to irrigate 150 acres with 2 inches of water? In this calculation 27,154 gallons (102,789 L) is one acre inch of water (2.54 cm). (Assume that the efficiency is 100%)

ANSWER:

$$\left(150 \text{ acres}\right)\left(\frac{2 \text{ in water}}{\text{acre}}\right)\left(\frac{27{,}154 \text{ gal}}{\text{in water}\big/\text{acre}}\right)\left(\frac{\text{Min}}{800 \text{ gal}}\right)\left(\frac{1 \text{ hr}}{60 \text{ min}}\right)=170 \text{ hours}$$

PROBLEM 8.14.

How long will it take to apply 1 inch of water to a 20 ft by 30 ft garden (6.1 by 9.1 m) with a 1 inch (2.54 cm) garden hose? (Assume the rate of flow is 1 minute per 5 gallons (18.9 L min^{-1}); 1 ft^3 contains 7.481 gallons of water (28.32 L). Further assume that the application efficiency is 100 %.)

ANSWER:

$$1.\text{Volume of water}=\left(20 \text{ ft}\right)\left(30 \text{ ft}\right)\left(1 \text{ in}\times\frac{1 \text{ ft}}{12 \text{ in}}\right)=50 \text{ ft}^3$$

$$2.\text{Gallons of water}=50 \text{ ft}^3\times\frac{7.481 \text{ gal}}{1 \text{ ft}^3}=374 \text{ gal}$$

$$3.\text{Flow rate:}\frac{1 \text{ min}}{5 \text{ gal}}$$

$$4.\text{Length of time}=\left(374 \text{ gal}\right)\left(\frac{1 \text{ min}}{5 \text{ gal}}\right)=74.8 \text{ min}$$

PROBLEM 8.15A.

Estimate the percent of plant available water that remains in the soil from the following information. For 21, 22, 23, and 24 July the estimated crop evapotranspiration (ET) is 0.40, 0.37, and 0.25 in. On 21 July, the soil contains 6 in water. It does not rain and irrigation is not applied. The soil can hold 9 in water. R1 refers to the plant growth stage.

ANSWER:

Date	Growth Stage	Temp (°F)	ET(in)	PAW(in)	%PAW
21 July	R1			6 (15.2 cm)	66.7 (100%×6/9)
22 July	R1	86 (36 °C)	0.4 (1.02 cm)	5.6 (13.2 cm)	62.2 (100%×5.6/9)
23 July	R1	82 (27.8°C)	0.37(0.94 cm)	5.23 (13.28 cm)	58.1 (100%× 5.81/9)
24 July	R1	76(24.4° C)	0.25(0.54 cm)	4.98 (12.65 cm)	55.3 (100%× 4.98/9)

If the farmer wants to keep the water in the profile at greater than 60% plant available water (PAW), when and how much irrigation should be applied?

ANSWER:

Based on data from 21 and 22 July, the farmer suspects that ET will be enough to take soil profile below the 60% PAW on 23 July, therefore he begins irrigating on 22 July. Additional information on crop water use and ET relative to crop growth stage is available in Al-Kaisi (2000).

PROBLEM 8.16.
How much water is removed from 1 acre for a two-foot drop of the water table if the drainable porosity is 15%?

ANSWER:

$$\left(\frac{2\ ft}{acre}\right) \times \left(\frac{0.15\ ft\ water}{foot\ soil}\right) = \frac{0.30\ ft\ water}{acre} \quad (22\ cm\ ha^{-1})$$

$$\left(\frac{0.30\ ft\ water}{acre}\right) \times \left(\frac{12\ in\ water}{ft\ soil\ water}\right) = \frac{3.6\ in\ water}{acre} \quad (22.6\ cm\ h^{-1})$$

$$\left(\frac{3.6\ in\ water}{acre}\right) \times \left(\frac{27{,}154\ gallons}{1\ in\ water/acre}\right) = 97{,}750\ gallons \quad (256{,}461\ L)$$

Additional Problems

8.17. If the soil has a bulk density of 1.15 g cm^{-3} and a particle density of 2.65 g cm^{-3}, what is the pore space?

8.18. If the soil bulk density is 1.3 g cm^{-3} and the gravimetric water content is 0.25 g g^{-1} what is the volumetric water content?

8.19. Estimate the amount of plant available water if the surface 6 in is a silt loam and the texture of the 6- to 36-in depth is loamy sand.

8.20. How might delaying the N rate increase the precipitation use efficiency? Hint, N stimulates growth which can reduce water during grain filling.

8.21. How might reducing the seeding rate increase the precipitation use efficiency? Hint consider the impact of skip row planting (Kline, 2017).

8.22. Calculate the water use efficiency for the following data. Yield is 8000 kg ha^{-1}, rainfall is 75 cm, the volumetric water at seeding is 8 cm and the volumetric water at harvest is 7 cm.

Summary

In summary, in many fields, both too much and too little water can be major factors influencing crop growth. In areas with too much water, drainage can be used to remove excess water. In areas with too little water, yields may be reduced 40 to 60% (Fig. 8.1) but yield reductions can be avoided with irrigation. However different irrigation strategies should be used for different crops and locations. Management strategies can also be used to improve precipitation use efficiency. For example, lowering plant populations may also reduce water stress and adopting no-tillage can reduce evaporation. Water stress may also produce unexpected impacts on plant growth and development. Hansen et al. (2013) reported that in response to water stress, corn downregulated genes associated with disease suppression and nutrient uptake, while up-regulating genes associated with saline and water stress. This chapter discussed techniques for estimating the amount of available water in the soil profile. Additional information is available in Cahoon et al., (1992), Blann et al. (2009), Evett et al. (2012), Irmak and Rudnick(2014), and Kumar et al.(2016).

ACKNOWLEDGMENTS

Support for this document was provided by South Dakota State University, South Dakota Soybean Research and Promotion Council, and the Precision Farming Systems communities in the American Society of Agronomy and the International Society of Precision Agriculture. Additional support was provided by the USDA-AFRI Higher Education Grant (2014-04572).

REFERENCES

Al-Kaisi, M., and J.F. Shanahan. 1999. Irrigation of winter wheat. Publication # 0.556. Colorado State University, Fort Collins, CO.

Al-Kaisi, M. 2000. Crop water use or evapotranspiration. Integrated Crop Management 484:84–85.

Blann, K., J.L. Anderson, G.R. Sands, and B. Vondracek. 2009. Effects of agricultural drainage on aquatic ecosystems: A review. Crit. Rev. Environ. Sci. Technol. 39:909–1001. doi:10.1080/10643380801977966

Cahoon, J., C.D. Yonts, and S.R. Melvin. 1992. Estimating effective rainfall. G92-1099-A. Univ. Nebr-Lincoln Extension Neb-Guide, Lincoln, NE.

Clay, D.E., J. Chang, S.A. Clay, J.J. Stone, R.H. Gelderman, C.G. Carlson, K. Reitsma, M. Jones, L. Janssen, and T. Schumacher. 2012. Corn yields and no-tillage affects carbon sequestration and carbon footprint. Agron. J. 104:763–777. doi:10.2134/agronj2011.0353

Clay, D.E., S.A. Clay, K.D. Reitsma, B.H. Dunn, A.J. Smart, C.G. Carlson, D. Horvath, and J. Stone. 2014. Does the conversion of grasslands to row crop production in semi-arid areas threaten global food supplies? Glob. Food Secur. 3:22–30. doi:10.1016/j.gfs.2013.12.002

Evett, S.R., R.C. Schwartz, J.J. Casanova, and L.K. Heng. 2012. Soil water sensing for water balance, ET, and WUE. Agric. Water Manage. 104:1–9. doi:10.1016/j.agwat.2011.12.002

Hatfield, J.L., T.J. Sauer, and J.H. Prueger. 2000. Managing soils to achieve greater water use efficiency: A review. Agron. J. 93:271–280. doi:10.2134/agronj2001.932271x

Hansen, S., S.A. Clay, D.E. Clay, C.G. Carlson, G. Reicks, J. Jarachi, and D. Horvath. 2013. Landscape features impact on soil available water, corn biomass, and gene expression during the late vegetative growth stage. Plant Genome 6:1–9. doi:10.3835/plantgenome2012.11.0029

Hay, C., and T. Trooien. 2016. Chapter 30: Managing high water tables in corn production. In: D.E. Clay, C.G. Carlson, S.A. Clay, and E. Byamukama, editors, iGrow corn: Corn best management practices. South Dakota State University, Brookings, SD.

Hay, C. 2016. Chapter 31: Reducing nitrate losses from drained land. In: D.E. Clay, C.G. Carlson, S.A. Clay, and E. Byamukama, editors. iGROW corn: Best management practices. South Dakota State University, Brookings, SD.

Heeren, D.M., T.P. Trooien, H.D. Werner, and N.L. Klocke. 2011. Development of deficit irrigation strategies for corn using a yield ratio model. Appl. Eng. Agric. 27:605–614. doi:10.13031/2013.38207

Irmak, S., and D.R. Rudnick. 2014. Corn irrigation management under water-limiting conditions. EC2007. Univ. Neb.-Lincoln Extension Circular, Lincoln, NE.

Kleine, R. 2017. Recommendations for implementing and fertilizing skip-row planting. Nebraska Crop Watch, Lincoln, NE.

Klocke, N.L., R.S. Currie, and T.J. Dumler. 2009. Water savings from crop residue management. Proceedings of the 21st Annual Central Plains Irrigation Conference. Colby, KS. 24-25 Feb. 2009. Kansas State University, Manhattan, KS.

Kranz, W.L., and J.E. Specht. 2012. Irrigating soybean. NebGuide G1367. University of Nebraska, Lincoln, NE.

Kumar, S., D.E. Clay, and C.G. Carlson. 2016. Chapter 14: Soil compaction impact on corn yield. In: D.E. Clay, C.G. Carlson, S.A. Clay, and E. Byamukama, editors, iGrow corn: Best management practices. South Dakota State University, Brookings, SD.

Mishra, U., D.E. Clay, T. Trooien, K. Dalsted, D.D. Malo, and C.G. Carlson. 2008. Assessing the value of using a remote sensing based evapotranspiration map in site-specific management. J. Plant Nutr. 31:1188–1202. doi:10.1080/01904160802134491

Ostrem, D., T. Trooien, and C. Hay. 2016. Chapter 33: Irrigating corn in South Dakota. In: D.E. Clay, C.G. Carlson, S.A. Clay, and E. Byamukama, editors, iGROW corn: Best management practices. South Dakota State University, Brookings, SD.

Sands, G. 2001. Agricultural drainage: Soil water concepts. University of Minnesota, St. Paul, MN.

Skaggs, R.W. 2007. Criteria for calculating drain spacing and depth. Trans. ASABE 50:1657–1662. doi:10.13031/2013.23971

Triplett, G.B., and W.A. Dick. 2008. No-tillage crop production: A revolution in agriculture. Agron. J. 100:S-153–S-165. doi:10.2134/agronj2007.0005c

Developing Prescriptive Soil Nutrient Maps

Richard B. Ferguson,* Joe D. Luck, and Rachel Stevens

Agronomic Basis for Fertilizer Recommendations

When developing a map to vary the rate of fertilizer across a field, the agronomic basis for applying fertilizer, and how much fertilizer, must be considered. A primary goal is assumed to be increased profit through increased yield, with a focus on crop production in the year of application. However, fertilizer decisions are more nuanced than it might seem at first glance. Factors such as land ownership or land tenure, economic trends in crop and fertilizer price, and environmental considerations and costs can influence fertilizer use decisions. In some cases, regulatory requirements may override other considerations. Management considerations for nitrogen (N), a very dynamic nutrient in the soil environment, will be quite different from phosphorus (P), potassium (K) or other relatively immobile nutrients.

In general, crop yield increases with fertilizer rate with a curvilinear response (Fig. 9.1). At some point along this curve, a point is reached where the added fertilizer just pays for itself; any additional application of fertilizer will cost more than the resulting increase in value from added yield. This point, the Economic Optimum Rate (EOR) is not easily predicted, given the many variables that influence crop response to fertilizer input, as well as fertilizer and crop selling prices. Chapter 11 (Fausti et al., 2017) provides more detail on EOR calculation; this chapter assumes that for a given location in a field fertilizer and crop-specific EOR can be estimated based on yield response to fertilizer application and the cost of fertilizer and the value of the crop. Figure 9.2 illustrates a typical yield response of corn to N and an example of the influence of fertilizer and N price on the EOR of N fertilizer (EONR) for a specific research site. In this case, the EONR ranges from 82 to 152 lb N acre⁻¹, depending on the price ratio of corn to N (Dobermann et al., 2010). Land grant university and soil test lab recommendations are generally based on research that estimates the profitability of fertilizer use. Nitrogen, as a dynamic nutrient, is not easily stored in soil; consequently, Fig. 9.2 represents EONR using a sidedress management approach, with a single fertilizer application around the V6 growth stage for corn (Abendroth et al., 2011). For N, multiple applications during the growing season could reduce the overall N rate needed, due to a potential reduction in climate-induced losses of N. The EONR for fertilizer input also is based on prior accounting for N supply to the crop from other sources, such as

Chapter Purpose

Understanding the process of developing variable rate fertilizer prescription maps is a critical component of precision agriculture. This chapter also describes some of the technological issues in prescription map development, though many technological factors will be software and hardware specific. Additional information about different fertilizer choices is available in chapter 10, determining the economic optimum fertilizer rate is available in Chapters 11 and 15, and identifying management zones is available in Chapter 7.

Key Terms

Grid soil sampling, management zone, crop removal, critical value, economic optimum nutrient rate, interpolation, kriging, inverse-distance, vegetation index.

Mathematical Skills

Interpolation, calculation of basic statistics such as mean, median, range.

R.B. Ferguson, Department of Agronomy, University of Nebraska—Lincoln, Lincoln, NE 68588; J.D. Luck and R. Stevens, Biological Systems Engineering Department, University of Nebraska—Lincoln, Lincoln, NE 68588. *Corresponding author (rferguson1@unl.edu)
doi:10.2134/practicalmath2016.0109

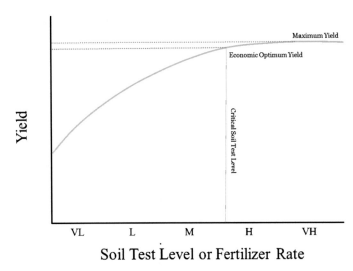

Fig. 9. 1. Typical crop yield response to soil test nutrient level or fertilizer rate (VL, very low; L, low; M medium; H, high; VH very high). The amount of fertilizer for very low to high test levels vary with specific crops. For additional information on calculating economic optimum rates see Chapter 11 (Fausti et al., 2017)

legume and manure credits, soil residual nitrate, and nitrate contained in irrigation water.

For relatively immobile nutrients such as P and K, options are different. Crop yield responses to these nutrients are still typically curvilinear in the year of application (Fig. 9.1). The shape of the response curve for a given field will be influenced by the nutrient soil test level and the crop being grown. Nutrients such as P and K typically have soil test critical values, derived from research, that indicate the soil test value for a specific lab analysis extraction method at which yield increase is unlikely with added fertilizer. If a soil test value is above the critical level, it will likely not be profitable in the year of application to apply fertilizer. However, there may be reasons one would choose to apply fertilizer in such a situation. For example, if fertilizer prices are relatively inexpensive, and one expects prices to increase significantly in the future, it may make economic sense over the long term to apply fertilizer even when there is little likelihood of a yield response in the year of application. If soil test values are near the critical value, one may choose to apply fertilizer just to maintain the

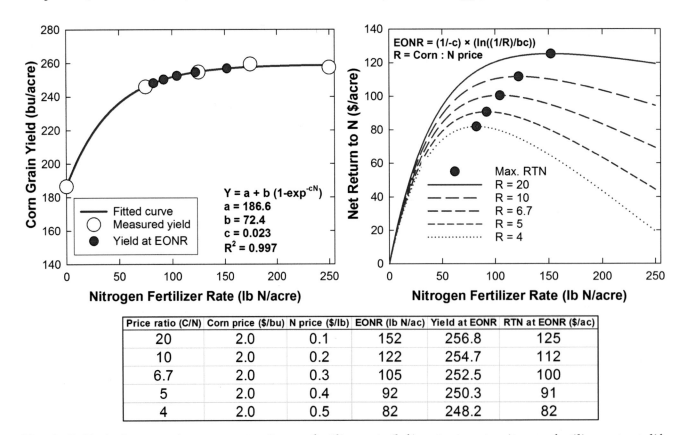

Price ratio (C/N)	Corn price ($/bu)	N price ($/lb)	EONR (lb N/ac)	Yield at EONR	RTN at EONR ($/ac)
20	2.0	0.1	152	256.8	125
10	2.0	0.2	122	254.7	112
6.7	2.0	0.3	105	252.5	100
5	2.0	0.4	92	250.3	91
4	2.0	0.5	82	248.2	82

Fig. 9. 2. Typical corn grain response to nitrogen fertilizer rate (left); net return to nitrogen fertilizer rate at different ratios (R) between corn grain price and nitrogen fertilizer price [EONR is the economic optimum nitrogen rate; Price ratio (R) (C/N) is the corn selling price divided by the nitrogen price; RTN is return on investment for nitrogen].

soil test value near the optimum, or to apply anticipated crop removal levels of nutrients for a couple of years (see nutrient budgets in Chapter 16; Clay, 2017), rather than each year, to reduce application costs. Such fertilizer storage in the soil assumes loss to runoff, leaching, or immobilization into plant unavailable forms to be minimal. If one owns land vs. rents a field for a single year, the decision to apply fertilizer, or not, and at what rate will depend on land tenure as much as likelihood of yield response. Different crops may have different soil test critical values. For example, in Nebraska the Bray-1 P critical value for soybean is 13 ppm, and for corn in a corn–soybean rotation the Bray-1 P critical value is 17 ppm. In a corn–soybean rotation, then, it makes sense to fertilize with P annually if the Bray-1 P level is 17 ppm or less, just to replace crop removal and maintain soil test level, though one would not expect significant yield increase in the soybean year.

These considerations suggest that a variable rate fertilizer map needs to reflect not only spatial variability, but also the temporal and management influences on the fertilization decision. For N, there may be more than one application map for a growing season, depending on management options and risk of N loss. For P and K, a variable rate map could represent inputs for more than 1 yr, or possibly the same map could be used for multiple years.

Spatial Approaches to Management

There are two primary approaches to developing variable rate technology (VRT) fertilization maps: grid soil sampling and management zone delineation based on spatial data layers. Both approaches have merit depending on one's management options, field history, and spatial information resources. Grid soil sampling is more commonly used if (i) there is no prior history or knowledge for the field; (ii) previous management has significantly altered soil fertility through confined livestock or manure application; (iii) small fields have been merged into a single field; or (iv) if an accurate base map of soil organic matter is desired (Ferguson and Hergert, 2009). A management zone approach may be preferable if several sources of spatial information are available from which yield potential or other types of zones can be derived. Such information can include multiple years of yield maps, aerial imagery, soil apparent electrical conductivity (EC$_a$), mapped soil series, or even grower experience.

Grid Soil Sampling

With a grid soil sampling approach, samples are collected at predetermined points using a regular or offset pattern. Typically, multiple soil samples collected in a tight radius (8–10 ft) around each grid point are composited into a single sample for lab analysis (Fig. 9.3). In this approach, the location of each sampling grid point should be identified. A regular grid has evenly spaced grid points horizontally and vertically across the field (Fig. 9.4). An offset grid pattern uses the same spacing, but offsets alternating grid points by half the grid interval every other pass (Fig. 9.3). Past research

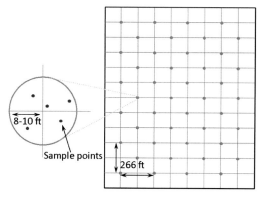

Fig. 9. 3. Suggested grid soil sampling pattern, with four to five soil samples collected in an 8- to 10-ft radius around each grid point.

PROBLEM 9.1.

A 160-acre field is grid sampled on a 2.5-acre grid. How many samples will there be for the field? What is the total cost for analysis if each sample costs $10 for the soil test lab?

ANSWER:

160 acre/2.5 = 64 samples   analysis cost = 64×$10 = $640

PROBLEM 9.2.

If the field is 0.5 miles in length, what are the dimensions of a 160-acre field?

ANSWER:

Total area = length × width; the total area is known and the length is given, therefore the width can be calculated.
160 acres = 43,560 ft²/acre×160 acres = 6969600 ft²
A half mile is 5280 ft/2 = 2640 ft Width of the 160 acre field = 6,969,600 ft²/2640 ft = 2640 ft

PROBLEM 9.3A.

The gridline starts 330 ft from the western edge of a field and 330 ft from the southern border. How many samples will be taken on this grid line?

ANSWER:

The field length is 2640 ft and the grid starts at 330 ft into the field. Typically we do not want to sample close to the edge of the field so we will end up 330 ft from the other edge as well =2640 – 330 = 2310 so there are 2310 ft left along the line. =2310/330 = 7 points

PROBLEM 9.3B.

How many grid lines should there be in this field?

ANSWER:

Each line has 7 points taken along a line, and there are 64 points desired, so number of grid lines = 64/7 = 9.1 lines. Since 9.1 is not a whole number, you decide to round this up to 10 lines.
This will produce 70 (10×7) samples. How far apart will you make these grid lines to have the last line 330 ft from the eastern field border? Line 1 starts at 330 ft from western border. Last line is 330 ft (at 2310) from eastern border. A total of 10 lines are needed. = 2640- 2(330) =1980 ft =1980/9 = 220 ft apart.

PROBLEM 9.4.

How many acres are represented by a 220- by 330-ft area?

ANSWER:

Acres = 220 ft ×330 ft/43,560 ft^2/acre = 1.6 acres

PROBLEM 9.5.

Another grower uses a 266- by 266-ft grid sampling area. How many grid lines, points along the grid, total grid samples, and area is represented by each point if the field is 160 acres - both width and length of 2640 ft?

ANSWER:

The number of grid lines = 2640/266 = 10. Number of points along each grid line = 2640/266 =10.
The area represented by the grid point = 266 ft× 266 ft/43,560 ft^2 = 1.62 acres
Total number of samples = 10×10 = 100

has found that a grid point approach using an offset grid may be preferable (Gotway et al., 1996). As always, more information is better. A VRT map developed from a fine-grid interval will be more accurate than a map developed from a coarse grid interval. However, the industry has standardized on a grid density around 2.5 acres per sample (a grid of about 330 ft by 330 ft); it is a good compromise between map accuracy and affordability. When developing a sampling protocol it is prudent to consider the limitations of the equipment that will be used to apply the treatment.

Management Zone Sampling

With a zone approach, one can expect that fewer soil samples will be collected than for a grid approach. With some management zone approaches, soil sampling may not be used. Once zones have been delineated (Chapter 7), individual soil samples can be randomly collected within each zone, then integrated into one sample for laboratory analysis. With this approach it is not necessary to identify the location of each soil sample point. However, recording soil sample location provides the ability to duplicate the sampling protocol in the future. One may also choose to analyze each soil sample separately rather than aggregating samples within a zone, but this will significantly increase the analysis cost. Normally one would develop fertilizer rate recommendations for each zone using mean soil test and yield expectations for that zone, resulting in a uniform fertilizer rate for each zone.

Converting Soil Analysis Into Fertilizer Recommendations

Eventually soil laboratory analysis results will need to be converted into a fertilizer application rate. The most straightforward approach is to use a soil test parameter to calculate a recommended fertilizer rate using a research-based formula. For example, the University of Nebraska-Lincoln equation for variable rate phosphorus fertilization for corn following corn (Shapiro et al., 2008) is shown in Eq. [9.1].

$$P_2O_5 \text{ rate (lb acre}^{-1}) = (25 - \text{Bray-1 P ppm}) \times 4 \qquad [9.1]$$

This calculation can be performed for every location in the field with a known soil test P value. For example, a location with a Bray-1 P value of 15 ppm would have a P_2O_5 recommendation of 40 lb acre^{-1}; a location with a Bray-1 P value of 23 ppm would have a P_2O_5 recommendation of 8 lb acre^{-1}. Locations with values of greater than 25 ppm would receive no fertilizer. In some cases, producers may wish to apply enough phosphorus fertilizer for several years. For a two-year application, the rates calculated using this equation would be doubled.

For a grid sampling approach, fertilizer rate recommendations can be calculated for each grid point. Estimates of fertilizer rate recommendations between sampling points are derived using spatial statistics to interpolate values between sampled points (Chapter 5; Gotway et al., 1996). For a management zone approach, interpolation is not necessary because the calculated fertilizer recommendation represents the entire zone.

Fertilizer Rate Recommendation Based on Crop Removal

Another approach for variable rate fertilization is based on crop removal. Using this approach, the producer attempts to replace nutrients removed in grain yield. For example, the International Plant Nutrition Institute (IPNI) estimates an average crop P removal rate for corn grain to be 0.38 lb P_2O_5 bu^{-1}; for potassium, crop removal for corn grain is 0.27 lb K_2O bu^{-1}. Thus for a corn grain yield of 180 bu acre^{-1}, crop removal is 68 lb P_2O_5 acre^{-1}; for a yield of

---

PROBLEM 9.6

Calculate the phosphorus fertilizer recommendation based on soil test values.
A transect of grid soil sample analyses across a field has the following Bray-1 P values in parts per million (ppm):

18	23	21	27	14	12

ANSWER:

Using Equation 9.1, the phosphorus fertilizer recommendation (P_2O_5 lb acre^{-1}) for each point above is:

28	8	16	0	44	52

PROBLEM 9.7A.

Calculate P Fertilizer Rate Based on Crop Removal. Given the corn grain yields below (bu acre^{-1}), calculate the P_2O_5 fertilizer rate in lb acre^{-1}, based on the IPNI estimate of 0.38 lb P_2O_5 removal bu^{-1}.

180	207	244	237	226

ANSWER:

68	79	93	90	86

PROBLEM 9.7B.

What if the last two yield observations were in an area close to a feedlot where manure was often applied, and consequently the average Bray-1 P soil test level is 85 ppm, while the first three yield observations were further away from the feedlot, where the average Bray-1 P soil test level is 23 ppm?

ANSWER:

In this case it would be most profitable to not apply P fertilizer in the high soil test area, even though crop removal is significant, until soil test levels decline. In this case, recommended P_2O_5 rates (lb acre^{-1}) would be:

68	79	93	0	0

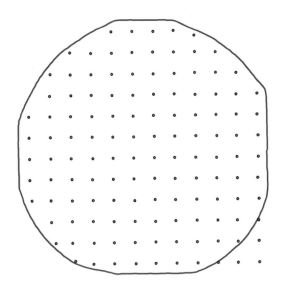

Fig. 9. 4. A high density, regular grid sample pattern for a center-pivot irrigated field.

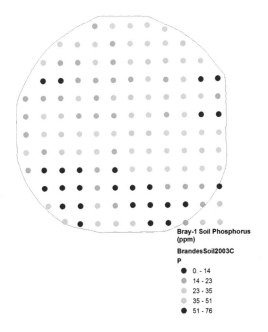

Bray-1 Soil Phosphorus (ppm)

BrandesSoil2003C
P

- 0. - 14
- 14 - 23
- 23 - 35
- 35 - 51
- 51 - 76

Fig. 9. 5. Bray-1 phosphorus soil test levels for a field site in Merrick County, NE.

245 bu acre^{-1}, crop removal is 93 lb P$_2$O$_5$ acre^{-1}. A straightforward way to use this approach is to use the yield map from a given year to generate the recommended fertilizer rate for the following year. However, this single year approach is discouraged for a couple of reasons. Grain yield and thus crop removal will vary year to year; averaging yield and thus crop removal over multiple years will more accurately predict actual trends in P fertilizer need. Also, a strict crop removal approach ignores the underlying soil resource. Applying fertilizer to areas of the field that have high soil test values will reduce overall profit if there is no corresponding increase in production. Realistically a crop removal approach should be combined with soil testing. Areas of the field with moderate soil test values may have fertilizer applied at crop removal rates. Areas with high soil test values may have fertilizer applied at rates below crop removal, or not at all. Additional information on this approach and examples of calculating crop nutrient removal are available in Chapter 16 (Clay, 2017) and Murrell (2008).

On-the-go Sensors for N Fertilizer Recommendations

A third approach for variable rate fertilization currently applies only to nitrogen. This is the use of on-the-go sensors, or remote sensing, to detect crop N status and develop spatial N fertilization recommendations (Barnes et al., 2000; Gitelson et al., 2005). In the case of an on-the-go crop canopy active sensor, reflectance from an internal light source is collected from multiple wavebands, and a vegetation index (VI) is calculated. (An active sensor has its own light source; a passive sensor uses the sun as a light source). An algorithm is used to convert vegetation index information into a N fertilizer rate as the applicator moves across the field. In such cases the prescription map is not developed ahead of time, but typically an as-applied map is created during application and retained for records. For aerial (manned or unmanned) or satellite imagery, crop reflectance will be collected as images in multiple wavebands using passive sensors, and these will be used to calculate a VI. The VI map then can be converted using an algorithm into a VRT N fertilization map. Due to the time required to collect and process aerial and satellite imagery, there is typically a delay of a few days between data collection and fertilizer application. Fertilizer rate recommendations developed from passive or active sensor information is such that the mathematics underlying the fertilizer rate calculations goes beyond the scope of this chapter. However, the basic principles are the same between on-the-go active sensors and passive sensors; crop canopy reflectance in multiple wavebands can be related to crop need for N and thus used to develop spatial fertilizer N recommendations.

Creating the Map

Example 1: Working with Grid Soil Test Data

For this example, the Bray-1 P values associated with sample locations in Fig. 9.4 are shown in Fig. 9.5. Soil test Bray-1 P in these 126 samples range from 6 to 76 ppm, with a mean of 24.8 ppm. For mapping purposes, values are grouped into five categories with equal counts of samples in each group. Such a map provides an idea of the range of soil test P values, and areas of the field which tend to be low or high. Using Eq. [9].1, soil test P values can be converted to a phosphorus fertilizer recommendation (Fig. 9.6). Here the recommended fertilizer rate ranges from 0 to 76 lb P_2O_5 acre^{-1}, with a mean recommended rate of 17 lb P_2O_5 acre^{-1}. However, such a map is of little use for variable rate fertilization. One has no information of what the fertilizer rate should be other than at sample points; there are large areas of the field which are unsampled. To generate a fertilizer application map, one must estimate values for unsampled points using the process of interpolation. The math underlying the interpolation process generally goes beyond the scope of this chapter and is discussed in Chapter 5 (Hatfield, 2017) and Gotway et al. (1996). However, the concept of interpolation is that there is a gradient of values between sampling points if the values are different. For example, if two values are the same along the grid, then values between the two points are the same. However, if the values differ, values for unmeasured points between will be weighted by proximity to known values. Thus, if two known points have values of 5 and 20, unknown points closer to the point with a value of 5 will have values closer to 5 than to 20. You can see that this is a simplified example if you examine the information provided in Fig. 9.6 and 9.7, which is typical of grid and interpolated data.

Today (2017) there are robust interpolation methods available designed to generate fertilizer application maps. Such interpolated maps provide fertilizer rate information for a fertilizer applicator anywhere in the field. Figure 9.7 is an interpolated map of P fertilizer rate for this field, using the kriging interpolation method. Kriging is generally considered the most accurate interpolation method when correctly applied, in that interpolated patterns more closely approximate those occurring naturally in fields. One outcome of the kriging process is the development of a contour map of fertilizer rate. However, kriging is a relatively complex interpolation method, and can generate inaccurate maps if improperly applied. In Fig. 9.7, a phosphorus application map is not produced outside the planned cropping area, but within the bounds of the center pivot irrigation

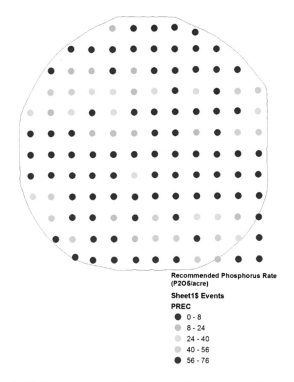

Recommended Phosphorus Rate (P2O5/acre)

Sheet1$ Events

PREC
- ● 0 - 8
- ● 8 - 24
- ● 24 - 40
- ● 40 - 56
- ● 56 - 76

Fig. 9. 6. Recommended phosphorus fertilizer rate (P_2O_5 lb acre^{-1}) resulting from soil test P levels in Fig. 9.5, using Eq. [9.1].

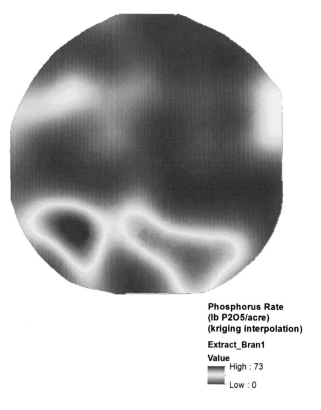

Phosphorus Rate (lb P2O5/acre) (kriging interpolation)

Extract_Bran1

Value
High : 73
Low : 0

Fig. 9. 7. Phosphorus fertilizer application rate map, using kriging interpolation from grid P rate values in Fig. 9.6.

PROBLEM 9.8.

Given that the Bray-1 P soil test for a point (x_0) that is 100 ft (x_0) into the field is 12 ppm (y_0), and the Bray-1 P soil test for a point (x_1) 200 ft into the field is 23 ppm (y_1), calculate the interpolated Bray-1 P value (y) for a point 125 ft into the field (x).

ANSWER:

The simplest form of interpolation is linear interpolation. For example, if soil test values for two points in the field are known, the estimate of a point halfway between the two points is simply the mean of values for the two known points. However, if one wanted to estimate the soil test value of a point in closer proximity to Point B than to Point A, the value of the estimate needs to be weighted according to the proximity to Point B. The formula for linear interpolation which generates a weighted average is:

$$y=\frac{y_0(x_1-x)+y_1(x-x_0)}{x_1-x_0}$$

$$y=\frac{12(200-125)+23(125-100)}{200-100}$$

$$y=14.75 \text{ ppm}$$

PROBLEM 9.9.

A field is soil sampled in an offset grid pattern, with spacing and known Bray-1 P values as shown in the diagram below. Calculate the inverse distance-squared interpolated value for Bray-1 P in the box: P concentrations and distances to known points are 23 ppm (150 ft), 14 ppm (206 ft), 11 ppm (50 ft) and 7 ppm (206 ft).

ANSWER:

The formula for inverse distance squared interpolation with these distances and Bray-1 P values is:

$$x=\frac{\dfrac{23}{150^2}+\dfrac{14}{206^2}+\dfrac{11}{50^2}+\dfrac{7}{206^2}}{\dfrac{1}{150^2}+\dfrac{1}{206^2}+\dfrac{1}{50^2}+\dfrac{1}{206^2}}$$

$$X = 12.03 \text{ ppm Bray-1 P}$$

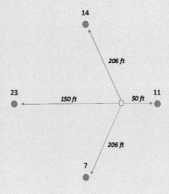

system. Note that Fig. 9.4 through 9.6 show samples collected in the southeast corner, which is outside the reach of the irrigation system.

Another common interpolation method is inverse distance weighted to a power (IDW), most often to a power of 2 (inverse distance squared). A phosphorus fertilizer application map using inverse-distance squared is shown in Fig. 9.8. In this map, areas of the field which have low and high P rate recommendations are generally the same as in Fig. 9.7, yet transitions from one region to another are not as smooth as in Fig. 9.7. Inverse distance interpolation is an exact interpolation process, in that values at known sample points are honored, rather than smoothed over. Thus IDW maps will typically have characteristic "bullseye" patterns around sample points, unless sampling is conducted at

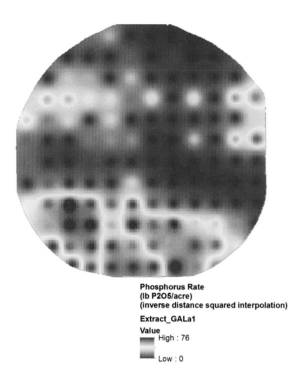

Phosphorus Rate
(lb P2O5/acre)
(inverse distance squared interpolation)

Extract_GALa1

Value
High : 76

Low : 0

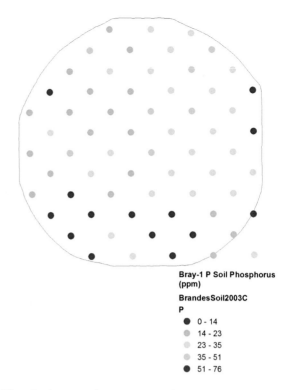

Bray-1 P Soil Phosphorus
(ppm)

BrandesSoil2003C

P
● 0 - 14
● 14 - 23
● 23 - 35
● 35 - 51
● 51 - 76

Fig. 9. 8. Phosphorus fertilizer application rate map, using inverse distance squared interpolation from grid P rate values in Fig. 9.6.

Fig. 9. 9. Data from Fig. 9.5 of Bray-1 P soil test levels, with every other point removed in an offset grid pattern.

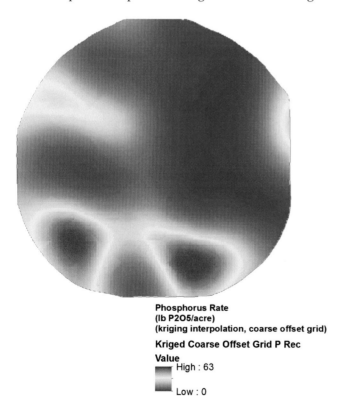

Phosphorus Rate
(lb P2O5/acre)
(kriging interpolation, coarse offset grid)

Kriged Coarse Offset Grid P Rec

Value
High : 63

Low : 0

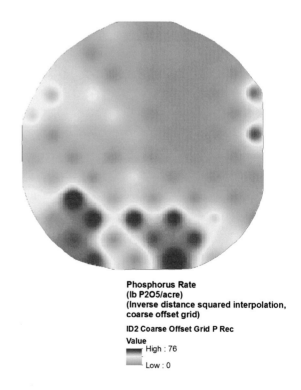

Phosphorus Rate
(lb P2O5/acre)
(Inverse distance squared interpolation,
coarse offset grid)

ID2 Coarse Offset Grid P Rec

Value
High : 76

Low : 0

Fig. 9. 10. Phosphorus fertilizer application rate map, using kriging interpolation from grid P rate values in Fig. 9.9.

Fig. 9. 11. Phosphorus fertilizer application rate map, using inverse distance squared interpolation from grid P rate values in Fig. 9.9.

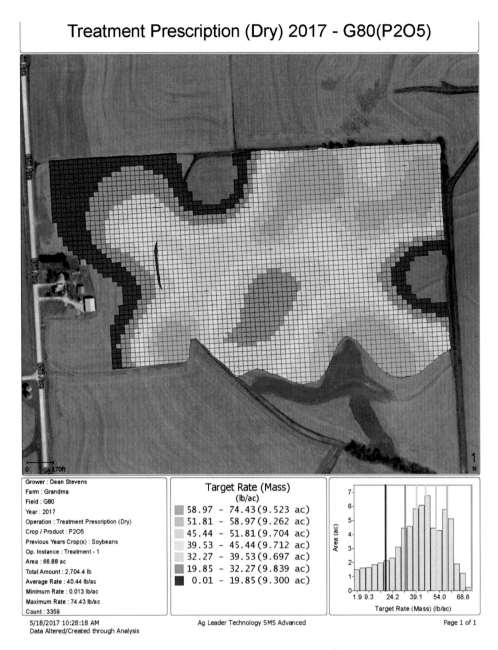

Fig. 9.12. Phosphorus fertilizer application rate map, developed from grid soil sampling, using a grid-based map approach in Ag Leader SMS Advanced software for a field in Richardson County, NE.

very high density. In practice fertilizer application from maps using IDW interpolation often do not differ much from those developed with kriging interpolation, and IDW interpolation is more easily conducted with less risk of error than kriging. For additional information on spatial statistics see Chapter 5.

The field in Fig. 9.4 through 9.8 was sampled at a density of 0.97 acres per sample–much higher than the typical 2 or 2.5 acres per sample most often practiced commercially. Figure 9.9 illustrates Bray-1 P soil test values for this field with half the sample points removed in an offset grid pattern. Figure 9.10 is the resulting P fertilizer application map using kriging interpolation, again for just the irrigated area of the field; Fig. 9.11 is the resulting P fertilizer application map using inverse distance squared interpolation. For this field, there are not major differences in the P application map with either sampling density or interpolation method. For fields with different patterns and trends in spatial variability, this may not be the case.

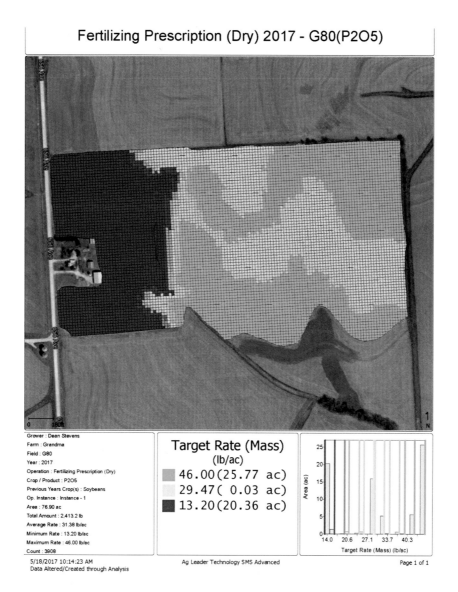

Fig. 9.13. Phosphorus fertilizer application rate map, using three management zones derived from previous yield history, using a grid-based map approach in Ag Leader SMS Advanced software for a field in Richardson County, NE.

Example 2: Comparing Grid and Management Zone Approaches

In this example, using a field in Richardson County, Nebraska, the variable rate prescription map for phosphorus fertilization was developed using both grid and management zone approaches. The grid-based prescription map (Fig. 9.12) used Eq. [9].1 with grid soil sample data to create a fertilizer rate for each grid point. This calculation was done using Ag Leader SMS Advanced software, which then developed a cell-based interpolated map of P fertilizer rate for the applicator. Using this approach, P fertilizer rates ranged from 0 to 74.4 lb P_2O_5 acre^{-1}; areas where the soil test level exceeded the critical value of 25 ppm P were left blank in Fig. 9.12, to receive no fertilizer. Figure 9.13 illustrates the same field segmented into three management zones for P based on previous crop yield history. In this case, soil samples were collected from within each of the three zones and averaged together to determine soil test P level. The mean soil test P within each zone was used with Eq. [9].1 to determine a P fertilizer rate for each zone. In this case the entire field is fertilized–there is no zone with an average soil test P level high enough that no P fertilizer is recommended. Note that the total amount of fertilizer used for the field with each approach is similar; 2704 lb material for the grid approach, 2413 lb material for the zone approach. The distribution is more varied with the grid approach, however, with rates ranging from 0 to 74 lb P_2O_5 acre^{-1}, while the zone approach has only three rates– either 13, 29 or 46 lb P_2O_5 acre^{-1}.

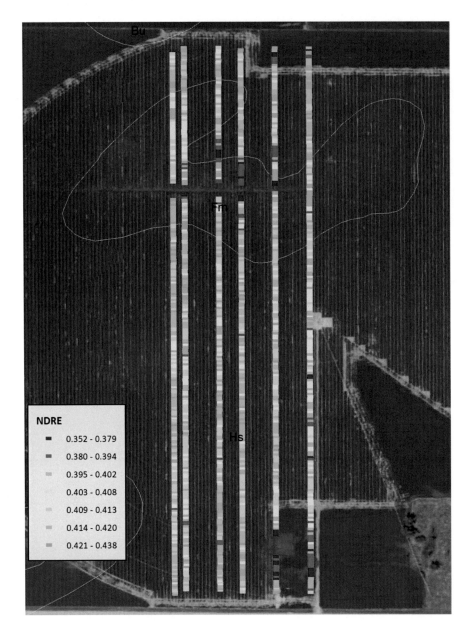

Fig. 9. 14. Normalized Difference Red Edge (NDRE) vegetation index (VI) for treatment strips in a York County, Nebraska field. Reflectance data collected at the V9 growth stage for corn.

Example 3: Working With On-the-Go Sensors

An alternative to a predictive prescription approach for spatial nutrient management is the approach in which a map of fertilizer rate is *not* developed ahead of time, and instead sensors are used in an "on-the-go" approach to vary fertilizer rate in real time. In its basic form, on-the-go sensor approaches to variable nutrient management do not use any prior mapped information about the field; instead, the producer relies on measurements collected by sensors as the fertilizer applicator traverses the field. The primary nutrient managed to date in this manner is N, using active crop canopy sensors to measure crop canopy light reflectance to vary fertilizer rate. Active crop canopy sensors use an internal light source which is modulated such that only canopy reflectance from the internal light source is detected by the sensor. This prevents other light sources (such as the sun) and varying lighting conditions from affecting canopy reflectance, allowing sensor-based variable rate fertilization to occur any time of day or night. Active canopy sensors typically use canopy reflection in visible and near-infrared wavelengths to inform the sensor of crop N status. Such sensing and in-season fertilization typically occurs with corn between V6 and V14 growth stages, and with wheat between Feekes growth stages 4 through 6.

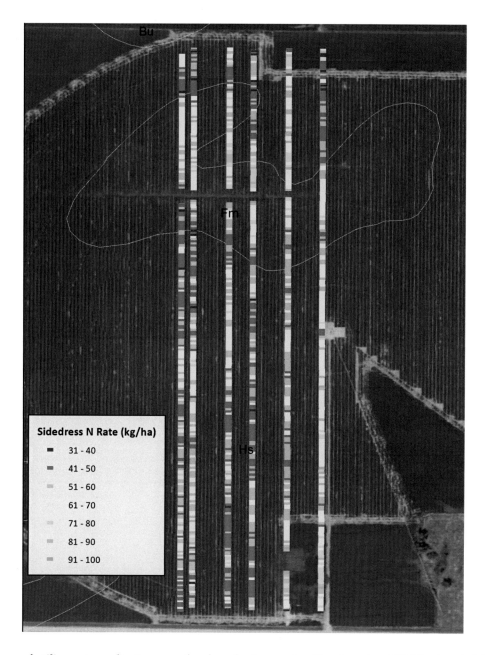

Fig. 9. 15. Sidedress nitrogen fertilizer rate application map developed using an algorithm based on NDRE values in Fig. 9.14.

PROBLEM 9.10.

A commercial crop canopy sensor (OptRx, Ag Leader) collects reflectance in red edge (730 nm) and NIR (780 nm) wavebands. Using Equation 9.3 above, calculate the NDRE vegetation index given the following measurements:

ANSWER:

Red edge reflectance: 0.209
NIR reflectance: 0.323

$$\text{NDRE} = \frac{(\text{NIR} - \text{Red edge})}{(\text{NIR} + \text{Red edge})}, \quad \text{NDRE} = \frac{(0.323 - 0.209)}{(0.323 + 0.209)} = 0.214$$

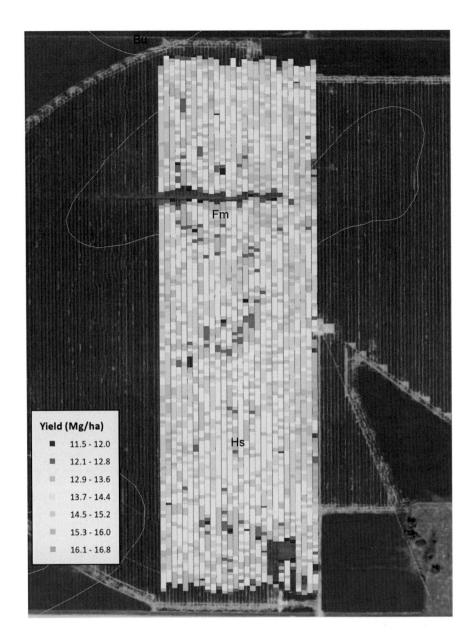

Fig. 9.16. Grain yield (Mg ha⁻¹) resulting from in-season N application in Fig. 9.15. Strips between those in Fig. 9.15 received a uniform in-season N rate based on the producer's standard practices.

Crop canopy sensors collect information in multiple wavelengths to create a vegetation index (VI) which can be useful in providing information about specific plant properties. The most well-known VI is the Normalized Difference Vegetation Index (NDVI) (Rouse et al., 1974). The Normalized Difference Vegetation Index uses the form in Eq. [9].2 below, with resulting values ranging from -1.0 to 1.0:

$$NDVI = \frac{(NIR - Red)}{(NIR + Red)}$$

[9.2]

Where: RED is the reflectance in red wavelengths (~ 670 nm), and NIR is the reflectance in near-infrared wavelengths (~ 780 to 800 nm).

Other commonly used VI with active canopy sensors are the Normalized Difference Red Edge index (NDRE) (Barnes et al., 2000, Eq. 9.3) which uses the same equation form as NDVI, but substitutes reflectance in the region of approximately 730 nm for reflectance in the red wavelength, and the Chlorophyll Index (CI) (Gitelson et al., 2005, Eq. [9].4) which uses reflectance in green (~590 nm) and NIR bands. The NDRE can be more sensitive to N status with high biomass crops due to reflectance saturation in the red region with leaf area index (LAI) greater than two to three (Gitelson et al., 2005).

$$NDRE = \frac{(NIR - Red\ Edge)}{(NIR + Red\ Edge)}$$ [9.3]

$$CI = (NIR \div Green) - 1$$ [9.4]

In Example 3, a commercial crop canopy sensor (OptRx, Ag Leader) was used to control the rate of in-season nitrogen application to a field in York County, Nebraska. This sensor uses the NDRE vegetation index to evaluate canopy N status for corn. Most algorithms (equations) which have been developed from research using active crop canopy sensors use an in-field calibration process, comparing a well-fertilized control area reflectance to that from the target areas of the field. By comparing a non-stressed area to a N stressed area, more accurate fertilizer recommendations can be generated. For example, the NDVI value for a well-fertilized control may be 0.85, while the NDVI value for a N-stressed area of the field may be 0.65. The difference between these two vegetation indices indicates that N fertilization is warranted. The relationship between the control and the target reflectance can be used to develop either a Sufficiency Index (Varvel et al., 2007) or Response Index (Raun et al., 2002), which can be used to develop equations to control N fertilizer application rates.

Figure 9.14 illustrates a map of NDRE readings from sensors on a high clearance applicator driving over the field in late June 2016. The strips with NDRE values are treatment strips which received sensor-based in-season N application, shown in Fig. 9.15. Treatment strips in between these sensor-based treatments received a uniform N rate determined by the cooperating producer. At this site, sensor-based N rate ranged from 31 to 100 kg N ha^{-1}. Grain yield for both sensor-based and uniform rate treatments is shown in Fig. 9.16, ranging from 11.5 to 16.8 Mg ha^{-1}. For this site, on average across all replications the uniform N rate was 188 kg N/ha (168 lb N acre^{-1}), while the sensor-based N rate was 191 kg N ha^{-1} (171 lb N acre^{-1}). The average grain yield was 11.5 Mg ha^{-1} (184 bu acre^{-1}) for the uniform treatment, whereas the sensor-based treatment yielded 13.1 Mg ha^{-1} (208 bu acre^{-1}). It should be noted that even with variable rate N application, yield still varied in variable rate strips, but the mean yield was greater for this application method. In addition, the total N applied per acre was somewhat higher than the grower's typical application, however, the distribution differed. For additional information on remote sensing see Ferguson and Rundquist, 2017.

Compatibility with the Application Controller

Once a prescription map has been developed, several items need to be addressed to ensure proper compatibility with the applicator rate control monitor or in-cab display. Most commercially available rate controllers will accept a generic shapefile (consisting of .shp, .prj, .dbf, and .shx files as a minimum) as an option for prescription file input. For additional information on GIS see Braise (2017). Such prescription maps could be created in any generic geographic information system (GIS) software. There are two primary considerations when using this approach. First, the projection of the shapefile should be defined during creation of the prescription map. In most cases, a geographic coordinate system (GCS) using the WGS 1984 projection will be acceptable to the rate controller (Chapter 4, Clay et al., 2017; Braise, 2017). It should be noted that some monitors will actually accept shapefiles that utilize a projected coordinate system (PCS); examples might include the UTM or State Plane systems.

A second consideration with generic shapefiles is the product units used in the attribute table. An example would be when developing a nutrient prescription map for N application. The units could be in either pounds per acre (lb acre^{-1}) of total N, or they could be converted into a product rate (e.g., 32–0–0) where the gallons per acre (gal acre^{-1}) of the 32% UAN product must be used in the attribute table. Some rate control monitors may be able to perform the conversion from lb/acre to gal/acre based on manual user input of the product form.

An additional note when developing prescription maps relates to how the product is metered on the application equipment. An example of this is with N in the form of anhydrous ammonia (NH_3). While some would designate N from NH_3 in lb acre^{-1} (i.e., a mass-based rate), this product is generally metered from application equipment in liquid form and is therefore measured via a flowmeter.

A second approach for prescription map development is to use farm management information system (FMIS) software. Examples of commercially available FMIS platforms include SMS, SST, FarmWorks, MapShots, and Apex, among others. FMIS software packages utilize a generic GIS software in the background for geospatial functionality, but have generally been developed to interface with the input and/or output file structure, which are not in basic shapefile format from most rate control monitors on the market. An example of this functionality can be seen in

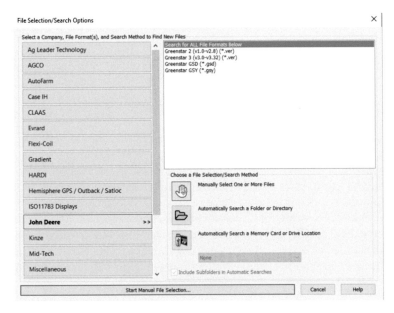

Fig. 9. 17. Ag Leader SMS Advanced screen for selection of the appropriate rate controller to use a previously developed fertilizer rate prescription map.

Fig. 9.17 where supported rate controllers from multiple manufacturers are shown for the current version of SMS software from Ag Leader. In Fig. 9.17, a John Deere monitor has been selected for interface; note that recent monitor firmware versions are available for selection at the right hand side. Firmware for in-cab displays is generally updated on an annual basis and should be checked before prescription maps are exported from the FMIS to the rate control monitor. Most FMIS software will allow the user to create a prescription map in either units of the nutrient (N, P, or K in lb acre^{-1}) or product (gal acre^{-1}). Once exported for a specific monitor configuration, these prescription files will be readable by that display only. One benefit of the FMIS method is that the grower, farm, and/or field file management structure may also be exported to contain prescriptions. This is often beneficial in the field as the applicator operator would be able to choose specific fields from that structure as opposed to uploading a generic shapefile. In addition, as-applied data from the rate control monitor would also be stored into the grower, farm, and/or field file management structure for easier import into the FMIS software.

A final consideration when developing prescription maps for in-field application is related to the number of specific rates contained in the attribute table of the shapefile or the proprietary database format for a particular rate control monitor. For example, some dry product rate controllers will only allow five different application rates while others may allow many application rates. This problem often occurs when an equation is used to generate a prescription map based on a gridded shapefile of soil samples. The equation might assign a unique rate to each grid; for a quarter section field with 2.5 acre grids, this could be up to 64 different rates. A solution to this, for example, might be to designate a minimum rate change of 15 lb acre^{-1} in a N rate prescription map to substantially reduce these rates.

Summary

This chapter addresses issues of collection and use of spatial information to prepare variable rate fertilizer application maps. This includes the use of soil test information, obtained through grid or zone sampling approaches, or crop yield maps for management zone delineation, and resulting calculations associated with developing fertilizer recommendations and map development. The chapter also describes issues to be aware of when importing prescription maps into controllers used with variable rate application equipment.

ACKNOWLEDGMENTS

Support for this document was provided by the Precision Farming Systems communities in the American Society of Agronomy and the International Society of Precision Agriculture. Additional support was provided by the USDA-AFRI Higher Education Grant (2014-04572).

Additional Problems

9.11. Calculate the potassium (K_2O) fertilizer rate recommendation for corn based on crop removal, using the IPNI crop removal value of 0.27 lb K_2O bu^{-1} corn grain yield (Murrell 2008), for corn yields of 197, 224, and 260 bu acre^{-1}.

9.12. Calculate the potassium (K_2O) fertilizer rate recommendation for corn based on the University of Nebraska–Lincoln algorithm of K_2O (lb acre^{-1}) = 125– soil test K (ppm) (Shapiro et al., 2008) for soil test values of 85, 100, and 150 ppm (85, 100, and 150 mg kg^{-1}).

9.13. Following the procedure in Problem 9.4, calculate the Normalized Difference Vegetation Index (NDVI) given Red reflectance of 0.310 and NIR reflectance of 0.44. (Hint: NDRE and NDVI formulas are the same; NDRE substitutes Red edge reflectance for Red).

9.14. Using the formula provided in Problem 9.3 for linear interpolation, calculate the estimated Bray-1 P soil test value for a point 520 feet into the field along a transect, where known Bray-1 P soil test values are 12 ppm at 400 ft and 20 ppm at 600 ft.

9.15. Using the formula provided in Problem 9.3 for linear interpolation, calculate the estimated Bray-1 P soil test value for a point 650 feet into the field along a transect, where known Bray-1 P soil test values are 20 ppm at 600 ft and 33 ppm at 800 ft. With this and information in Problem 4, what is the recommended P fertilizer rate (P_2O_5 lb acre^{-1}) for points at 400, 520, 600, 650, and 800 ft into the field, using Eq. [9].1?

9.16. A grower is only willing to pay for 40 samples across a 160-acre field. He has an 80-ft boom. Set up a sampling grid that would optimize the placement of the 40 samples, based on his equipment.

REFERENCES

Abendroth, L.J., R.W. Elmore, M.J. Boyer, and S.K. Marlay. 2011. Corn growth and development. Iowa State University Extension, Ames, IA. https://store.extension.iastate.edu/Product/Corn-Growth-and-Development (24 July 2017).

Barnes, E.M., T.R. Clarke, and S.E. Richards., P.D., Colaizzi, J. Haberland, M. Kostrzewski. 2000. Coincident detection of crop water stress, nitrogen status and canopy density using ground-based multi-spectral data [CD-ROM]. In: P.C. Robert et al., editor, Proceedings 5th International Conference on Precision Agriculture, Bloomington, MN. 16–19 July 2000. ASA, CSSA, and SSSA, Madison, WI.

Braise, T. 2017. Geographic information systems (GIS). In: K. Shannon and D.E. Clay, editors, Precision agriculture basics. ASA, Madison, WI. (In press).

Clay, D.E. 2017. Assessing a fertilizer program and determining the impact of management on soil health using soil and yield monitor benchmarks. In: D.E. Clay, S.A. Clay, and S. Bruggeman, editors, Practical mathematics for precision farming. ASA, CSSA, and SSSA, Madison, WI.

Clay, D.E., T.A. Brase, and G. Reicks. 2017. Mathematics of latitude and longitude. In: D.E. Clay, S.A. Clay, and S. Bruggeman, editors, Practical mathematics for precision farming. ASA, CSSA, and SSSA, Madison WI.

Dobermann, A., C.S. Wortmann, R.B. Ferguson, G.W. Hergert, C.A. Shapiro, D.D. Tarkalson, and D.T. Walters. 2010. Nitrogen Response and Economics for Irrigated Corn in Nebraska. Agron. J. 103:67–75. doi:10.2134/agronj2010.0179

Fausti, S., B.J. Erickson, D.E. Clay, and C.G. Carlson. 2017. Deriving and using equations to calculate the economic optimum fertilizer and seeding rates. In: D.E. Clay, S.A. Clay, and S. Bruggeman, editors, Practical mathematics and agronomy for precision farming. ASA, Madison WI.

Ferguson, R.B. and D. Runquist. 2017. Remote sensing for site-specific crop management. In: K. Shannon and D.E. Clay, editors, Precision agriculture basics. ASA, Madison WI. (In press).

Ferguson, R.B., and G.W. Hergert. 2009. Soil sampling for precision agriculture. Extension Circular 154, University of Nebraska Extension, Lincoln, NE. (http://extensionpublications.unl.edu/assets/pdf/ec154.pdf (verified 24 July 2017).

Gitelson, A.A., A. Viña, V. Ciganda, D.C. Rundquist, and T.J. Arkebauer. 2005. Remote estimation of canopy chlorophyll content in crops. Geophys. Res. Lett. 32:L08403. doi:10.1029/2005GL022688

Gotway, C.A., R.B. Ferguson, G.W. Hergert, and T.A. Peterson. 1996. Comparison of kriging and inverse-distance methods for mapping soil parameters. Soil Sci. Soc. Am. J. 60:1237–1247. doi:10.2136/sssaj1996.03615995006000040040x

Hatfield, G. 2017. Spatial statistics. In: D.E. Clay, S.A. Clay, and S. Bruggeman, editors, Practical mathematics and agronomy for precision farming. ASA, Madison WI.

Murrell, T.S. 2008. Average nutrient removal rates for crops in the North Central Region. International Plant Nutrition Institute, editors, Plant Nutrition Today, Fall 2008, No. 4. International Plant Nutrition Institute. Norcross, GA. http://www.ipni.net/publication/pnt-na.nsf/0/8A409B10BB8A4ECE85257CD600736675/$FILE/PNT-2008-Fall-04.pdf (verified 24 July 2017)

Raun, W.R., J.B. Solie, G.V. Johnson, M.L. Stone, R.W. Mullen, K.W. Freeman, W.E. Thomason, and E.V. Lukina. 2002. Improving nitrogen use efficiency in cereal grain production with optical sensing and variable rate application. Agron. J. 94:815–820. doi:10.2134/agronj2002.8150

Rouse, J.W., R.H. Haas, J.A. Scheel, and D.W. Deering. 1974. Monitoring vegetation systems in the Great Plains with ERTS. In: S.C. Freden, E.P. Mercanti, and M.A. Becker, editors, Proceedings, 3rd Earth Resource Technology Satellite (ERTS) Symposium, Vol. 1: Technical Presentations Washington, D.C. 10-14 Dec. 1973. NASA, Washington, D.C. p. 48-62.

Shapiro, C.A., R.B. Ferguson, G.W. Hergert, C.S. Wortmann, and D.T. Walters. 2008. Fertilizer suggestions for corn. Extension Circular 117, University of Nebraska–Lincoln, Lincoln, NE.

Varvel, G.E., W.W. Wilhelm, J.F. Shanahan, and J.S. Schepers. 2007. An algorithm for corn nitrogen recommendations using a chlorophyll meter based sufficiency index. Agron. J. 99:701–706. doi:10.2134/agronj2006.0190

Essential Plant Nutrients, Fertilizer Sources, and Application Rates Calculations

10

Jiyul Chang, David E. Clay,* Brian Arnall, and Graig Reicks

Chapter Purpose

In precision farming, even though the fertilizer rates may be varied across the landscape, the calculations for converting a recommendation into an application map are identical. For many farmers, purchasing fertilizers is a major expense for their operation and understanding the products and associated calculations is required to optimize productivity. This chapter provides a background on the nutrients needed for plant growth, fertilizer types, manure, fertilizers used for organic agriculture, and how to convert nutrient recommendations from a soil testing laboratory value into a recommendation. Practical problems and their associated answers are provided. A critical component associated with fertilizers is understanding that all fertilizers should be applied following the 4R protocol (right rate, right source, right time, and right place).

Key Terms

Fertilizers, manure, organic agriculture, soil nutrients.

Mathematical Skills

Converting nutrient recommendations to fertilizer rates.

Introduction

In precision farming, variable fertilizer application rates often are used across the field; however, the application rates are dependent on the site-specific recommendations. Optimizing the fertilizer recommendation by location requires an understanding of the nutrients needed for plant growth and the timing of nutrient uptake (Table 10.1). Since the advent of agriculture, the techniques used to meet the needs of the plant have changed. Old world civilizations, including the Egyptians, the Babylonians, and the Romans used minerals and manure to enhance soil and crop productivity, whereas Native North Americans inserted small fish or a nitrogen fixing plant adjacent to corn seedlings to provide N and other nutrients (Clay et al., 2017). The South American Andean peoples harvested guano that was used as a soil amendment approximately 1000 yr ago. The Andeans valued this resource, and if a person was caught hurting seabirds, the guano source, or harvesting these materials

J. Chang and D.E. Clay, South Dakota State University, Brookings, SD 57007-2201; N. Kitchen, Division of Plant Sciences, University of Missouri, Columbia, MO; B. Arnall, Oklahoma State University, Oklahoma City, OK 74078; G. Reicks, South Dakota State University Extension, Brookings, SD 57007.*Corresponding author (david.clay@sdstate.edu).

doi:10.2134/practicalmath2017.0026

Table 10.1 A list of commonly used inorganic fertilizers, fertilizer grades, densities, and S and Cl contents. For all fertilizers, the fertilizer grade is equivalent to the unit of a nutrient in 100 units of that fertilizer. For example, 100 lbs of 46–0–0 contains 46 lbs of N. Similarly, 100 kg of 46–0–0 contains 46 kg of N. All fertilizers contain soluble salts which can reduce seed germination when applied with the seed. Information for determining the amount of fertilizer that can be applied with the seed is available at https://igrow.org/up/resources/03–5000–2016–26.pdf.

	N	P_2O_5	K_2O	Density	S	Cl
Solid fertilizers	%	%	%	lb gal^{-1}	%	%
Ammonium nitrate (NH_4NO_3)	33	0	0			
Urea ($(NH_2)_2CO$)	46	0	0			
Potassium nitrate (KNO_3)	13	0	44			
Calcium nitrate ($Ca(NO_3)$)	15					34% CaO
Ammonium sulfate ($(NH_4)_2SO_4$)	21	0	0		24	
Diammonium phosphate (DAP, $(NH_4)_2HPO4$)	18–21	46–53	0		0–1.5	
Monoammonium phosphate (MAP, $NH_4H_2PO_4$)	10–12	48–55	0		0–1.5	
Triple superphosphate ($Ca(H_2PO_4)(H_2O)_2$	0	46	0			
Nitrophosphates ($H_3PO_4+Ca(NO_3)_2$)	21	7	14			
	28	14	0			
	12	24	12			
Potassium chloride (KCl)	0	0	62			47
Potassium sulfate (K_2SO_4)			50–53		17	
Potassium nitrate (KNO_3)	13	0	44			
Elemental S (S)	0	0	0		90	
Liquid fertilizers						
Urea ammonium nitrate (UAN, urea + NH_4NO_3)	28–32	0	0	10.6–11.0		
Ammonium polyphosphate (chain of NH_4PO_3–NH_4PO_3–	10	34	0	11.6		
(polyphosphates)	10	37	0	11.9		
Thiosulfates ATS [$(NH_4)_2S_2O_3$]	12	0		11.0–11.2	26	
KTS ($K_2S_2O_3$)	0	0	250	1.46	18.7	
Gas fertilizer						
Anhydrous ammonia (NH_3)	82	0	0			

without approval, they could be sentenced to death. After the fall of the Incan Empire, guano was harvested and shipped around the world as a fertilizer source. In fact, before synthetic fertilizer was produced, excrements of seabirds, seals, and bats (guano) was a major source of fertilizer (COHA, 2011). Quano can contain high concentrations of N, P, and K.

Crop rotations have been used to reduce pests and use N-fixing plants to provide N to following crops in the rotation. Since the turn of the 20th century, the ability to maintain the soil's nutrient capacity has changed from reliance on rotations and manure, which provided organic forms of nutrients, to inorganic fertilizers, where P and K fertilizers were mined and ammonia (NH_3) was produced through the Haber–Bosch process. More recently, other fertilizer products are being developed for specific applications (Chien et al., 2009). For example, some N fertilizers have incorporated inhibitors into the formulation to slow urea hydrolysis and volatilization, whereas in others, nitrification inhibitors are used to slow the release of nitrogen from the fertilizer granule.

Fertilizers can be solids, liquids, or gases. Each fertilizer type has unique strengths and weaknesses. For example, a weakness of urea is that it is generally not recommended for use as a seed treatment, as urea can be toxic to seedlings, whereas a strength is that it is easy to store. The most common fertilizers applied to soils contain the elements N, P, and K (Table 10.1). These nutrients are called macronutrients and are reported in the fertilizer grade, % N, % P2O5, and % K2O. A series of fact sheets on these nutrients, prepared by the International Plant Nutrition Institute, are available at https://www.ipni.net/nutrifacts-northamerican. When fertilizers are added to soil, the individual nutrient availability is dependent on pH (Fig. 10.1). Many nutrients are most available between pH values of 5.5 and 7.5. In addition, detailed information on the nutrients and appropriate application techniques are presented in Roy et al. (2006) and Havlin et al. (2013). The fourth percentage in a fertilizer grade (if present) generally represents the amount of sulfur (S) contained in the fertilizer (e.g., 12–0–0–26S). A compound fertilizer is

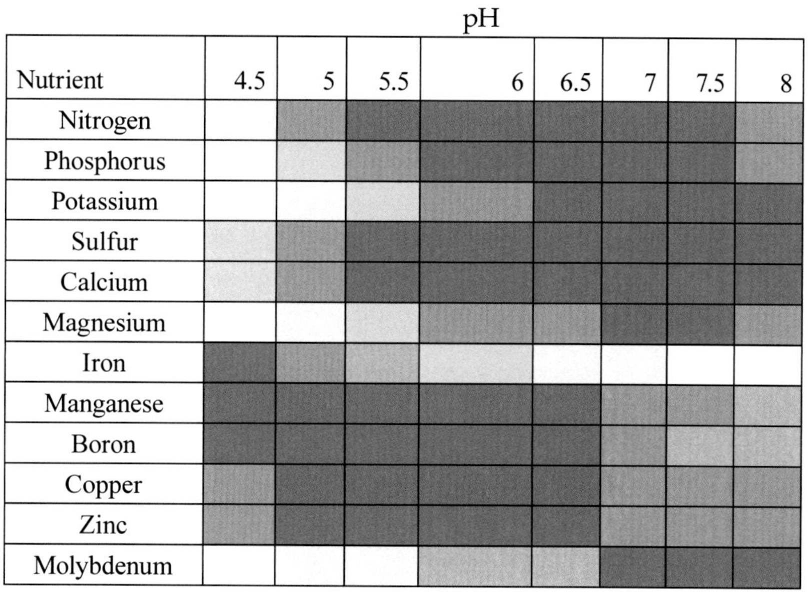

pH

Nutrient	4.5	5	5.5	6	6.5	7	7.5	8
Nitrogen								
Phosphorus								
Potassium								
Sulfur								
Calcium								
Magnesium								
Iron								
Manganese								
Boron								
Copper								
Zinc								
Molybdenum								

Fig. 10.1. Relationship between pH and soil nutrient availability. The darker the color within a nutrient, the higher the availability at that pH. Most nutrients have differential availability across soil pH values.

Fig. 10.2. In-season N liquid fertilizer applicator (Courtesy of John Fulton, The Ohio State University).

PROBLEM 10.1.

Convert 10 kg K ha^{-1} to kg K_2O ha^{-1}.

ANSWER:

The molecular weight of K and O are 39.1 and 16 g, respectively. Based on these weights, one mole of K_2O weighs 94.2 g.

$$\frac{10,000 \text{ g K}}{\text{ha}} \times \frac{94.2 \text{ g K}_2\text{O}}{78.2 \text{ g K}} \times \frac{1 \text{ kg}}{1000 \text{ g}} = 12.1 \text{ kg K}_2\text{O/ha (10.8 lb N/acre)}$$

When doing these calculations, check the units. In this case, g K in the numerator are cancelled by g K in the denominator. If the units are not correct the answer is not correct. Second, does the answer make sense. For example, should the amount of K_2O be more or less than the amount of K. The answer has to be greater than 10 kg, since K_2O also contains two molecules of K and a molecule of O.

PROBLEM 10.2.

Convert 20 kg P_2O_5 (44.2 lb) to kg P.

ANSWER:

The molecular weight of P is 30.98 and the molecular weight of oxygen is 16. Since the molecular weight of P_2O_5 equals to 141.96 g (30.98×2+16×5) and each mole contains two atoms of P weighing 61.96 g (2*30.98), then

$$20 \text{ kg P}_2\text{O}_5 \times \frac{2(30.9 \text{ kg P})}{141.96 \text{ kg P}_2\text{O}_5} = 8.7 \text{ kg P}$$

Check your answer by cancelling the units. In this example, kg can be converted to pounds by multiplying kg by 2.20462 (44.1 lb).

PROBLEM 10.3.

Convert 50 g nitrate (NO_3) to nitrate—N (NO_3—N).

ANSWER:

The nitrate molecule contains three atoms of oxygen and one atom of N. When nitrate is reported as NO_3—N, only the N is considered. This means that we only consider the 14 g of N in the nitrate molecule.

When N is reported as NO_3 both the nitrogen and the oxygen is considered and the molecular weight is 62 g (14+16×3).

$$50 \text{ g NO}_3 \times \frac{14 \text{g N}}{62 \text{g NO}_3} = 11.3 \text{ g N}$$

Check your answer by cancelling the units.

typically a solid product that contains multiple nutrients within each granule. A fertilizer blend, however, is created when different granules with different compositions are mixed together. In general, compound fertilizers are more expensive than blended fertilizers.

The oxide percentages of P_2O_5 (molecular weight = 141.9) and K_2O (molecular weight = 94.3) can be converted to elemental amounts of P (molecular weight = 30.97) and K (molecular weight of 39.1) by multiplying the percentage found in the fertilizer grade by their elemental molecular weights. For both the P and K calculations, the oxide contains two molecules of the element, so that the molecular weight of the element is multiplied by 2. For example, if a fertilizer grade is 0–52–65, the conversion equations to calculate the elemental amount of P and K are:

$$52\% \text{ P}_2\text{O}_5 = \frac{52 \text{ lb P}_2\text{O}_5}{100 \text{ lb fert.}} = \frac{(30.97 \times 2) \text{ lb P}}{141.9 \text{ lb P}_2\text{O}_5} \times \frac{52 \text{ lb P}_2\text{O}_5}{100 \text{ lb fert.}} = \frac{22.7 \text{ lb P}}{100 \text{ lb fert.}}$$

$$65\% \ K_2O = \frac{65 \text{ lb } K_2O}{100 \text{ lb fert.}} = \frac{(39.1 \times 2) \text{ lb } K}{94.3 \text{ lb } K_2O} \times \frac{65 \text{ lb } K_2O}{100 \text{ lb fert.}} = \frac{53.89 \text{ lb } K}{100 \text{ lb fert.}}$$

In addition, when a value is reported as NO_3–N, only the N in the nitrate molecule is considered, whereas when reported as NO_3 both the N and O are considered. Nitrate-N (NO_3–N) is converted to nitrate (NO_3) by multiplying nitrate N by 4.43 (62/14).

Essential Plant Nutrients

Nitrogen (N) is an essential element and it is contained in all plant and animal proteins. Because many soils cannot provide enough N to meet the plant requirement, fertilizers are routinely applied. Nitrogen fertilizer typically is applied at planting or after the crop has emerged. The application may be the entire rate, or may be applied as a split application, with some applied at planting, and the rest applied later in the season. Nitrogen fertilizer can be applied by broadcasting granules onto to the soil surface, injecting the fertilizer into a band below the soil surface as a gas, spraying a liquid fertilizer onto the soil or leaf surface, or applying a liquid formulation through a streamer nozzle bar close to the soil surface to reduce leaf burn (Fig. 10.2).

Even though most of the soil N is contained in organic matter, only the inorganic forms (NO_3^- and NH_4^+) are taken up by plants. Many N fertilizers are NH_4–N based, and when added to soil, the ammonium is nitrified to NO_3^-. The nitrification process produces H^+ ions, which over time, lowers the soil pH. Information on the impact of fertilizer source on soil acidification is available in Table 6.7 (Clay et. al, 2017). The nitrate molecule, having a negative charge, can move (leach) with percolating water or be denitrified. Ammonium nitrification rate can be reduced by treating the fertilizer with a nitrification inhibitor. The rate that urea is hydrolyzed can be slowed by treating the urea molecule with a urease inhibitor.

Fig. 10.3. Nitrogen deficiency in corn. In this image, the deficicies are first observed in lower leaves (Source: Felix Francis, Univeristy of Delaware, bugwood.org).

Nitrogen is considered mobile in plant tissue, meaning that the nutrient can move from one part of the plant to another. Deficiency symptoms are first seen in older leaves, as these nutrients are moved to newer, developing tissue that require the nutrient for growth. Nitrogen deficient plants may have pale green leaves with symptoms that first appear as a V pattern along the central vein in older tissues leaves (Fig. 10.3), whereas young tissue remains green. Additional information and images for identifying nutrient deficiencies are available in Hosier and Bradley (1999), McCauley et al. (2011), and Sawyer (2004).

Phosphorus (P) is needed in relatively large amounts by plants and it has a critical role in energy storage and use (Beegle and Durst, 2002). Adequate P levels have been linked to improved water use efficiency, disease resistance, and helps manage abiotic stress. The maximum availability of P occurs in soils with pH values between 6.5 and 7, and in many soils P uptake is most efficient when it is band applied (surface, point injection, or placed within an inch or two of the seed). Numerous sources of P exist, including commercial fertilizers, or contained within animal wastes and biosolids. Phosphorus is generally mobile in the plant and deficiencies (purple on the leaf edges) will appear first in older leaves (Fig. 10.4).

Unlike P and N, potassium (K) is not integrated into complex organic molecules in the plant, but is integrally involved in water stress regulation and balancing electrical charges in the cell. Potassium helps the plant withstand stress, reduces lodging, reduces insect damage, and helps to mitigate diseases. Potassium can be added to the soil through irrigation water, fertilizers, and manure. The release of K from clay minerals is also another K source. Potassium is considered mobile in the plant and deficiencies of necrosis will appear first along the older leaf margins (Fig. 10.5).

Fig. 10.4. Phosphorus deficiency in corn. Purple along the leaf-edges is a common symptom (Source: South Dakota State University).

Fig. 10.5. Potassium deficiency in corn. (Courtesy, International Plant Nutrition Institute, crop nutrient deficiency image collection, image number IPNI12014hs101–1348, M.K. Rakkar)

Calcium (Ca) is a component of several soil amendments (gypsum and lime), is a crucial regulator of many plant processes, and is required in the plant cell wall and membranes. Deficiencies are not common; however, they have been observed in acidic soils. Gypsum ($CaSO_4$) is often added to soil as a remediation strategy for sodic soils, whereas lime ($CaCO_3$) is added to raise the soil pH. Calcium is considered immobile in plants, meaning that this nutrient does not move in the plant. Deficient plants may be stunted and have necrotic areas on the margins of the leaves. Additional information for identifying deficiencies are available in Hosier and Bradley (1999), McCauley et al. (2011), and Sawyer (2004).

Sulfur (S) is a component of several amino acids, and due to reduced S emissions and subsequent down-wind deposition of S from coal-burning power plants, sulfur fertilization in some areas of the U.S. has become increasingly important. Sulfur in the soil is primarily found in organic matter and becomes plant available during organic matter mineralization. There are many different sources of fertilizer sulfur that can be applied to soil, including gypsum, ammonium sulfate, and elemental S. When added to soil, $(NH_4)_2SO_4$ (ammonium sulfate) rapidly dissolves in soil to the ammonium (NH_4^{+1}) and sulfate (SO_4^{-2}) components. In the microbial process of converting elemental S also to SO_4^{-2}, the soil pH is lowered. In sodic soils that contain $CaCO_3$, elemental S has been used as a chemical amendment to help rebuild the soil structure. In plants, sulfur is not mobile, and deficiency symptoms may include younger leaves that are pale green.

Boron (B) fertilization must be managed very carefully because toxicity and deficiency levels occur within a very narrow range of concentrations. Boron is involved in cell differentiation and deficient plants may be stunted and deformed. Boron deficiencies or excesses can limit alfalfa, clover, and trefoil growth (Kelling, 1999). Because boron can be toxic to germinating seeds, borated fertilizers should not be placed with corn, soybeans, or oats seeds during planting (Kelling, 1999). For these crops, B should be broadcast applied and thoroughly mixed with the soil. Boron fertilizers include, borax, boric acid, sodium pentaborate, sodium tetraborate, solubor®, and boron frits. Boron is not mobile in the plant and deficiency symptoms might include stunted growth and dead growing tips.

Magnesium (Mg) has many roles, including being a central atom in the chlorophyll molecule and it is involved in ATP synthesis. Magnesium is a mobile nutrient and the deficiency symptoms may include yellowing between leaf veins (chlorosis) in older leaves. Plant deficiencies are most often found in acidic sandy soils. Magnesium can be provided by irrigation water and the weathering of soil minerals. It is also contained in many soil amendments (dolomite, magnesium chloride, and magnesium sulfate). Additional information for identifying deficiencies are available in Hosier and Bradley (1999), McCauley et al. (2011), and Sawyer (2004).

Chloride (Cl) is involved in many plant processes including nutrient transport, stomata activity, and water movement. It helps increase stem strength, and in wheat, may reduce the severity of take-all, tan spot, strip rust, and Septoria. In plants, Cl is a mobile nutrient, and deficiency symptoms can result in wilting and leaf mottling. A common Cl⁻ application approach is to apply potassium chloride (KCl) as the K source. Other chloride sources are magnesium chloride ($MgCl_2$), ammonium chloride (NH_4Cl), and calcium chloride ($CaCl_2$).

Copper (Cu) has an essential role in both plants and animals and, in plants, it is not a mobile nutrient. Copper is needed for photosynthesis and respiration, and deficient plants may appear stunted with leaves that are pale green. Cu deficiencies may be observed in high pH sandy soils that contain high amounts of available P.

Because copper can be tightly bound to organic matter, deficiencies may also be observed in soils with > 8% organic matter. Alfalfa, oats, rice, and wheat are most responsive to copper, whereas canola, potato, and soybean are least responsive. Common fertilizer sources are copper sulfate, copper chelate (EDTA), copper sulfate monohydrate, copper acetate, copper ammonium phosphate, cupric oxide, and animal manures.

Zinc (Zn) deficiencies are a worldwide problem. Zinc is involved in protein synthesis and energy production and it is not a mobile nutrient. Deficiency symptoms may include delayed maturity, short internodes, and reduced leaf sizes. Deficiencies may be found in organic, eroded, and sandy soils containing low soil organic C levels. Zinc availability from soil decreases with increasing pH. Zinc fertilizers can be grouped into synthetic chelates, inorganic compounds, and organic materials. Organic materials such as manure can provide a substantial amount of Zn to the soil. Corn, rice, and flax are most responsive to zinc, whereas oats, rye, and lettuce are least responsive.

Manganese (Mn) is a plant enzyme activator and is essential for root growth. Manganese is immobile in plant tissues with deficiencies appearing first in young tissues. Soybeans, small grains, and peanuts are considered Mn sensitive plants. Symptoms might include interveinal chlorosis and necrotic spots (dead spots) on the leaves. Manganese deficiencies may be observed in sandy soils with low organic matter and high pH. Mn concentration decreases as the pH increases, whereas the concentrations become toxic when soil pHs are very low. Manganese sulfate ($MnSO_4$) can be broadcast (10–15 lbs Mn acre^{-1}), banded into soil (3–5 lbs Mn acre^{-1}), or applied as a foliar (1–2 lbs Mn acre^{-1}) spray. In high pH soils, banding maybe more efficient than broadcast applications.

In plants, iron (Fe) is a critical component of many enzymes that have numerous biological functions. Iron is not a mobile nutrient in the plant with yellowing first appearing on young leaves. In most soils, only a small portion of the total Fe contained in the soil is available to plants. In high pH soils, Fe^{+3} availability can be reduced by Ca and bicarbonate ions (HCO_3^-), whereas in poorly drained soils, Fe^{+3} may be reduced to Fe^{+2}, which is more water soluble. Deficiencies can be reduced by growing tolerant varieties, seed treatment, or applying foliar or soil-based Fe fertilizers. In soybeans, an oat companion crop has been successful in reducing iron chlorosis deficiencies (Kaiser et al., 2014).

Molybdenum (Mo) is needed to produce enzymes that are involved in N fixation and it is a mobile nutrient in plants. In legumes, Mo deficiency will have a similar appearance as N deficiency. Molybdenum availability increases with pH and its concentration can be reduced by adding sulfate fertilizers. Acid sandy soils that were formed under humid conditions may be Mo deficient. Molybdenum fertilizer can be banded or broadcast at rates ranging from 0.5 to 2 lbs acre[19 if] and is often mixed with other fertilizers. It can also be applied as foliar application or seed treatment. Common fertilizer sources are sodium molybdate ($Na_2MoO_4 \cdot 2H_2O$), ammonium molybdate [$(NH_4)_2 MoO_4$], and molybdenum trioxide (MoO_3).

Nickel (Ni) is a component of the urease enzyme, which converts urea to ammonia. Deficiency symptoms may include necrosis (dead tissue) starting at the leaf tip. Nickel helps the plant protect itself against pathogens, and its availability decreases with increasing pH. Due to limited information, soil testing for Ni availability is not routinely conducted. Trace amounts of Ni in some fertilizers may provide adequate Ni for the plant requirements. When deficient, foliar sprays can be applied.

Cobalt (Co) has a critical role in the synthesis of vitamin B_{12}, a protein involved in N_2 fixation. Deficiencies are more likely to occur in animals consuming low Co biomass than plants. Animals grazing on plants growing in soils with low cobalt concentrations may experience vitamin B_{12} deficiencies. These deficiencies can be reduced by applying a small amount of cobalt to the plants. In plants, low Co concentrations may include reduced seed germination, and in leguminous plants, N_2 fixation. Co deficiencies can occur in highly weathered, coarse textured soils.

Silicon (Si) adds strength to cell walls and stalks, and assists the plant in resisting fungal pathogens. Shortages can result in weak stems and roots and increased diseases. A commonly applied silicon fertilizer is calcium silicate ($CaSiO_4$). The effects of Si fertilizers has had mixed outcomes, however there have been situations where corn, wheat, rice, and sugarcane have responded positively to silicon fertilizers. Problems can be reduced by not harvesting the crop residue.

Selenium (Se) is an essential element required by humans. Plants can uptake selenate (SeO_4^{-2}) and selenite (SeO_3^{-2}). Selenate may be dominant in well-drained soils with neutral pH (7), whereas selenite may be dominant in well-drained soils with a neutral to acid pH values (pH 5–7). Soils from Australia or New Zealand may contain low Se concentrations. Selenate is less soluble than selenite, and in poorly-drained soils, selenide (Se^{-2}) may be the dominant species. Plants may contain selenium concentrations that can be either toxic or deficient to grazing animals. When Se is less than the animal's requirement, Se concentrations can be increased by foliar or seed treatments. Plants containing toxic Se levels may be stunted and the leaves may be pale green or yellow. Ammonium sulfate and superphosphate fertilizers can contain substantial amounts of Se.

Slow Release Fertilizers

Slow release fertilizers can be characterized as uncoated slow release, coated slow release, or bioinhibitors. The primary goal of a slow-release fertilizer application is to apply a single application that will satisfy the plant's nutritional needs by making the nutrient available only when needed ('spoon feed') throughout the season. Slow release fertilizers should, therefore, minimize nutrient leaching, volatilization, and denitrification losses. This is different from most

PROBLEM 10.5A.

How much N is contained in a gallon of liquid fertilizer that is 28% N? The density is 10.8 lb gal^{-1}.

ANSWER:

$$1 \text{ gallon fertilizer} \times \frac{10.8 \text{ lb}}{1 \text{ gallon}} \times \frac{0.28 \text{ lb N}}{1 \text{ lb fertilizer}} = 3.02 \text{ lb N gal}^{-1}$$

PROBLEM 10.5B.

How many gallons of 28-0-0 are needed to apply 100 lb N acre^{-1}?

ANSWER:

$$\frac{100 \text{ lb N}}{acre} \times \frac{1 \text{ lb fertilizer}}{0.28 \text{ lb N}} \times \frac{gal}{10.8 \text{ lb fertilizer}} = \frac{33.1 \text{ gal}}{acre}$$

commercial fertilizers, where nutrient availability is dependent on the nutrient solubility, and when and where they are applied. An advantage of slow release fertilizers is that multiple applications should not be required.

Coated Fertilizers

Two general types of coated fertilizers are available. The first type is sulfur coated products and the second type is polymer coated products. In sulfur coated fertilizer, the nutrients become available following oxidation of an elemental sulfur coating. One of the first coated fertilizers that was commercially available was sulfur coated urea.

The effectiveness of a coating can be enhanced by adding an organic polymer creating a polymer-coated type of fertilizer. Polymer coated products include osmocote, polyon, duratin, and ESN. These products are effective in slowing the nutrient release into the soil and they are most effective when the fertilizer release and plant uptake patterns are synchronized. For polymer coated urea, the coating slows N diffusion through the porous polymer membrane. Different coatings have different release patterns. Additional information is available in Chien et al. (2009).

Uncoated Slow Release Fertilizer

In uncoated slow release fertilizers, the nutrient is in a form that is slowly made available to the plant. Examples include urea-formaldehyde or isobutylindine diurea. An advantage of uncoated materials over the coated ones is that individual uncoated particles can be smaller and more homogenous. Urea formaldehyde is produced by reacting urea with formaldehyde, and the resulting length of the chain that forms influences its release pattern. These products may be marketed under ureaform, methylene ureas, and UF solutions.

Isobutylindene diurea (IBDU) is produced by reacting urea with isobutyraldehyde. This product is approximately 30% N. When placed in water, isobutylindene diurea is hydrolyzed to urea and isobutyraldehyde. The hydrolysis rate is dependent on particle size, the product's water solubility, pH, and temperature.

Bioinhibitors

Bioinhibitors can be used to slow the nutrient release from solid, liquid, and gas fertilizers. These products work by slowing and inhibiting microbial processes such as nitrification ($NH_4 - NO_3$) or hydrolysis (urea$-NH_4$).

The three most widely-used nitrification inhibitors are nitrapyrin, DCD (dicycandiamide), and DMPP (3,4-dimethylpyrazole phosphate). Nitrapyrin can be injected into the soil with anhydrous ammonia, mixed with manure, or coated onto the fertilizer granule. DCD is often coated onto the solid fertilizer granule, and can remain active for several months. DMPP is generally preblended with the fertilizer and it can remain active for several months.

Urease inhibitors slow the conversion of urea to ammonium, and the most widely known urease inhibitors are NBPT [N-(n-Butyul) thiophosphoric tramide], PPD (phenylphosphorodiamidate), hydroquinone, and ammonium thiosulfate. NBPT can be added to urea or mixed with UAN. Urease inhibitors can reduce volatilization losses of NH_3, and they can remain active for several weeks. Additional information is available in Chien et al. (2009).

Table 10.2. The influence of manure type on the amount of organic N, inorganic N, and P2O5 contained in liquid and solid manures (Modified from Clay et al., 2009).

	Liquid manure			Solid manure		
	Organic N	Inorganic N	Phosphorus	Organic N	Inorganic N	Phosphorus
	lb N per 1000 gal	lb N per 1000 gal	lb P_2O_5 per 1000 gal	lb N per ton	lb N per ton	lb P_2O_5 per ton
Swine	7–21	8–39	18–42	4–11	3–6	5–9
Dairy	22–26	5–21	14–22	4–7	2–5	3–4
Beef	13–21	7–8	16–18	4–7	3–4	4–7
Poultry	17–50	5–37	15–52	13–39	4–12	51–53

Organic Farming

Many of the organic farming standards were defined by the International Federation of Organic Agriculture Movement (IFOAM), and in the United States, certification is available through the USDA (https://www.ams.usda.gov/about-ams/programs-offices/national-organic-program). Other information is also available in Coleman (2012). Organic foods are produced using ecological principles that are designed to mimic natural systems. To be classified as an organic product, the farm must first meet, and then continue to follow, strict guidelines. Following these guidelines reduces producer and consumer exposures to insecticides, transgenic crops, and antibiotics. Premium prices are often paid for organic foods.

In organic farming, commercially produced fertilizers generally are prohibited, whereas naturally occurring products are allowed. Therefore, the N sources are plant- or animal-derived and not made through industrial processes. Plant-derived N sources might result from using cover crops, compost, and legume rotations that have the capacity to fix N_2. Animal-derived N sources might include cattle, poultry, pig, and sheep manure (Table 10.2). However, the use of sewage sludge or biosolids is not allowed. An acceptable inorganic P source is rock phosphate, whereas acceptable organic sources are bone meal, fish meal, wood ash, manure, and compost. Acceptable K sources include potassium minerals such as Sul-Po-Mag and Polyhalite, wood ash, manure, and composts.

When applying manure, it is important to consider the variability of the product in the area and what portion of nutrient is, or is not readily, available. In manure, the inorganic sources of nutrients are generally readily available, whereas the organic nutrient sources must be converted to inorganic forms before they become plant available (Table 10.2). The amount of available N from organic N materials depends on the source. Rosen and Bierman (2005) reported that the amounts of N available in the first year of application from swine, beef with bedding, beef without bedding, poultry with litter, and poultry without litter manures, were about 50, 25, 35, 45, and 50% of the total N present, respectively, whereas P availability in the first year often ranges from 80 to 100% (Beegle and Durst, 2002; Rosen and Bierman, 2005).

Converting Nutrient Recommendations to Fertilizer Rates

To correctly apply site-specific fertilizer recommendation, the fertilizer grade must be converted to a rate. This conversion is different for different fertilizer sources. For liquid fertilizers, the fertilizer grade is based on weight, whereas the application rate is based on volume. The density of the liquid fertilizer is required to convert a recommendation in pounds of N $acre^{-1}$ to gallons $acre^{-1}$. If the density of the fertilizer is not known, this can be calculated by filling a five-gallon bucket and weighing the bucket of liquid fertilizer. Remember to know the empty bucket weight and subtract this amount off OR tare the bucket and then fill.

EXAMPLE 10.1.

Calculating N and P from manure (Modified from Rosen and Bierman, 2005).

1. Determine the amount of a nutrient required by the crop. For example, this season's crop needs 120 lb N acre^{-1} and 60 lb P_2O_5 acre^{-1}.
2. Determine the amount of nutrient in the organic amendment (compost or manure). This may involve sampling the amendment and sending the sample to a lab for analysis. If analysis was not conducted, then the nutrient concentrations can be estimated based on data in Table 10.2.
3. In this example, the selected manure per ton was found to contain 13 lb ammonium—N ton^{-1}, 20 lbs total N ton^{-1}, and 16 lb P_2O_5,

   a. Ammonium—nitrogen is an inorganic N form and is considered 100% available,

   b. Total N contains both the ammonium—N and organic N. In one year it is estimated that 45% of the organic N is made available and 90% of the P_2O_5 is available.
4. Determine the amount of nutrient available,

   a. Total amount of available N: $[(20-13)\times0.45]+13= 16.2$ lb N ton^{-1}

   i. In this calculation, 20 is total N, 13 is ammonium N, and 45% percent of the organic N is mineralized in a year.

   b. Total available P: $16\times0.9=14.4$ lb P_2O_5 ton^{-1}

   i. In this calculation, 16 is lb P_2O_5 ton^{-1} and assumes that 90% of the P is made available in one year.
5. Calculate the manure rate needed to supply the needed nutrients,

$$\text{N:} \quad \frac{120 \text{ lb N}}{\text{acre}} \times \frac{\text{ton manure}}{16.2 \text{ lb N}} = \frac{7.4 \text{ ton manure}}{\text{acre}}$$

$$\text{P:} \quad \frac{60 \text{ lb } P_2O_5}{\text{acre}} \times \frac{\text{ton manure}}{14.4 \text{ lb } P_2O_5} = \frac{4.17 \text{ ton manure}}{\text{acre}}$$

Note: If 7.4 tons of this manure are applied to meet the N requirement, the P amount is over applied. The overapplication of P is of environmental concern because P can be transported to stream and rivers with run-off. Therefore, the lower amount (4.2 tons) should be applied and another N source with no P should be found OR a different manure with a lower P value should be used.

Summary

In precision farming, even though the rates may be varied across the landscape, the calculations for converting a recommendation into application maps are identical. This chapter provides a background on the nutrients needed for plant growth, fertilizer types, fertilizers used for organic agriculture, and how to convert nutrient recommendations from a soil testing laboratory into application rates. Practical problems and their associated answers are provided.

Fertilizers are used to increase crop yields. However, if too much fertilizer is applied, the nutrient is not utilized by the crop, which wastes money and can produce adverse environmental problems. By understanding differences among the different fertilizer sources, the costs can be calculated and both cost and environmental problems can be minimized. In addition, this chapter provided background information on important soil nutrients, a list of commonly used fertilizers, a discussion of slow release fertilizers, and sample problems. Additional problems are in Clay et al. (2011).

ACKNOWLEDGMENTS

Support for this document was provided by South Dakota State University, Precision Farming Systems community in the American Society of Agronomy, International Society of Precision Agriculture, and the USDA-AFRI Higher Education program (2014-04572).

PROBLEM 10.6.
You are planning on applying 10 lb acre^{-1} of N to a field. If the fertilizer source is 28-0-0 how much 28% should be applied?

ANSWER:

$$\frac{10 \text{ lb N}}{\text{acre}} \times \frac{1 \text{ lb UAN}}{0.28 \text{ lb N}} \times \frac{1 \text{ gal UAN}}{10.8 \text{ lb UAN}} = \frac{3.30 \text{ gal}}{\text{acre}}$$

PROBLEM 10.7.
How much P is contained in 1500 lb of DAP?

ANSWER:
DAP has a fertilizer grade of 18-48-0.

$$1500 \text{ lb fertilizer} \times \frac{0.48 \text{ lb P}_2\text{O}_5}{1 \text{ lb fertilizer}} = 720 \text{ lb P}_2\text{O}_5 \qquad 720 \text{ lb P}_2\text{O}_5 \times \frac{61.96 \text{ lb P}}{141.6 \text{ lb P}_2\text{O}_5} = 315 \text{ lb P}$$

PROBLEM 10.8.

How much N is contained in 1500 kg of diammonium phosphate (DAP)?
Note that in Table 10.1 the N can range from 18-21 and the P$_2$O$_5$ can range from 46 to 53, therefore always find out the true fertilizer grade prior to working the problem.

ANSWER:
In this case, the DAP has a fertilizer grade of 18-48-0.

$$1500 \text{ kg fertilizer} \times \frac{0.18 \text{ kg N}}{1 \text{ kg fertilizer}} = 270 \text{ kg N}$$

PROBLEM 10.9.
DAP (18-48-0) is selling for $550 ton^{-1}, what is the cost per pound of P$_2$O$_5$? In this problem assume that the price of N is $0.326 lb^{-1}.

ANSWER:
DAP has a fertilizer grade of 18-48-0
Calculate the value of the N in fertilizer.

$$2000 \text{ lb DAP} \times \frac{0.18 \text{ lb N}}{1 \text{ lb DAP}} \times \frac{\$0.326}{1 \text{ lb N}} = \$117$$

Subtract the value of N from cost of DAP.
$550-$117= $433
Calculate the value of P$_2$O$_5$.

$$\frac{\$433}{2000 \text{ lb}} \times \frac{1 \text{ lb}}{0.46 \text{ P}_2\text{O}_5} = \frac{\$0.47}{\text{lb P}_2\text{O}_5}$$

PROBLEM 10.10.

How much P_2O_5 should be applied to increase a soil test P value from 3 to 11? The fertilizer grade DAP is 18-46-0.

ANSWER:

As a general rule, it takes from 10 to 30 lbs of P_2O_5 to increase the Bray soil test 1 ppm (1 mg kg^{-1}). Using a median of 20 lbs P_2O_5 per 1 ppm, to change a soil test from 3 to 11 (Olsen P) will require the addition of (11-3)×20 = 160 lbs P_2O_5 or 348 lb of DAP acre^{-1}. (= 160/0.46)

$$160 \text{ lb actual } P_2O_5 \times \frac{1 \text{ lb DAP product}}{0.46 \text{ lb } P_2O_5} = 348 \text{ lb DAP}$$

PROBLEM 10.11.

A farmer needs 150 lbs of N acre^{-1} and 100 lbs P_2O_5, how much DAP (18-46-0) and urea (46-0-0) should be applied?

ANSWER:

1. Calculate DAP added: $\dfrac{100 \text{ lb } P_2O_5}{\text{acre}} \times \dfrac{100 \text{ lb fert}}{46 \text{ lb } P_2O_5} = \dfrac{217 \text{ lb Fert}}{\text{acre}}$

2. Calculate N in DAP: $\dfrac{217 \text{ lb Fert}}{\text{acre}} \times \dfrac{0.18 \text{ lb N}}{1 \text{ lb fert}} = \dfrac{39 \text{ lb N}}{\text{acre}}$

3. Subtract N in DAP from requirement: $\dfrac{150 \text{ lb N}}{\text{acre}} - \dfrac{39 \text{ lb N}}{\text{acre}} = \dfrac{111 \text{ lb N}}{\text{acre}}$

4. Calculate urea N needed: $\dfrac{111 \text{ lb N}}{\text{acre}} \times \dfrac{100 \text{ lb urea}}{46 \text{ lb N}} = \dfrac{241 \text{ lb urea}}{\text{acre}}$

Apply 217 lbs DAP acre^{-1} and 241 lb urea acre^{-1}.

PROBLEM 10.12A.

Calculate the P removal rate for a soybean and corn crop (See table 16.1). In these calculations assume that 1 bu of corn removes 0.38 lb of P_2O_5 and 1 bu of soybeans removes 0.84 lb acre^{-1} of P_2O_5. In the calculations only consider the nutrients contained in the grain. In this example, the corn yield is 175 bu of corn and the soybean yield is 55 bu.

ANSWER:

Calculate the P_2O_5 removed in the corn and soybean grain.

$$\frac{175 \text{ bu corn}}{\text{acre}} \times \frac{0.38 \text{ lb } P_2O_5}{\text{bu corn}} + \frac{55 \text{ bu beans}}{\text{acre}} \times \frac{0.84 \text{ lb } P_2O_5}{\text{bu beans}} = 112.7 \text{ lb } P_2O_5$$

Note: a table and examples of nutrient removal is available in table 16.1.

PROBLEM 10.12B.

How much DAP would need to be applied to replace the amount removed?

ANSWER:

Calculate the amount of DAP (18-46-0) needed to replace this P.

$$\frac{112.7 \text{ lb } P_2O_5}{\text{acre}} \times \frac{100 \text{ lb DAP}}{46 \text{ lb } P_2O_5} = 245 \text{ lb DAP}$$

PROBLEM 10.12c.

How much N is in this DAP fertilizer?

ANSWER:

Calculate the amount of N from the DAP that is contained in the DAP fertilizer.

$$\frac{245 \text{ lb DAP}}{\text{acre}} \times \frac{18 \text{ lb N}}{100 \text{ lb DAP}} = \frac{44 \text{ lb N}}{\text{acre}}$$

PROBLEM 10.12D.

If you want to apply 150 lb N acre^{-1} how much additional N should you apply?

ANSWER:

Subtract the N in the DAP from the overall N recommendation

N recommendation = 150 lb N acre^{-1} – 44 lb N from DAP = 106 lb N.

Additional Problems

10.13. How many lbs of N are contained in 1000 lb of MAP?

10.14. How many lbs of N are contained in 100 gal of 28–0–0? (density = 10.8 lb gal^{-1})

10.15. If the goal is to apply 100 lb of N acre, how much urea should be purchased?

10.16. If urea sells for $300 ton^{-1}, what is the price per pound of N?

10.17. If the recommendation is to apply 200 lb of N and 60 lb of P_2O_5, how much anhydrous ammonia (82–0–0) and MAP (10–48–0) would you apply?

10.18. If the N recommendation for the summit and footslope areas are 95 and 150 lb N acre^{-1}, how many gallons of 28–0–0 (density = 10.8 lbs gal^{-1}) should you apply to these areas?

10.19. You do not know the density of a liquid fertilizer, how might you determine it?

REFERENCES

Beegle, D.B., and P.T. Durst. 2002. Managing phosphorus for crop production. Agronomy Facts 13, Penn State Extension, State College, PE.

Chien, S.H., L.I. Prochbnow, and H. Cantarella. 2009. Chapter 8. Recent developments of fertilize production and use to improve nutrient efficiency and minimize environmental impacts. In: D. Sparks, Advances in Agronomy. Vol. 102. Academic Press, Burlington, VT. pp. 267-322.

Clay, D.E., C. Robinson, and T.M. DeSutter. 2017. Soil Sampling and Understanding Soil Test Results for Precision Farming. In: D.E. Clay, S.A. Clay, and S. Bruggeman. Practical Mathematics for Precision Farming. ASA, CSSA, SSSA, Madison, WI.

Clay, D.E., S.A. Clay, and K. Reitsma. 2009. South Dakota corn best management practices. South Dakota State University, Brookings, SD.

Clay, D.E., S.A. Clay, C.G. Carlson, and S. Murrell. 2011. Mathematics and science for agronomists and soil scientists. International Plant Nutrition Institute, Norcross, GA.

Clay, D.E., T.M. DeSutter, S.A. Clay, and C. Reese. 2017. From plows, horses, and harnesses to precision technologies in the north American Great Plains. Oxford Press. Oxford, UK. doi:0.1093/acrefore/9780199389414.013.196

COHA (Council of Hemispheric Affairs). 2011. The great Peruvian guano bonaza. Rise, fall, and legacy. Council of Hemispheric Affairs. http://www.coha.org/the-great-peruvian-guano-bonanza-rise-fall-and-legacy/ (accessed June 2017).

Coleman, P. 2012. Guide for organic crop producers. National Center for Appropriate Technology. USDA Organic. Fayetteville, AK. https://www.ams.usda.gov/sites/default/files/media/Guide%20for%20Organic%20Crop%20Producers_0.pdf (verified 29 June 2017).

Havlin, J.L., S.L. Tisdale, W.L. Nelson, and J.D. Beaton. 2013. Soil fertility and fertilizers: An introduction to nutrient management 8th ed. Pearson, New York.

Hosier, S., and L. Bradley. 1999. Guide to symptoms of plant nutrient deficiencies. Publication AZ1106. Arizona Cooperative Extension. https://extension.arizona.edu/sites/extension.arizona.edu/files/pubs/az1106.pdf (verified 29 June 2017).

Kaiser, D.E., J.A. Lamb, P.R. Bloom, and J.A. Hernandez. 2014. Comparison of field management strategies for preventing iron deficiency chlorosis in soybean. Agron. J. 106:1963–1974.

Kelling, K.A. 1999. Soil and applied boron. University of Wisconsin, Extension A2522, Madison, WI. http://www.soils.wisc.edu/extension/pubs/A2522.pdf (verified 29 June 2017).

McCauley, A., C. Jones, and J. Jacobsen. 2011. Plant nutrient functions and deficiencies and toxicity symptoms. Montana State University Extension. 4449-9, Bozeman, MT. http://landresources.montana.edu/nm/documents/NM9.pdf (verified 29 June 2017).

Rosen, C.J., and P.M. Bierman. 2005. Nutrient management for fruit and vegetable crop production. University of Minnesota Extension Service. St Paul. MN. http://www.extension.umn.edu/garden/fruit-vegetable/using-manure-and-compost/docs/manure-and-compost.pdf (verified 29 June 2017).

Roy, R.N., A. Finck, G.J. Blair, and H.L.S. Tandon. 2006. Plant nutrition for food security: A Guide for Integrated Nutrient Management. Bulletin 16. FAO Fertilizer and Plant Nutrition. Rome, Italy.

Sawyer, J. 2004. Nutrient deficiency and application injuries in field crops. IPM 42. Iowa State University Extension. Ames, IA. http://extension.missouri.edu/scott/documents/Ag/Agronomy/Nutrient-Deficiencies-Row-Crops-ISU.pdf (verified 29 June 2017).

Deriving and Using an Equation to Calculate Economic Optimum Fertilizer and Seeding Rates

11

Scott Fausti, Bruce J. Erickson, David E. Clay, and C. Gregg Carlson*

Chapter Purpose

Producing a profit requires that returns are greater than the costs. However, optimization for individual decisions requires an analysis where the return for a specific investment is compared with the input costs for that investment. Seeding and fertilizer rates are treated differently than many other costs, because these rates can be varied. Rates for many pesticides cannot be varied, because they must follow the labeled rates. This chapter derives an equation that can be used to calculate economic optimum seeding and fertilizer rates. This solution is based on the use of a second order polynomial equation. If a different model is used, this solution will not be appropriate. Even though the derivation of this equation requires calculus, the use of the equation does not require calculus. Examples are provided.

Key Terms

Economic optimum rate, site-specific seeding rates, 2nd order polynomial equation, derivative.

Mathematical Skills

Deriving the economic optimum rate equation for a 2nd order polynomial equation, determining economic optimum rates for an input with diminishing returns such as nitrogen fertilizer rate or seeding rate, and developing a site-specific seeding algorithm.

Calculating Site-Specific Economic Optimum Rates

The economic optimum rate (EOR) is the point where an incremental change in input costs equals an incremental change in the value of product produced (Bullock and Bullock, 1994; Cerrato and Blackmer, 1990; Kyveryga et al., 2007; Fig. 11.1). Traditionally, this condition defines profit maximization where marginal revenue equals marginal cost of a production activity. Mathematically, EOR is the ratio where the change in revenue (from production) equals the change in input cost. Or where marginal revenue equals marginal cost, where Δ denotes incremental change:

$$1 = \frac{\Delta \text{ value of product}}{\Delta \text{ input cost}} \qquad [11.1]$$

In agriculture, the input or total costs (TC) are defined as the unit input cost multiplied by total input units per unit area [TC = price per unit purchased × number of units per acre], and the value of the product, or total revenue (TR)

S. Fausti, California State Monterey Bay, Seaside, CA 93933; B.J. Erickson, Purdue University, Agronomy Department, West Lafayette, IN 49707; D. Clay and G. Carlson, South Dakota State University, Brookings, SD 57007. *Corresponding author (gregg.carlson@sdstate.edu)
doi:10.2134/practicalmath2016.0027

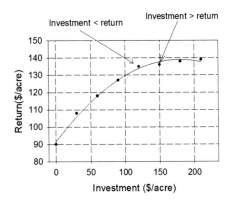

Fig. 11.1. Relationship between investment (cost) and return. The economic optimum rate is the point where the marginal investment and equals the marginal return. On the chart, two points are identified, investment < return and investment > return. This chapter discusses how to identify the point where investment = returns. The location of the economic optimum rate depends on the cost of the inputs and value of the products.

per unit area. Per unit area is defined in dollars per unit of product or the price per unit of output multiplied by total units produced per unit area [TR = price per unit sold × number of units per acre]. In discussion below, it is assumed that the crop is corn, the inputs are nitrogen (N) and corn seed (S), and the unit area is one acre. In this solution, the input costs for multiple inputs are not considered simultaneously, and therefore N application rate costs are considered separately from the seeding rate calculations on a per acre basis. A solution that considers seeding and fertilizer rates simultaneously is available in Clay et al. (2011). To simplify the discussion the unit price of corn is denoted P_C, unit cost of N is P_N, and the unit cost of corn seed is P_S.

For example, if the yield is 100 bushels per acre and the return on each bushel is $4.00 ($P_C$), then the value per unit area is $400 acre^{-1} (TR). Mathematically, total return (TR) is defined as:

$$TR = P_C \times \text{bushels per acre.}$$

The total cost (TC) of the input is calculated by multiplying the unit cost of product purchased times the number of products purchased per unit area. Mathematically TC is defined as:

$$TC = \text{Unit input price} \times \text{total input usage per unit area}$$

Or for N fertilizer

$$TC = P_N \times \text{pounds applied per acre.}$$

For example, if fertilizer cost $341 per ton and a pound of fertilizer contains 0.46 pounds of N, then the cost of N per pound is $0.37 or $\left(\dfrac{\$341}{2000\ \text{lb}} \times \dfrac{1\ \text{lb}}{0.46\ \text{lb N}}\right)$. Assume the application rate is 150 lb N per acre, then the cost per unit area is $55.50 per acre $\left(\dfrac{\$0.37}{\text{lb of N}} \times \dfrac{150\ \text{lb N}}{\text{acre}} = \dfrac{\$55.50}{\text{acre}}\right)$. Based on these definitions and the mathematical relationships discussed above, the cost and revenue components of the EOR per acre are defined by the equation,

$$1 = \frac{\Delta(\text{value of product})}{\Delta(\text{cost of inputs})} = \frac{\Delta(\text{TR})}{\Delta(\text{TC})} = \frac{P_C}{P_N} \qquad [11.2]$$

Eq. [11.2] states that the ratio of value of the products (corn grain) to the cost of the inputs (N fertilizer) are the marginal revenue and marginal cost, respectively. Problems 11.1a and 11.1b provide simple examples of how to calculate TR and TC per acre.

The TR and TC values are used to determine the economic optimum rate (EOR) and they are used to determine the return per acre, which is equal to the TR minus TC. Mathematically, the return or profit per acre is defined as:

$$\text{Return} = P_C \times Y(N) - P_N \times N \qquad [11.3]$$

Where, Y is corn yield per acre and is defined as a function of nitrogen inputs Y(N). Notice that yield (Y) is defined as a function of N, the sole input cost for corn production. A reasonable assumption is that the more N applied to an acre of corn the higher the yield, however as the amount of N applied increases the return decreases

---

PROBLEM 11.1A.
If P_N is $0.50 per pound and you apply 100 lb N per acre, what is the TC per acre?

ANSWER:
TC = 100 lb × $0.5 = $50 per acre.

PROBLEM 11.1B.
If P_C = $4 per bushel and you harvest 150 bushels per acre, what is the TR per acre?

ANSWER:
150 bu per acre × $4 per bushel = $600 per acre

(Fig. 11.1). This is known as the "Law of Diminishing Returns." The basic economic principle of the Law of Diminishing Returns can be mathematically incorporated into an equation that predicts outputs based on inputs (Fig. 11.1). The solution for determining the EOR that is discussed below, requires the use of the second-order polynomial equation, which has the form,

$$\text{Yield} = a + b \times N^1 + c \times N^2 \qquad [11.4]$$

Where, "a" is the y-intercept and represents the yield when no N is applied, and coefficients "b" and "c" represent the nonlinear relationship between nitrogen input and corn yield. The slope of the second-order yield equation describes the relationship between the nitrogen inputs and yield. The slope of the second order polynomial equation is the first derivative of the Yield with respect to the inputs (N). In the first derivative, the symbol dY/dN, represents the incremental change in yield resulting from an incremental change in N rate. For example, if the yield when 0 lb of N is applied is 100 bushels per acre, and the yield when 0.000001 lb N per acre are applied is 100.00001, the change in yield (0.00001 bushels per acre) per change in N (0.000001), represents the incremental change in yield per incremental change in N. The first derivative of Eq. [11.4] is defined as:

$$\frac{dY}{dN} = b + 2cN \qquad [11.5]$$

The derivate dY/dN is called the marginal product of the input, N. The economic interpretation of this equation is that it provides the transformation rate of N per acre into the yield per acre. Next, by combining equations [11.3] and [11.4] by substituting Eq. [11.4] for Y(N) in Eq. [11.3], and taking the derivative of return with respect to N will result in Eq. [11.6]:

$$\frac{dReturn}{dN} = P_C (b + 2cN) - P_N \qquad [11.6]$$

Note, that P_C is the marginal revenue (selling price of the corn) and P_N is the cost of the N fertilizer ($ per lb). Equation [11.6], alters the EOR condition depicted in Eq. [11.1] In Eq. [11.1], ΔTR is equal to P_C. Now $\Delta TR = P_C(b + 2cN)$ where P_C denotes marginal revenue and $b + 2cN$ is the marginal product of N. Incorporating the yield—N relationship into the return function Eq. [11.3] allows the defining of the economic optimum rate for N in the following equation,

$$(b + 2cN) = \frac{P_N}{P_C} \qquad [11.7]$$

where, b and c are based on the use of the yield—N production model (Eq. [11.4]) discussed above. Notice that the EOR condition now requires that the transformation rate of N per acre into the yield per acre be equal to the ratio of corn price per bushel to the cost per pound of N. The solution value for N, which is the optimal N rate per acre, can be found by solving for N in Eq. [11.7]. The resulting solution is,

$$\left(\frac{P_N}{P_C}\right)\left(\frac{1}{2c}\right) - \left(\frac{b}{2c}\right) = \text{Optimal Application Rate for N} \qquad [11.8]$$

In this solution, the b and c values are the coefficients in Eq. [11.4], P_N is the cost of each unit of N fertilizer and P_c is the selling price of each bushel of corn.

PROBLEM 11.2.

Assume the following estimated quadratic equation values for a corn return function includes the following values: $b = 0.78$ and $c = (-0.0024)$, $P_C = \$4$ and $P_N = \$0.55$. First step is to solve Eq. [11.7] for N. Then substitute values given to solve for the EOR application rate for N per acre.

ANSWER:

$$N = \left(\frac{P_N}{P_C}\right)\left(\frac{1}{2c}\right) - \left(\frac{b}{2c}\right) = 133.8 \text{ lb per acre}$$

PROBLEM 11.3.

What happens if the cost of N increases to $0.60 per pound?

ANSWER:

Note that if the corn price remains the same and the N price increses, the return has decreased (133.8 vs. 131.2).

Problem 11.4.

Using the b and c values provided in Problem [11.3], calculate the return if N is $0.60 per pound, and the price of corn increases from 4 to $5 per bushel?

Answer:
137.5 pounds per acre
In this example, more N is economical because the price for corn increased.

Problem 11.5.

Using values in Problem 11.3, determine the application rate of N if corn is selling at $6.00 per bushel and N costs $0.70 per pound.

Answer:
138.2 pounds per acre

Problem 11.6.

Determine the economic optimum rate using the quadratic model, yield = $106+0.96x-0.00343x^2$. First take the derivative of this equation and then use the information to determine the EOR of N.
The derivative is $0.96-2(0.00343)x = 0.96 -0.00686x$,

$b= 0.96$ and
$2c = -0.00686$

In this problem, N is the nitrogen rate in pounds per acre, 106 is a, 0.96 is b, and -0.00343 is c. The fertilizer is selling for $0.50 per pound ($P_N$) and corn is worth $3.50 per bushel ($P_C$).

Answer:

$$N=\left(\frac{P_N}{P_C}\right)\left(\frac{1}{2c}\right)-\left(\frac{b}{2c}\right)=119.1 \text{ lb per acre}$$

Problem 11.7.

What is the quadratic equation for the hypothetical relationship between N rate and corn yield?

N rate	yield
pounds per acre	bushels per acre
0	100
60	130
120	160
180	170
240	171

Answer:
This equation can be derived in Microsoft Excel 2013. However, the equation is determined differently in different versions. In Excel 2013, highlight the data, select insert, select recommended charts, click on scatter and select OK. At this point the chart should be created, left click on the first point, select add "trendline", select polynomial, select display equation on chart and display R^2 on chart. The resulting equation is, Yield = $98.7 + 0.67$ (N rate) $– 0.00155$ (N rate)^2 The R^2 value is 0.99 which means that 99% of the yield variability was defined by this equation. In this equation the b can c coefficients are 0.67 and -0.00155.
The equation can also be derived using regression under Data and Data Analysis. To use the Data Analysis program it must be activated. In Microsoft Excel 2013 select Developer, Add-Ins, and click on Analysis ToolPak, and click on OK. If you use the data set above, the program will only provide a linear equation. To produce a 2nd order polynomial equation, you must add an additional column containing the N rate squared values. When using the data analysis tool, both columns (N rate and N rate squared must be selected). The data must be provided in columns. If provided in rows, the program may not work.

Problem 11.8.

What is the economic optimum nitrogen rate for the quadratic model where yield = $108+0.948x-0.00356x^2$, where N is the nitrogen rate in pounds per acre? In this calculation N is selling for $0.50 per pound and corn is worth $3.50 per bushel.

Answer:

$$N=\left(\frac{P_N}{P_C}\right)\left(\frac{1}{2c}\right)-\left(\frac{b}{2c}\right)=113.08 \text{ lb per acre}$$

PROBLEM 11.8B.

Next derive yield, N cost, and corn revenue per acre at the economically optimum N rate in this example.

ANSWER:

Corn Yield = $106 + 0.96 \times (113.08) - 0.00343 \times (113.08)^2$ = 169.68 bushels per acre.

N Cost = 0.5×113.08 = $56.54

Total Revenue (profit) = $3.50 \times 169.68 - $56.54 = $537.33

PROBLEM 11.9.

In the summit and footslope of a field, what is the economic optimum N rates for the two landscape positions? Corn is selling for $4 per bushel and N is selling for $0.55 per pound. Calculate profit for footslope and Summit for EOR N rate.

N rate	Footslope yield	Summit yield
pounds per acre	bushels per acre	bushels per acre
0	100	60
60	130	90
120	160	120
180	170	130
240	171	131

ANSWER:

First, use excel to estimate two regression equations for both landscape positions as shown in Problem 11.7. The second order polynomial equations are:

Footslope: yield = $98.7 + 0.67$ (N rate) $- 0.00155$ (N rate)^2

Summit: yield = $58.7 + 0.67$ (N rate) $- 0.00155$ (N rate)^2

Note that both equations have identical derivatives,

$\dfrac{dy}{dx} = 0.67 - (2 \times 0.00155)x$, thus they have identical b and c terms. The N rate per acre is given by the EOR equation:

Foot Slope yield = 168 bu.
Foot Slope Profit = TR − TC
$4 \times 168 - $0.55 \times 171.78 = $577

Summit yield = 128
Summit Slope Profit = TR - TC
$4 \times 128 - $0.55 \times 171.78 = $417

Economic Optimum Seeding Rate

The next topic is estimating the EOR optimal seeding rate. The methodology for estimating the optimal seeding rate (S) is identical to the N example above. Therefore Eq. [11.3] and [11.4] can be modified as follows:

$$\text{Return} = P_C \times Y(S) - P_S \times S \qquad [11.9]$$

$$\text{Yield} = a + b \times S^1 + c \times S^2 \qquad [11.10]$$

The EOR equation (Eq. [11.8]) is modified as follows (Eq. [11.11]) so that it can be applied to deriving the optimal seeding rate that will maximize returns. Note the only change in Eq. [11.11] is that P_S (cost of seed per 1000 seeds) has been substituted for P_N.

$$S = \left(\frac{P_S}{P_C}\right)\left(\frac{1}{2c}\right) - \left(\frac{b}{2c}\right) = \text{Optimal Seeding Rate} \qquad [11.11]$$

Developing Site-specific Seeding Rate Algorithms

Since the advent of agriculture, farmers have always been interested in reducing their input costs. To help identify appropriate rates, agricultural companies have conducted numerous seeding rate experiments across the world. For a specific cultivar, data from multiple experiments can be aggregated to create site-specific seeding rate algorithms.

The following discussion assumes that water stress is the primary factor limiting yield. This assumption is based on experimentation that has shown that water plays a major role in impacting yields especially across landscapes with undulating topography (Mishra et al., 2008; Carlson et al., 2011). With corn, optimum yields can be achieved by increasing the seeding rates in high yielding areas and reducing the rate in low yielding areas (Clay et al., 2009). Site-specific algorithms

PROBLEM 11.10.

You conduct an experiment and measure the resulting corn yields. In this experiment, populations of 18000, 24000, 30000, and 36000 plants per acre yielded 128, 150,162, and 167 bushels per acre, respectively. The resulting relationship between the number of seeds (×1000) and yield were, yield =12.95 + 8.525(seed) -0.1181(seed)^2 (derived from Excel polynomial curve fit equation). The cost of the seed is $160 per 80,000 kernel unit (1 bag of corn seed contains 80,000 seeds), and the corn selling price (P_c) is $3.00 per bushel. What is the optimum seeding rate per acre, using 1000 seeds as a unit per acre? When solving these problems the units used to create the 2nd order polynomial equation must be the same as those used in the calculations below.

ANSWER:

First, find the cost of 1000 seeds (80/$160=$2 per 1000 seeds). So it costs $2 per 1000 seed unit or P_S=$2.00. The following calculations are based on 1000 seed per seed unit:

$$S=\left(\frac{P_S}{P_C}\right)\left(\frac{1}{2c}\right)-\left(\frac{b}{2c}\right)=\frac{2}{3}\times\frac{1}{2\times(-0.1181)}-\frac{8.525}{2(-0.1181)}=33.27 \text{ rounded to } 33.30 \text{ seeds units per acre}$$

Remembering that a unit of seed was 1000 seeds, the optimum seeding rate is 33,300 seeds per acre. Convert this to seeds ha^{-1}. (1 ha = 2.47 acres)

$$\frac{33,300 \text{ seeds}}{\text{acre}}\times\frac{2.47 \text{ acre}}{\text{ha}}=\frac{82,200 \text{ seeds}}{\text{ha}}$$

PROBLEM 11.10B

At a seeding rate of 33,300 seeds, what is the expected yield? What is the profit?

ANSWER:

Corn yield: 12.95+8.525(33.3) -0.1181(33.3)^2 = 165.9 bushels per acre.
Profit for optimal seeding rate: = TR – TC= $3×165.9 - $2×33.3 = $431.10.

PROBLEM 11.10C.

The planter was not calibrated correctly and seeding occurred at 27,000 seeds. What is the expected yield and profit?

ANSWER:

Corn yield: 12.95 + 8.525(27)-0.1181(27)^2 = 157 bushels per acre
Profit = $3×157 – ($2×27) = $417; this means that there would be $14 per acre lost

PROBLEM 11.10D

Assume the seeding rate was 40,000 per acre. What is the expected yield and profit?

ANSWER:

Based on the calculations above the yield would be 165 and profit would be $415 or a $16 per acre loss. Although yields at the optimum and higher seeding rates are similar the cost of seed is higher, resulting in less profit than if seeded at a lower rate with lower yield.

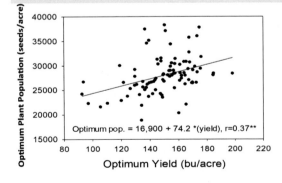

Fig. 11.2. The calculated site-specific corn seeding algorithm, based on Northern Great Plains experiments conducted between 1998 and 2002 (Carlson et al., 2011). A second order polynomial equation is not used because, each individual point was identified using the techniques described above.

should account for this variability. However, because plants with different genetic capacity have different response functions, the data must be sorted appropriately (Cox, 1997). A program that converts data from a number of studies into a site specific algorithm is available in Carlson et al. (2011). The data and the algorithm (shown in Fig. 11.2) is based on the economic optimum rate (EOR) discussion above. Based on this equation, the optimum population level for a yield of 140 and 200 bu acre^{-1} yield are 27,300 (= 16,900 + 74.2 × 140) and 31,700 (= 16,900 + 74.2 × 200) plants per acre, respectively.

PROBLEM 11.11.

Based on the calculations above, you determine that the desired live plant population is 35,200 plants acre[-1] and what is the expected yield? If the germination rate is 95% how many seeds should be planted?

ANSWER:

$$\text{Seeding rate (seeds per acre)} = \frac{\text{Desired population at harvest (plants per acre)}}{\dfrac{\%\text{ emergence of planted seeds}}{100}}$$

Expected yield: $35{,}200 = 16900 + 74.2\,(\text{yield}) = 35200 - 16900 = 74.2(\text{yield}) = 18300/74.2 = \text{yield}$ yield = 246 bushels per acre

For this calculation % emergence of planted seeds with the germination

$$\text{Seeding rate (seeds per acre)} = \frac{35{,}200\text{ plants per acre}}{\dfrac{95\%\text{ germination rate}}{100}} = 37{,}100 \text{ seed per acre}$$

PROBLEM 11.12.

What is the relationship between the economically optimum plant populations for the following data? In this calculation, corn is selling for $3.00 per bushel and a bag of corn (80,000 seeds) is selling for $250 per bag. Footslope areas are located in lower landscape positions and summit areas are located in upland areas.

Plant (×1000) per acre	Footslope bushels per acre	Summit bushels per acre
20	125	95
25	140	112
30	160	125
35	175	123
40	170	120

Estimated footslope and summit yield equations are:

Footslope: yield= -30.3+10.2(population)-0.129(population)^2; r^2=0.96, b= 10.2, and c= -0.129
Summit: yield = -55.2 + 10.65(population) – 0.157(population)^2; r^2=0.98, b=10.65, and c = -0.157

Seed costs are $3.125 per 1000 seeds and plant populations are in 1,000 seeds per acre.

ANSWER:

Footslope:
The optimum seeding rate in the footslope is 1000 time 35.50 or 35,500 seeds per acre.
Summit:
The optimum seeding rate in the summit is 1000 time 30.60 or 30,600 seeds per acre.
Next: what is the cost difference between the two seeding rates?
$250/80 = $3.125/1000 seeds 35.5 × $3.125 = $111 for footslope and 30.6 × $3.125 = $95.6 for the summit.
Then, what the Yield and Profit for the footslope and summit?

Foot Slope yield = 169.2 bu.	Summit yield= 123.7 bu
Foot Slope Profit = TR – TC	Summit Slope Profit = TR – TC
$3×169.22 - $3.125×35.49 = $396.75	$3×123.68 - $3.125×30.6 = $275.42

Summary

In summary, this chapter derives an equation for determining the economic optimum rate for a second order polynomial equation and provides examples for determining the EOR for N and seeding rates. Based on the analysis of individual experiments, an analysis across experiments was conducted to derive a precision farming seeding rate algorithm.

ACKNOWLEDGMENTS

Support for this document was provided by South Dakota State University, South Dakota Soybean Research and Promotion Council, and the Precision Farming Systems communities in the American Society of Agronomy and the International Society of Precision Agriculture. Additional support was provided by the USDA-AFRI Higher Education Grant (2014-04572).

Additional Problems

11.13. Will the equation provided in this chapter do equally as well for determining economic optimum rates using with a linear or exponential model?

11.14. What does this equation, $1 = \dfrac{\Delta \text{ value of product}}{\Delta \text{ input cost}}$ mean?

11.15. Why does the N or seeding rate decrease with increasing cost?

11.16. How would landscape position influence the N recommendation model?

11.17. If seed cost $300 a bag (80,000 seeds) and corn is selling for $3.50 per bushel, what is the economic optimum seeding rate?

Plant	Yield
(×1000) per acre	bu per acre
20	130
25	145
30	162
35	178
40	175

Calculate the profit using the optimum seeding rate and then 20% higher and lower than that rate. Would overseeding or underseeding result in less loss?

11.18. Given the information on N rate application for footslope and summit yields, calculate the optimal N rate, yield, and profit associated with optimal N rate for footslope and summit. Then calculate yield and returns if the application rate was 15% lower or higher than the economic optimum. Would over or under application result in less loss compared with the optimum?

N	Nsq	Footslope yield	Summit yield
0	0	90	75
60	3600	120	95
120	14400	155	120
180	32400	170	130
240	57600	171	131

REFERENCES

Bullock, D.G., and D.S. Bullock. 1994. Quadratic and quadratic-plus-plateau models for predicting optimum nitrogen rate of corn: A comparison. Agron. J. 86:191–195. doi:10.2134/agronj1994.00021962008600010033x

Carlson, G., D.E. Clay, and J. Schefers. 2011. A case study for improving nutrient management efficiency by optimizing the plant population. In: D.E. Clay and J. Shanahan, editors, GIS in agriculture: Nutrient management for improved energy efficiency. CRC Press, New York. p. 157–172.

Cerrato, M.E., and A.M. Blackmer. 1990. Comparison of models for describing corn yield response to nitrogen fertilizer. Agron. J. 82:138–143. doi:10.2134/agronj1990.00021962008200010030x

Clay, D.E., S.A. Clay, C.G. Carlson, and S. Murrell. 2011. Mathematics and calculations for agronomists and soil scientists. International Plant Nutrition Institute, Peachtree Corner, Georgia.

Clay, S.A., D.E. Clay, D.P. Horvath, J. Pullis, C.G. Carlson, S. Hansen, and G. Reicks. 2009. Corn (Zea mays) responses to competition: Growth alteration vs limiting factor. Agron. J. 101:1522–1529. doi:10.2134/agronj2008.0213x

Cox, W.J. 1997. Corn silage and grain yield response to plant density. J. Prod. Agric. 10:405–410. doi:10.2134/jpa1997.0405

Kyveryga, P.M., A.M. Blackmer, and T.F. Morris. 2007. Disaggregating model bias and variability when calculating economic optimum nitrogen fertilization for corm. Agron. J. 99:1048–1056. doi:10.2134/agronj2006.0339

Mishra, U., D.E. Clay, T. Trooien, K. Dalsted, D.D. Malo, and C.G. Carlson. 2008. Assessing the value of using a remote sensing based evaportranspiration map in site-specific management. J. Plant Nutr. 31:1188–1202. doi:10.1080/01904160802134491

Cost of Crop Production

12

Scott Fausti* and Tong Wang

Chapter Purpose

Cost of production is the dollar value of all your inputs for growing a specific crop. For example, to produce an acre of corn, these inputs would include all costs including seed, fertilizer, chemicals, insurance, labor, machinery and land. Each of these inputs has a dollar value, and the cost of corn production is determined by summing these costs. To accurately estimate the cost of production, detailed recordkeeping is necessary to ensure that all expenses are recorded. This chapter provides an analysis on how to determine the cost structure and its relationship to profitability. The cost reduction and reduced yield variability benefits of implementing variable rate seeding and fertilizer technologies are demonstrated. Details are provided on estimating ownership and operating costs while developing the enterprise budget.

Key Terms

Economic cost structure of crop production, Heterogeneous soil and field slope characteristics, site-specific seeding and fertilizer rates, second order polynomial equation, Profit Maximization, Variable (operating) cost, Fixed (ownership) cost, spreadsheet applications to determine cost of production.

Mathematical Skills

Calculating the cost of production, conducting cost and corn yield analysis for multi-rate seeding and fertilizer application rates, cost classification and estimation using enterprise budget.

The Cost Structure of Crop Production

Profitability of an agricultural enterprise is dependent on total revenue earned from the sale of agricultural crops and livestock, and the cost structure of production. The agricultural producer competes in a market environment that economists describe as a perfectly competitive market. This market structure has the unique feature where the grain producer has no influence over price (e.g., price per bushel of corn). Thus, total revenue (TR) is defined as P×Q, where P is the price of the product and Q is the level of output produced (bushels per acre). For example, if the price of corn is $3.00 and yield per acre is 150 bushels, TR = $450.00 per acre.

The cost structure of a farming operation refers to both fixed costs (FC) and variable cost (VC) of production. Economists often define the total cost (TC) as TC = FC + VC. In this equation, the fixed cost refers to the cost that will not change depending on the level of production, for example machine depreciation cost, whereas variable cost refers to the cost that changes with production level, which include the seed, fertilizer, fuel, labor, and pesticide costs. While TR is a linear function of the level of production, the TC function is nonlinear due to the basic economic principle of diminishing returns to production. This principle states that as you attempt to extend the productive activity, it becomes increasingly more difficult to do so. The implication for crop production is

S. Fausti, California State Monterey Bay, Seaside, CA 93933; T. Wang, South Dakota State University, Brookings, SD 57007. *Corresponding author (sfausti@csumb.edu)

doi:10.2134/practicalmath2017.0032

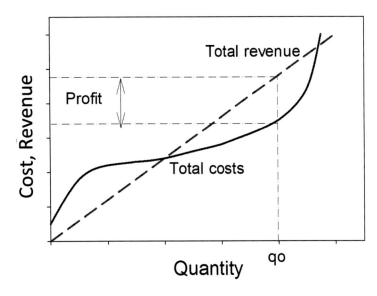

Fig. 12.1. The relationship between TR, TC, Profit per acre as influenced by cost and quantity. The point q_0 is the point where profit is maximized.

that as you try to increase yield, each additional bushel becomes more expensive to produce.

Per-acre profit (Π) is defined as: $\Pi = TR-TC$. Figure 12.1 illustrates the relationship between TR, TC, and Π. The intersection of the TR and TC curves represents the breakeven point. Fixed cost is graphically depicted at the point where the TC curve intersects the vertical axis. Profit is maximized at q_0, where the distance between TR and TC is greatest.

Cost of Growing Corn in the United States

Table 12.1 provides estimates of the revenue and cost associated with growing an acre of corn in the United States for 2015 and 2016. These estimates account for the cost of materials (seed, chemicals, etc.), land, labor (excluding operator labor hours), and capital (estimated as depreciation). The value of production less total costs was based on a corn yield of 167 bushel per acre that

Table 12.1. The average revenue and cost per acre of corn grown in the United States in 2015 and 2016 (USDA-ERS 2017).†		
Item	2015	2016
Gross value of production		
Primary product: Corn grain	611.22	581.00
Secondary product: Corn silage	1.38	1.27
Total, gross value of production	612.60	582.27
Operating costs		
Seed	101.62	98.64
Fertilizer	137.33	115.76
Chemicals	27.95	28.80
Custom operations	19.04	19.42
Fuel, lube, and electricity	21.28	19.37
Repairs	26.18	26.44
Purchased irrigation water	0.12	0.14
Interest on operating capital	0.28	0.71
Total, operating costs	333.80	309.28
Other variable costs per acre		
Hired labor	3.28	3.41
Capital recovery of machinery and equipment	102.63	103.13
Land Rent	179.15	167.40
Taxes and insurance	11.01	11.00
General farm overhead	19.83	19.85
Total, allocated overhead	315.90	304.79
Total, costs listed	**649.70**	**614.07**
Value of production less total costs listed	-37.10	-31.80
Value of production less operating costs	278.80	272.99
Supporting information		
Yield (bushels per planted acre)	167	175
Price (dollars per bushel at harvest)	3.66	3.32
Enterprise size (planted acres) 1/	280	280

†https://www.ers.usda.gov/data-products/commodity-costs-and-returns/. Corn data Excel spreadsheet 2015–2016.

were sold for $3.66 per bushel. Thus, TR = 611.22, adding the value of corn silage, TR = $612.60. Total cost in 2015 equaled operating cost plus other variable cost, TC = $649.70. Thus, the average profit per acre in 2015 was TR-TC = -$37.10 per acre, indicating an operating loss for each acre of corn produced. These calculations indicate that to improve the return, the costs must be reduced, the price of corn must increase, or, to break even, about 11 more bushels of corn must be produced at the same cost.

PROBLEM 12.1.

In Problem 11.12, the data below was used to estimate a quadratic yield equation based on seeding rate. For problem 12.1 the assumptions of corn selling for $3.00 per bushel and a bag of seed-corn (80,000 seeds) selling for $250/bag will be continued. Thus, the price per 1000 seeds is $3.125. Microsoft Excel generated the quadratic relationships R1 and R2 below. Examples of footslope areas are located in lower landscape positions and summit areas are located in upland areas (Fig. 11.1)

R1 (Footslope):   yield= -30.3+10.2(population)-0.129(population)^2; r^2=0.96
b= 10.2 and c= -0.129

R2 (Summit): yield = -55.2 + 10.65(population) – 0.157(population)^2; r^2=0.98
b=10.65 and c = -0.157

Optimal seeding rate calculated in Problem 11.12 for footslope gradient was 35,500 seeds per acre. For the summit region the optimal seeding rate was 30,600 seeds per acre. Assuming that instead of varying the seeding rate based on field gradient, the producer plants the average seeding rate for the field or 33,050 seeds per acre. The question is: How does the producer's decision of not applying a variable seeding rate based on field gradient affect per acre yield and profitability?

ANSWER:

First step is to calculate footslope and summit yield per acre at a seeding rate of 33,050 seeds per acre using the yield equations R1 and R2 above. The footslope yield is estimated at 165.9 bushels per acre and summit is estimated at 125.3. The footslope profit per acre is $394.43 and the summit profit per acre is estimated at $272.59. Comparing these yield estimates to optimal seeding rate yield estimates derived in Chapter 11, (Fausti, 2017), we find that using a single (average) seeding rate results in a footslope–yield decline of 3.32 bushels and a summit increase in yield of 1.62 bu per acre. This decline in yield translates into a $2.33 per acre decline in profit per acre for the footslope gradient and increased seed cost resulted in a $2.80 decline in profit for the summit gradient (See chapter 11 for optimal yield and profit calculations). Based on the slim margins for profit, it would be best to plant populations in a site-specific manner.

PROBLEM 12.2.

The table below provides data on N application rates per acre and corn yield per acre. Microsoft Excel is used to estimate the quadratic yield equations (discussed in chapter 11). Equations R1 and R2 are the estimated equations for the non-linear relationship between N application rates per acre and yield per acre for different field gradients. Assume the price of N is $0.70 per pound and the selling price of corn is $4 per bushel. Calculate optimal N application rates, yield, and profit for each landscape position. Next calculate the average N application rate. Recalculate yield and profit for each gradient and the difference in yield and profit associated with variable N application versus a single rate.

R1 (Footslope):   yield= 80.774 +1.121 (N)-0.00324(N)^2; r^2=0.95
b= 1.121 and c= -0.00324

R2 (Summit):   yield = 50.786 + 1.012(N) – 0.00238(N)^2; r^2=0.97
b=1.012 and c = -0.00238

From Chapter 11, the optimal N application rate can be calculated using the following formula:

$$\left(\frac{P_N}{P_C}\right)\left(\frac{1}{2c}\right)-\left(\frac{b}{2c}\right)= \text{Optimal Application Rate for N per acre.}$$

For example, optimal N application rate per acre for the footslope gradient can be calculated as: [(0.7/4)(1/{2*[-0.00324)]}- {1.121/(2*[-0.00324)]} = 146 (footslope).
Next calculate the optimum for the summit and then the average N rate.

ANSWER:

Optimal N application rate per acre for summit it is 175.84 lbs. The average N application rate is 160.92 lb.

PROBLEM 12.2B

How does the producer's decision to not apply a variable N rate based on field gradient effect per acre yield and profitability?

ANSWER:

First step is to calculate yield and profit per acre for both footslope and summit using equations R1 and R2 with the optimal N application for each gradient. Next, recalculate yield and profit using the average N application rate of 160.92 lb. Results indicate that the profit will decrease $2.63 per acre in the footslope area (too much N applied) and profits will decrease $2.09 in the summit area (too little N applied).

This is a per acre calculation. To examine the field returns it would have to be known how many acres are in a 'summit' position and how many are in a 'footslope' position. In addition, between the summit and footslope (the two extremes), is a 'backslope' position and this may need a calculation for a different N rate to analyze the cost and returns across the entire field.

Reducing Variable Cost Using Precision Farming

Precision farming provides an opportunity to reduce the production costs by optimizing the seeding and fertilizer costs. Problems 12.1 and 12.2 address the economic implications for failing to account for landscape variability when deciding the optimal amount of inputs. For these calculations, it is not necessary to determine the cost of production.

What we learned in Problem 12.1: Marginal revenue versus marginal cost

In problem 12.1, if a single seeding rate of 33,050 was used instead of the optimal rate for the footslope and summit, there was a negative effect on profit. The question is why did profit decline when yield increased? The answer has to do with the economic relationship between revenue and cost when the producer decides to change input levels.

For the footslope a lower seeding rate reduced yield (3.325 bushels), revenue ($3 per bushel ×3.325 bushels = $9.975) and seed cost ($3.125 × 2.45 = $7.65). In this case, the marginal decline in revenue exceeded the marginal decline in seed cost. Thus, footslope profit declined as a result of using an average seeding rate.

For the summit, a higher than the optimal seeding rate of 30,600 per acre resulted in an increase of 1.62 bushels or an increase in total revenue of $4.86. However, the increase in revenue was the result of increased use of seed per acre of 2450 seeds which cost $7.65. Thus the marginal increase in revenue was less than the increase in the marginal cost of the additional seed. As a result, increasing the seeding rate above the optimal level resulted in a decline in profit. For additional discussion on the topics of revenue, cost, and profit see Graham (2013).

Determining the Cost of Production

To develop a profitability map, it is necessary to determine the cost of production. This value can be determined with an online enterprise budget template. An enterprise budget template organizes all costs by categories using a computer spreadsheet. Many of calculators are targeted for different crop, problems, and areas. These calculators can be used to determine the estimated costs and returns of your production system. Most of calculators use known or estimated production costs and expected returns to estimate expected returns. Many of the calculators provide many rotational options. Most of the models require information on:

1. Expected market prices,
2. Planted acres,
3. Variable costs,
4. Rotation and planted crop, and
5. Variable costs,
6. Fertilizer and lime costs,
7. Herbicide costs,
8. Insecticide costs,
9. Seed costs,
10. Drying costs,
11. Grain hauling costs,
12. Machinery costs,
13. Labor costs,
14. Insurance costs, and
15. Irrigation costs.

Determining these costs are discussed in this chapter. Crop production cost calculators are available at the following websites:

Iowa State: https://www.extension.iastate.edu/agdm/crops/html/a1-20.html

Kansas State: http://www.agmanager.info/decision-tools

North Dakota: https://www.ag.ndsu.edu/farmmanagement/crop-budget-archive

Texas A &M: https://agecoext.tamu.edu/resources/crop-livestock-budgets/

The Ohio State University: https://aede.osu.edu/research/osu-farm-management/enterprise-budgets

DuPont Pioneer: https://www.pioneer.com/home/site/ca/agronomy/tools/production-cost-calculator/

Variable Costs

Herbicides, fertilizers, insecticide, and seed costs are considered as variable costs. Typical costs for these expenses (2015 and 2016) are provided in Table 12.1. Costs of preharvest and harvest machinery, including tractors, planters, and combines, are more difficult to estimate. On many farms, these costs are incurred by several enterprises, therefore you need to determine the appropriate percentage allocation for each expense item to each enterprise. For example, if a tractor is likely to be used for all of the crops grown on a farm, not just for the corn enterprises, then the costs of the tractor should be divided between all enterprises. Suppose that a tractor is used a total of 800 hours in a year, and 400 hours are in corn fields, then half of the tractor expenses should be allocated to the corn crop.

Fixed costs

Machinery costs can be divided into ownership costs (also called fixed costs), which occur no matter if the machine is used or not, and operating costs (also called variable costs), which vary with the amount of machine use. Ownership costs include depreciation, interest, taxes, insurance, and housing and maintenance facilities (Edwards, 2015).

The annual depreciation amount can be calculated using a straight-line formula as:

$$\text{Annual depreciation}(\$) = \frac{\text{Purchase cost} - \text{salvage value}}{\text{Years of machine use}} \qquad [12.1]$$

In Eq. [12.1], salvage value is the sales or trade-in value of the machine at the end of its economic life, which is typically between 10 and 15 yr for most farm machines. If you plan to keep the machine until it is worn out, then its salvage value will be zero. For example, if a combine is purchased at $200,000 and has an economic life of 15 yr

PROBLEM 12.3A.

Suppose you purchased a 180-PTO horsepower diesel tractor at the price of $200,000. At the end of its economic life, which is 15 years, the tractor have a salvage value estimated as 20% of its purchase price. Estimate: (i) the tractor's annual depreciation using the straight line method; (ii) annual interest payment, given the annual interest rate is 7%; (iii) taxes, insurance and housing cost for the tractor.

ANSWER:

1. Salvage value = $200,000 × 20%= $ 40,000.
Annual depreciation = (purchase price - salvage value)/ No. of years machine will be used
= ($200,000 - $40,000)/15
= $10,667
2. Annual interest payment = ($200,000 + $ 40,000)/2 × 7%= $8,400.
3. TIH = ($200,000 + $ 40,000)/2 × 1%= $1,200.

PROBLEM 12.3B.

Suppose the above tractor is used for a total of 800 hours in a year, and 400 hours are in the 800-acre corn fields. To the corn enterprise, what is the ownership cost of the tractor on a per acre basis?

ANSWER:

Total ownership cost = annual depreciation + interest payment + TIH
= $10,667 + $ 8,400 + $1,200 = $20,267
Ownership cost for the corn enterprise =$20267 × 400/800 = $10,134.
Ownership cost per acre = $10,134/800= $12.67

PROBLEM 12.4A.

Continue with the 180-PTO horsepower diesel tractor in Problem 12.3. It operates for an average of 800 hours per year, 50% of the time in the 800-acre soybean field, and 50% of the time in the 800-acre corn field. Assume its total annual maintenance and repair cost during its economic life averages at 3% of its purchase price, $200,000. The average amount of diesel fuel required to operate in the corn field is $18.00/hour. For the corn enterprise, calculate the following annual costs of the tractor on a per acre basis: (i) maintenance and repair cost; (ii) fuel cost; (iii) lubrication cost, assuming it to be 15% of the fuel cost; (iv) labor cost, assuming labor hour equals to 110% of the tractor's operating hour and labor cost is $15 per hour.

ANSWER:

1) Annual maintenance and repair cost for the corn enterprise
= $200,000 × 3%× 50%= $ 3,000
Per acre maintenance and repair cost = $ 3,000/800 = $3.75
2) Annual fuel cost = $18.00 per hour × (800 hr × 50%) = $7,200.00
Per acre fuel cost = $ 7,200/800= $9.00
3) Per acre lubrication cost = $9.00 × 15% = $1.35
4) Total labor cost = 800 hr × 50% x 1.1 x $15 per hour= $6,600.
Per acre labor cost = $ 6,600/800 = $8.25 per acre

PROBLEM 12.4B.

What are the two components of machinery cost? Using the example from Problem 12.3b and 12.4a, calculate the total machinery cost on a per acre basis for the corn enterprise.

ANSWER:

Machinery cost comprises of two categories, ownership cost and operating cost.
Based on 12.3b, ownership cost per acre = $10,134/800= $12.67
Based on 12.4a, operating cost per acre = repair cost + fuel cost + lubrication cost + labor cost = $3.75 + $9.00 + $1.35 + $8.25 = $22.35.
Machinery cost per acre = ownership cost + operating cost = $12.67 + $22.35 = $35.02.

Based on these figures, you can now decide if owning and operating a piece of equipment is reasonable for your operation OR if custom planting, combining, and/or pesticide application would make more financial sense. Remember that custom crews may not be as timely as you many want, and this also should be taken into account.

with a salvage value of $50,000, then based on the straight-line formula in Eq. [12.1], the depreciation each year will be ($200,000- $50,000)/15 = $10,000.

Annual interest payment is calculated based on the average value over the economic life of the machine:

$$\text{interest} = \frac{\text{purchase cost} + \text{salvage value}}{2} \times \text{annual interest rate} \qquad [12.2]$$

Assuming an annual interest rate of 7%, then in our example, the interest of the machine is calculated as: ($200,000 + $50,000)/2 × 0.07 = $8,750. For calculations in Eqs. [12.2–12.3] it is important to convert the interest percent to the decimal value. For example 5% is 0.05.

Compared with depreciation and interest, taxes, insurance and housing (TIH) costs account for an insignificant amount (usually 1% or less) of the total machinery cost. For simplicity they can be lumped together and calculated as:

$$\text{TIH} = \frac{\text{purchase cost} + \text{salvage value}}{2} \times 1\% \qquad [12.3]$$

For our combine example, the annual TIH costs are: ($200,000 + $50,000)/2 ×1% = $1,250. Total ownership cost of the combine, which includes depreciation, interest and TIH, can be found by adding these three cost items together: $10,000 + $8,750 + $1,250 = $ 20,000. If 50% of the combine time is spent in the corn enterprise of 800 acres, then the annual per acre ownership cost is $20,000 × 50%/800 = $12.5 per acre.

Calculating machine operating costs

Next, we will calculate the operating cost of the machine, which includes repairs and maintenance, fuel, lubrication and operator labor. Labor cost is included in machinery analysis but different machines have different

PROBLEM 12.5.

Based on the following information on corn production, calculate operating cost (variable cost) and ownership cost (fixed cost) on a per acre basis. Assuming corn yield is 160 bushels per acre, calculate the operating cost and ownership cost on a per bushel basis.

Item	Cost per acre
Seed	$40.50
Machinery ownership cost	$25.30
Land ownership cost	$80.00
Fertilizer and chemicals	$59.00
Fuel and lubrication	$10.00
Machinery labor	$8.00
Irrigation labor	$5.00
Repairs	$15.50
Machinery insurance	$7.50
Interest over operating cost	$5.35
Machinery interest	$3.50
Crop insurance	$6.00
Custom harvest	$25.50
Other ownership costs	$15.00
Other operating costs	$12.00

ANSWER:

Operating Costs	Cost per acre
Seed	$40.50
Fertilizer and chemicals	$59.00
Fuel and lubrication	$10.00
Machinery labor	$8.00
Irrigation labor	$5.00
Repairs	$15.50
Interest over operating costs	$5.35
Crop insurance	$6.00
Custom harvest	$25.50
Other operating costs	$12.00
Total operating costs	$186.85

On a per bushel basis, operating costs account for $186.85/160=$1.17 per bushel, ownership costs account for $131.30/160=$0.82 per bushel.

Ownership costs	Cost per acre
Machinery ownership costs	$25.30
Land ownership costs	$80.00
Machinery insurance	$7.50
Machinery interest	$3.50
Other ownership costs	$15.00
Total ownership costs	$131.30

labor requirements. The labor cost is an important factor to consider when comparing machine ownership to custom hiring.

To estimate repair and maintenance costs, the best data are records of your own past repair expenses. Suppose the average repair and maintenance costs of a combine amount to $6,400 per year, and 50% of its time is spent on the 800-acre corn field, then the annual repair cost would be $6,400 × 50%/800 = $4.00 acre per acre. You can calculate fuel cost by multiplying the fuel usage by the fuel cost per gallon. For example, if the average amount of diesel fuel required to harvest 800-acre of corn field is 1160 gallons, at a cost of $2.40 per gallon, then the fuel cost per acre is 1160 x $2.40/800 = $3.48 per acre.

Lubrication cost on most farms averages about 15% of the fuel cost. With the previous example, lubrication cost per acre can be estimated as $3.48 × 15% = $0.52 per acre. Suppose the combine harvests at a rate of 8 acres per hour, then the total combine time is 100 hours for the 800-acre corn field. Labor hours can be estimated by multiplying the combine time by a factor of 1.1 to 1.2, due to the extra time required for maintenance and lubrication. Taking the factor of 1.1, then the labor time would be 100 × 1.1 = 110 h. At a labor cost of $15 per hour, the labor cost per acre is estimated as: $15 × 110/800 = $2.06 per acre. To sum up, the combine per acre operating cost is $4.00 (repairs and maintenance) + $3.48 (fuel) + $0.52 (lubrication) +$ 2.06 (labor) = $10.06 per acre.

Now that the ownership cost and operating cost of the machine has been estimated, they can be added up together to obtain the total cost of the machine. In our combine example, the total cost is $12.5 + $10.06 = $22.56 per acre.

Estimating interest costs

Interest reflects the amount of money you pay on borrowed money or that amount you could have earned had your money not been tied up in production and invested in alternative uses in the market. Spreadsheet in Problem 12.5 includes interests in "other" category. Interest for operating expenses can be calculated as follows: Total operating expense × $(1 + \text{annual interest rate}/12)^{(\text{The number of months the expense is tied up in crop production})}$ - Total operating expense.

Example: Given that total operating expense = $6000; interest rate 5%; number of months in crop production = 8. Interest can be calculated as Interest = $6000*[1+(0.05/12)]^8$- $6000 = $202.94.

The number of months that expense is tied up in crop production starts when the operating capital is invested and ends when it is recovered, usually referring to the harvesting period or sale month for the crop. For example, assuming your fertilization and weed control operations are done in early May and your harvest time is late October, then the interest will be charged for six months.

After all of the expenses have been sorted and allocated correctly, they can be added up to obtain the total expenses. Expenses per acre, per bushel, et cetera can be calculated by using the total expenses divided by the total planted acres and expected yields.

Summary

This chapter provides examples on the importance of determining the optimal input level to maximize profitability. Profitability of the system can be maximized by choosing the optimal level of a specific input based on land characteristics. Variable rate technology examples are provided to demonstrate the value of precision agricultural technology to farm profitability. For simplicity, the examples in this chapter assume that the profitability is maximized by optimizing the individual decisions. The cost of production requires keeping detailed input record and a careful analysis with an enterprise budget spreadsheet. Formulas are provided to determine difficult cost items such as machinery and interest costs. Examples are provided. Additional information is provided at Doye and Sahs, 2017, Fausti et al., 2017, and Lessley et al., 2015.

Additional Problems

12.8. Redo problem 12.2 assuming price of N is 90 cents and price of corn is $3.00 per bushel

12.9. Redo problem 12.2 assuming price of N is 50 cents and price of corn is $6.00 per bushel

12.10. Compare your results from Problems 12.8 and 12.9. What can you conclude about how the change in the price of N and the change in the price of corn effects N application rates and profitability?

12.11. Using the results for Problem 12.2 calculate the marginal change in revenue and cost associated with using the average rate of N application relative to the optimal rate of N application for the footslope and summit.

ACKNOWLEDGMENTS

Support for this document was provided by South Dakota State University, South Dakota Corn Utilization Council, Precision Farming Systems community in the American Society of Agronomy, International Society of Precision Agriculture, and the USDA-AFRI Higher Education Grant (2014-04572).

REFERENCES AND ADDITIONAL INFORMATION

Graham, R. 2013. Managerial economics for dummies. John Wiley & Sons. New York.

Doye, D., and R. Sahs. 2017. Using Enterprise Budgets in Farm Financial Planning. AGEC-243. Oklahoma Cooperative Extension Service. http://pods.dasnr.okstate.edu/docushare/dsweb/Get/Document-1658/ (Accessed 2 June 2017).

Edwards, W. 2015. Estimating farm machinery costs. Ag Decision Maker, File A3-29. Iowa State University, Extension and Outreach. Ames, IA. https://www.extension.iastate.edu/agdm/crops/pdf/a3-29.pdf (Accessed 2 June 2017).

Fausti, S., B.J. Erickson, D.E. Clay, and C.G. Carlson. 2017. Chapter 11: Deriving and using equations to calculate the economic optimum fertilizer and seeding rates. In: D.E. Clay, S.A. Clay, and S. Bruggeman, Practical Mathematics and Agronomy for Precision Farming. ASA, Madison, WI.

Lessley, B.V., D.M. Johnson, and J.C. Hanson. 2015. Using the partial budget to analyze farm change. Fact Sheet 547. Maryland Cooperative Extension. Leonardtown, MD. https://www.arec.umd.edu/sites/arec.umd.edu/files/_docs/Using%20the%20Partial%20Budget_0.pdf (Accessed 2 June 2017).

United States Department of Agriculture, Economic Research Services. 2017. Commodity costs and returns. https://www.ers.usda.gov/data-products/commodity-costs-and-returns/commodity-costs-and-returns/ (Accessed on 14 May 2017).

Mathematics Associated with Seed Emergence, Plant Population, Stand Uniformity, and Harvest Losses

## 13

Stephanie A. Bruggeman,* Sharon A. Clay, Cheryl L. Reese, and C. Gregg Carlson

Collecting Plant Health Benchmarks

To assess the effectiveness of a precision farming treatment, it is important to collect adequate plant health benchmarks to ensure that the appropriate treatments were applied (Carlson and Clay, 2013; 2016). A benchmark is a reference point or starting point from which comparisons can be made. Specific benchmarks might include soil test values, a record of weed distributions and population levels, and seed emergence. A good record keeping system with accurately measured and recorded benchmarks is a cornerstone of precision farming, and is critical for calculating the success of any given treatment.

To assess the effectiveness of a management *practice* (the overall effects of all the treatments), the collected information should be stored for future reference. Many Integrated Pest Management programs recommend measuring yield and tracking the extent of pest populations and distribution; whereas to assess soil nutrient improvements, soil samples should be collected and analyzed. The list of benchmarks should include plant emergence, plant variability, the extent of pest populations and distribution, soil nutrient levels, remote sensing, and yield data (Nielsen, 2001; Liu et al., 2004; Carlson et al., 2013). Seed rates, plant emergence, stand uniformity, and harvest losses are discussed below.

Calculating the Seeding Rate

The desired plant population at harvest should not be used as the seeding rate. Seed germination rates (the percentage of seed that will germinate after planting), seed purity (actual percentage of crop seed in a bag of seed), and percentage of seedling emergence all need to be taken into account. To assess the effectiveness of a variable rate seeding plan, measurements and evaluations of plant populations and emergence at many different seeding rates should be performed. Seeding and emergence rate are calculated with the equations,

$$\text{Seeding rate} = \frac{\text{Desired population at harvest}}{\dfrac{\text{\% emergence of planted seeds}}{100}}$$

Chapter Purpose

Accurate benchmarks on plant health, seed germination, and harvest losses are needed to assess if precision treatments are successful. This chapter discusses and provides examples how to conduct these assessments.

Key Terms

Benchmarks, seeding rate, plant emergence rate, harvest loss, seed germination standard deviation.

Mathematical Skills

Comparison of measured values with appropriate benchmarks, calculating emergence of germinated seeds, calculating stand uniformity and harvest losses.

South Dakota State University, Brookings, SD 57007. *Corresponding author (Stephanie.Bruggeman@sdstate.edu)
doi: 10.2134/practicalmath2016.0029

PROBLEM 13.1

Determine the percentage of germinated seed that emerged from the soil. If the seed germination rate is 95%, the seeding rate is 35,000 seeds acre^{-1}, and the postemergence counted plant population is 33,000 plants per acre, what is the percentage emergence of germinated seeds (EGS)?

ANSWER:

$$\%EGS = \dfrac{\dfrac{\text{\# seed emergence}}{\text{Planted rate}}}{\dfrac{\%\text{germinated seeds}}{100}} = \dfrac{\dfrac{33,000 \text{ plants acre}^{-1}}{35,000 \text{ plants acre}^{-1}}}{0.95} = 0.992$$

This calculation suggests that 99% of the germinated seeds emerged from the soil.

PROBLEM 13.2

If the seedling plant population is 33,000 plants acre^{-1} and the plant population at harvest is 31,000 plants acre^{-1} what is the survival of seedlings to harvest.

ANSWER:

$$\% \text{ survival} = 100 \times \frac{31,000}{33,000} = 93.9\%$$

The seed emergence rate and the emergence of germinated seeds are two different values. The % emergence of planted seeds (emergence rate) is based on two values, % germinated seed and % emergence of germinated seed and it can range from 90% to 95%. These values are related by the equation below.

$$\frac{\% \text{ emergence of planted seeds}}{100} = \frac{\% \text{ germinated seeds}}{100} \times \frac{\% \text{ emergence of germinated seeds}}{100}$$

In this equation, the germination rate (% germinated seeds) is provided by the seed seller, whereas the % of germinated seeds that emerged from the soil is not known and therefore must be estimated or calculated. This value is important because it can reveal planter problems. Sample calculations for these values are provided in problems 13.1, 13.2, and 13.3.

Seed Emergence of Planted and Germinated Seeds

Seed emergence is influenced by many factors including seedbed preparation, crusting, and diseases, and it is calculated with the equation:

$$\% \text{ Seed emergence} = 100\% \times \left(\frac{\text{Plant population after emergence}}{\text{Seeding rate}} \right)$$

The plant population can be measured as soon as the plants emerge by counting the number of plants in a specified area. Many agronomists recommend measuring the plants in 1/1000 of an acre (Table 13.1; Carlson et al., 2013). The distance along a row that represents 1/1000 of an acre depends on the row width. For a 30-inch row, the length of the row for 1/1000 of an acre is 17 feet and 5.1 inches (Table 13.1).

Determining Stand Uniformity

Increasing the plant population may increase the yield per unit area, however it can also reduce the yield per plant (Nielsen, 2001; Carlson et al., 2002). Yields are reduced because adjacent plants compete for water, nutrients, and light. Lower or higher yields may be a function of stand uniformity, as opposed to cultivar. The field variability of a stand by location of individual plants along a tape measure (Fig. 13.1).

Variability is determined by recording the location of the plants, followed by determining the standard deviation of the distances between the plants (Table 13.2). This process should be repeated at a number of locations in the field. A standard deviation of 2 inches is excellent, and there is approximately a 4 bu acre^{-1} yield loss per inch for standard deviations greater than 2 inches. Landscape, soil, and seedbed differences may influence stand variability.

Table 13.1 The distance along a row representing 1/1000 of an acre. On the row, the number of plants should be counted. The plant population is 1000 times the number of plants in 1/1000 of an acre. "Feet" and "Inches" = distance in feet and inches of row to measure.

	Row width (inches)									
Distance	6	7	8	10	14	15	20	21	28	30
Feet	87	74	65	52	37	34	26	24	18	17
Inches	1.4	7.1	4.1	3.3	4	10.2	1.6	10.7	8	5.1

PROBLEM 13.3.

Calculate the seeding rate if the purity is 96%, and 90% of the planted seeds emergence rate (emerged seeds/planted seeds). The desired live population is 135,000 plants acre^{-1}.

ANSWER:

$$\frac{135,000 \text{ plants } @V2}{\text{acre}} = \left(\frac{\text{seeding rate}}{\text{acre}}\right) \times \text{Purity} \times \text{Emergence rate}$$

$$\frac{135,000 \text{ plants } @V2}{\text{acre}} = \left(\frac{\text{seeding rate}}{\text{acre}}\right) \times 0.96 \times 0.90 = \frac{156,250 \text{ seeds}}{\text{acre}}$$

PROBLEM 13.4.

If the % germination is 94%, the expected survival of germinated seed to harvest is 92.2%, and the target plant population is 34,000 seed acre^{-1}, what is the seeding rate?

ANSWER:

$$\text{Seeding rate} = \frac{\text{Target population at harvest}}{\dfrac{\% \text{ germinated seed}}{100} \times \dfrac{\% \text{ emergence of germinated seeds}}{100}}$$

$$\text{Seeding rate} = \frac{34,000/\text{acre}}{0.94 \times 0.922} = 39,230 \text{ plants acre}^{-1}$$

PROBLEM 13.5.

Determine the seed emergence if the seeding rate is 38,000 plants acre^{-1}.
Measure the row width, and if your row width is 30 inches, count the number of plants in a row that is 17 feet and 5.1 inches long. If 35 corn plants are contained in the row, then your plant population is 35,000 plants acre^{-1} (35×1000).

$$\text{Seed emergence} = 100\% \times \left(\frac{35,000}{38,000}\right) = 92.1\%$$

In a second example, you plant corn in 15-inch rows, what is the length of row to produce 1/1000th of an acre? Based on data in Table 34.1, count the number of plants in a row that is 34 feet and 10.2 inches long.

The calculations in Table 13.2 are used to determine the standard deviation of the distance between adjacent plants. As shown in B12, the average distance is determined with the command = *average (start, end)*. As shown in B13, the standard deviation is determined with the command = *stdev(start, end)*. The standard deviation is a measure of variability. A high standard deviation value has a high yield loss and is the result of a poorly calibrated planter or poor seed bed. The harvest loss in B12 is calculated with the equation, *harvest loss = (standard deviation -2)×4*. This is based on a standard deviation of three, which has a yield loss of 4 bu acre^{-1}.

Harvest Losses

Increasing the population has the potential to reduce the size of the individual kernels, which in turn can impact harvest losses. Harvest losses are unavoidable; however, they can be minimized by careful management (Humburg

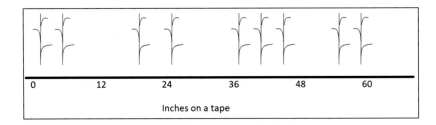

Fig. 13.1. The number of corn plants along a transect within a single row. In this example, corn plants are located at 2, 6, 18, 25, 37, 43, 46, 56, and 58 inches.

Table 13.2. Sample spreadsheet showing how to calculate plants acre^{-1} and yield losses due to variable seeding (for information in Fig 13.1). The tables below show the locations on a tape measure. In the table on the right, the equations behind the values in column B are shown. The row spacing is 30 in.

	A	B	C	A	B
	Measured location of each corn plant (inches)	Spacing distance between each pair of plants	Row spacing (inches)		Equations
1	0		30		
2	2	2			= A2-A1
3	6	4			= A3-A2
4	18	12			= A4-A3
5	25	7			= A5-A4
6	37	12			= A6-A5
7	43	6			= A7-A6
8	46	3			= A8-A7
9	56	10			= A9-A8
10	58	2			= A10-A9
					= A11-A10
12	Average	6.44			= average(B2:B11)
13	Standard deviation	4.07			= stdev(B2:b11)
14	Bu acre^{-1} in estimated yield loss	8.26			= (B13–2)*4
15	Plants acre^{-1}	32,467			= (1/(C1*B12))*144*43,560

2016a, 2016b). Harvest losses are determined by counting the number of kernels on the ground before and after harvest. If a combine is not properly adjusted, losses can be as high as 4 to 5% of the crop. Different calculations are required for different crops. For corn, two types of losses are possible, through kernels and/or ears, whereas for soybeans, the beans may fall on the ground. To assess preharvest losses, the field must be inspected prior to combining. When checking the field, identify any factor that may have contributed to the loss. For example, is the yield loss a result of a nutrient deficiency or pests?

Corn, wheat, and soybean seeds on the soil surface can be counted using an open circle or square with a known area, such as a square foot. Many agronomists use a circle or square that contains one square foot. The circle is placed on the soil, and number of seeds contained within the designated area are counted. To improve the estimate, counts should be conducted at a number of locations. If the field contains multiple cultivars or seeding rates, counts should be made in each areas. The yield loss is dependent on the number and size of seeds (Table 13.3).

Losses associated with the combine are associated with shattering, stubble, loose stems, separator, and machine leakage. Shattering loss is associated with cutting and gathering processes at the head. Stubble losses result from pods that are located below the cutting bar. Loose stem losses are the pods that remain after the combine has stripped the plant. Separator losses are associated with soybeans that are not threshed. Machine leakage losses occur from machine wear. To fix the problem, the type of loss must be identified (Humburg 2016a, 2016b). The calculations for converting the number of soybeans ft^2 to bu acre^{-1} loss are provided in Table 13.4.

As with corn and soybeans, the weight of the individual wheat kernels are variable (Table 13.5). A bushel of wheat can contain over 1 million kernels and the number of seeds lb^{-1} can range from 16,000 to over 22,000 kernels (Lyon and Klein, 2001; Nielsen, 2004; Carlson et al., 2013). If a bushel contains 1 million kernels, then 23 kernels per 1 ft^2 would represent 1 bu acre^{-1}.

PROBLEM 13.6.

What is the estimated corn yield loss if five kernels that are medium in size are found in a 1-ft^2 area? Use 43560 ft^2 per acre and 90,000 medium kernels per bushel.

ANSWER:

$$\frac{5 \text{ kernels}}{\text{ft}^2} \times \frac{43560 \text{ ft}^2}{\text{acre}} \times \frac{\text{bu}}{90,000 \text{ kernel}} = \frac{2.4 \text{ bu}}{\text{acre}}$$

This calculation assumes that kernels are medium in size (Table 13.3). If the kernels were small, then 110,000 kernels bu^{-1} would be used.

Table 13.3. The relationship between corn kernel size and kernels ft^2 equal to a 1 bu acre^{-1} yield loss. These values are used to determine yield losses using the number of kernels found on the ground following harvest.

Kernel Size	Weight bu^{-1} (15.5% moisture)	Kernels bu^{-1}	Kernels lb^{-1} (15.5% moisture)	Kernels ft^2 equivalent to 1 bu acre^{-1}
	lb			
Large	56	70,000	1250	1.6
Medium	56	90,000	1607	2.1
Small	56	110,000	1964	2.5

Table 13.4. The relationship between soybean seed size and the number of beans ft^2 equal to a yield loss of 1 bu acre^{-1}. The number of beans per ft^2 that is equivalent to one bu acre-1 is determined by dividing the number of soybeans bu^{-1} by 43,560. The data in Table 13.4 is based on size of the soybeans, which can be determined by weighing 200 soybeans. The weight is then converted to beans per bushel. As with corn, harvest losses can be measured by collecting the amount of kernels contained within a specified area (1 ft^2). The harvest loss is dependent on the weight of the kernels and the number of kernels collected. Loss increases with kernel weight.

	Weight bu^{-1} (13% moisture)	Bean bu^{-1}	Beans lb^{-1} (13% moisture)	Soybeans in beans ft^2 that are equivalent to 1 bu acre^{-1}
	lb			
Small soybeans	60	218,000	3633	5
Medium soybeans	60	174,000	2900	4
Large soybeans	60	130,800	2180	3

PROBLEM 13.7.

How many beans are contained in a bushel if 200 soybeans weight 0.95 ounces?

ANSWER:

$$\frac{beans}{bushel} = \left(\frac{200\ beans}{0.95\ ounces}\right) \times \frac{16\ ounces}{1\ lb} \times \frac{60\ lb}{bushel} = \frac{202,100\ beans}{bushel}$$

PROBLEM 13.8.

There are 15 soybeans ft^2 in a ground loss count. What is the estimated loss if there are 240,000 beans bu^{-1}?

ANSWER:

$$\frac{bushel}{acre} = \left(\frac{15\ beans}{ft^2}\right) \times \frac{43,560\ ft^2}{acre} \times \frac{bushel}{240,000\ beans} = \frac{2.72\ bushels}{acre}$$

These calculations are based on a bushel of soybeans containing 240,000 soybeans (Table 13.4) If the soybeans were medium in size a bushel would contain 174,000 soybeans.

PROBLEM 13.9.

Behind a combine harvesting wheat, 23 kernels ft^{-2} were found. What is the loss if there are 1,000,000 kernels bu^{-1}?

ANSWER:

$$\left(\frac{23\ wheat\ kernels}{ft^2}\right) \times \frac{43,560\ ft^2}{acre} \times \frac{bushel}{1,000,000\ kernels} = \frac{1.00\ bushel\ wheat}{acre}$$

Table 13.5. Approximate seeds gram^{-1}, seeds lb^{-1} and the standard weight and moisture content of selected annual crops.

Plant	Seeds gram^{-1}	Seeds lb^{-1}	lb bushel^{-1}	Standard moisture %
Barley	30	12,000–15,000	48	14.5%
Corn	3	1000–1,500	56	15.5%
Field pea	8	1500–3,600	60	15.0%
Lentil	20	9000	60	13.5%
Oats	30	13,000	32	14.0%
Rye	40	18,000	56	14.0%
Rice	65	29,000	45	12.0%
Soybean	9	2500–3,000	60	13.0%
Wheat	35	12,000–15,000	60	13.5%

Summary

In summary, to assess the effectiveness of a precision treatment, benchmarks for comparison are needed. In some situations, these benchmarks might include soil chemical analysis, whereas in other situations seed emergence and plant stand variability should be measured. This chapter provides examples on how to establish benchmarks for a seed emergence, stand uniformity, and harvest losses. These measurements provide clues on how to increase agronomic efficiency.

ACKNOWLEDGMENTS

Support for this document was provided by South Dakota State University, South Dakota Soybean Research and Promotion Council, and the Precision Farming Systems communities in the American Society of Agronomy and

Additional Problems

13.10. If the plant population is 120,000 plants acre^{-1} and the seeding rate is 125,000 what is the emergence rate?

13.11. What is plant population if the row width is 30 in, and there are 22 plant in a row that is 17 ft and 5.1 in long?

13.12. What is the plant population if the row width is 15 in and there are 25 plants along row that is 37 feet and 4 in long?

13.13. You measure the distance of corn plant on a tape, plants are located at 0, 3, 6, 15, 22, 23, 30, 37, 42, 47, 56, and 57 in. What is your standard deviation and what is the yield loss due to variability?

13.14. Why are benchmarks important?

13.15. Behind a combine, you find 5 medium sized corn kernels ft^{-2}, what is the estimated loss?

13.16. Behind a combine in a strip trial study, you find, 15 corn kernels ft^{-2} (small size) for Variety 1 and 5 corn kernels (medium size) for Variety 2, what is the estimated loss and what do you conclude?

the International Society of Precision Agriculture. Additional support was provided by the USDA-AFRI Higher Education Grant (2014-04572).

REFERENCES AND ADDITIONAL INFORMATION

Carlson, C.G., and D. Clay. 2013. Estimating the soybean plant population and seed emergence rate. Chapter 13. In: D.E. Clay, C.G. Carlson, S.A. Clay, L. Wagner, D. Deneke, and C. Hay, editors. iGrow soybean: Best management practices. South Dakota State University, Brookings, SD.

Carlson, C.G., and D. Clay. 2016. Chapter 34: Estimating the corn plant population and seed emergence rate. In: D.E. Clay, C.G. Carlson, S.A. Clay, and E. Byamukama, editors, iGrow corn: Best management practices. South Dakota State University, Brookings, SD.

Carlson, C.G., D. Clay, K. Reitsma, and K. Gustafson. 2013. Estimating soybean yield. Chapter 43. In: D.E. Clay, C.G. Carlson, S.A. Clay, L. Wagner, D. Deneke, and C. Hay, editors, iGrow soybean: Best management practices. South Dakota State University, Brookings, SD.

Carlson, C.G., T. Doerge, and D.E. Clay. 2002. Estimating corn yield losses from unevenly spaced corn. SSMG 37. In: D.E. Clay, et al., editors, Site Specific Management Guidelines. Potash and Phosphate Institute, Norcross, GA.

Humburg, D. 2016a. Chapter 36: Profitability can be enhanced by reducing corn harvest losses. In: D.E. Clay, C.G. Carlson, S.A. Clay, and E. Byamukama, editors, iGrow corn: Best management practices. South Dakota State University, Brookings, SD.

Humburg, D. 2016b. Chapter 37: Combine adjustment to reduce harvest losses. In: D.E. Clay, C.G. Carlson, S.A. Clay, and E. Byamukama, editors. iGROW corn: Best management practices. South Dakota State University, Brookings, SD.

Liu, W., M. Tollenaar, G. Stewart, and W. Deen. 2004. Response of corn grain yield to spatial and temporal variability in emergence. Crop Sci. 44:847–854. doi:10.2135/cropsci2004.8470

Lyon, D.J., and R.N. Klein. 2001. Estimating winter wheat grain yield. Neb Guide G1429. Nebraska Extension Service, Lincoln, NE.

Nielsen, R.L. 2001. Stand establishment variability in corn. AFRY- 91-01. Purdue University, West Lafayette, IN.

Nielsen, R.L. 2004. Estimating corn grain yields prior to harvest. Corny News Network. Purdue University. West Lafayette, IN.

Estimating Weed and Insect Development, In-Season Yield Losses, and Economic Thresholds

14

Sharon A. Clay* and Adam Varenhorst

Chapter Purpose

This chapter discusses and demonstrates how to estimate the impact of weed and insect pests on yield losses and how to calculate economic thresholds. Techniques for converting pest population into anticipated yield loss are discussed. The models used calculate yield losses and economic thresholds are pest specific, with the economic threshold, the point where the cost of the control equals the estimated cost of yield loss. For weeds, a hyperbolic model is used to estimate how weed density impacts crop yields. For weeds and insect development, growing degrees have been used to assess phenological development. Examples are provided.

Key Terms

Degree days; growing degree days; insect development; biofix; weed interference; linear model; curvilinear model, economic thresholds.

Mathematical Skills

Degree day calculations; linear regression; nonlinear regression.

Why Control Pests

Weed and insect pests can reduce crop yields though many different mechanisms including: (i) competition for water, nutrients, and light; (ii) downregulation of genes associated with photosynthesis, and (iii) tissue damage that can expose the plant to diseases, reduce leaf area, or reduce root growth with subsequent reduction in nutrient and water uptake (Pedigo and Rice, 2009; Clay et al., 2009). When managing pests, it is important to remember that if best management practices are not followed, minor problems can transition from a nuisance to a major problem relatively quickly. For example, under the correct weather conditions, small populations of insects can become full field infestations in a few days. To minimize these risks, it is important to watch weather patterns and understand the invasive species biology. In addition, if chemical resistant pests are suspected, proactive management may be required to prevent an expansion of the problem.

Collecting Information

Pest Management Inputs

Integrated pest management starts with collecting information and using science-based techniques to match inputs to the problem. Pest management

South Dakota State University, Department of Agronomy, Horticulture, and Plant Science, Brookings, SD 57007-2201.*Corresponding author (sharon.clay@sdstate.edu).

doi: 10.2134/practicalmath2017.0106

decisions should consider many diverse options that could be available, including long-term solutions, such as crop rotations, and short-term management choices, such as applying chemical controls. Whenever possible, the control costs should be balanced with the expected returns. If the costs are too high, the risk and outcome assessment of a 'no management (or inputs) at this time' decision should be considered. If the risks are too high, then cost may only play a small role in the final management decision(s).

Pest management (e.g., weeds, insects etc.) inputs include the costs associated with genetic enhancements, seed treatments, cultivation, rotational changes, and chemical applications. In many fields, these costs can exceed the investment for seed or fertilizer and, if not carefully evaluated, they can exceed the value of the crop. Therefore, the cost of managing pests must be balanced with the calculated yield loss and commodity price. For example, weed presence can result in 100% crop loss if not managed, but even with herbicides, weeds can still result in 1 to 15% crop loss (Bridges, 1992). Yield losses can be separated into short-term and long-term impacts. Short-term impacts may represent yield gains or losses within the given year, whereas the long-term impacts may be associated with increased weed seeds in the soil or pest resistance (Egley and Williams, 1990; Toole and Brown, 1946).

For short term calculations, information is needed to compare the management costs with the expected returns on the investment. However, making these calculations often is limited by the lack of information on specific pests located in a specific environment and climate. This problem can be partially solved by extrapolating pertinent data from related species. Yield loss estimates still may be inaccurate due to the many variables that influence yield. Yield potential, pest pressure, and the amount of loss are functions of yearly climate, abiotic variables, such as heat, nutrient, and water stress, and biotic variables, such as predator numbers or disease extent, all of which influence responses.

For some species, such as corn leaf aphid, black cutworm, and potato leaf hopper, which are transported from one region of the country to another by wind, considering a one-year time frame for control strategies may be appropriate. However, if the pest overwinters in the area, a long-term management plan would be more fitting.

Effective pest management must consider the timing of the infestation and associated management strategy. For example, five to 10 greenbugs, *Schizaphis graminum*, (Homoptera: Aphididae), per plant on seedling spring wheat may cause 60% yield loss compared with no yield loss if 100 greenbugs per plant are found on wheat plants at the dough to mature plant growth stages (Kieckhefer and Kantack, 1980) (Table 14.1). Contrast this aphid scenario with a weed example. If a weed goes to seed, the plant may produce hundreds, if not thousands of seeds that will remain in the soil for many years. The economics of managing a weed the first year may not be favorable, but future problems may be avoided by doing so. This chapter deals with predicting both yield losses and economic considerations for different types of pests.

Determining the Extent and Magnitude of the Problem

Before yield loss estimates can be calculated, the extent of the problem and the success of the treatment should be estimated. Field scouting is used to identify the boundaries of problem, the density of the infestation by field area, and the crop growth stage. Obtaining this information requires the use of pest appropriate scouting strategies.

Scouting methods vary depending on the pest and crop growth stage. Soil sampling is needed for nematode and belowground insect infestations, whereas sweep nets, sticky traps, plant sampling, or visual observations of crop damage provide information about aboveground insects. Additional information on scouting for pests is available in Deneke and Johnson (2016). Visual observations are needed to examine weed species and densities present and, where possible, all pest species present in the area should be identified (Clay, 2016a, 2016b; Strunk and Byamukama, 2016). Some pests are highly mobile and area-wide observations, not just observations from a specific field, may be needed. If a field is not uniform, based on topography, soil types, or other characteristics, site specific information may be gathered and carefully analyzed. Drier hilltops, side hills, and wetter toeslope areas may be considered individual areas that should be scouted and treated differently when appropriate. Soil type, (sandy vs. clay loam for instance) may also influence pest occurrence and severity and should be considered during scouting.

Using the Degree-day Method to Estimate Insect Development

The occurrence risk of some insect problems can be estimated using the degree-day method. This approach is effective because insects are cold-blooded (i.e., poikilothermic) and therefore depend on appropriate temperatures to complete their life-cycle. Insect activity can be estimated using the degree-day method because temperatures are positively related to insect development (Pedigo and Rice, 2009). In this method, insect developmental stages are predicted based on the summation of the average daily temperature (Eq. [1]). Degree days are calculated with the equation,

PROBLEM 14.1.

The following temperature data has been collected over a seven day period. Calculate the degree days based on (i) an insect with a development threshold of 50 °F; (ii) developmental threshold of 40 °F; and (iii) a biofix model when the insect infestation was first observed on day 3 and has a developmental threshold of 55°F and a developmental optimum of 87°F. A biofix model starts accumulating degree days when the first individual of the species of interest that is observed in the current season.

	Min. temp.	Max. temp.	Case 1 Min temp = 50	Case 2 Min temp = 40	Case 3 Biofix day 3 observed then base temp = 55 and max temp = 87
Day 1	44	55	2.5	9.5	
Day 2	60	75	17.5	27.5	
Day 3	49	77	13.5	23	insect observed
Day 4	42	80	15	21	12.5
Day 5	52	84	18	28	14.5
Day 6	60	90	25	35	18.5
Day 7	65	92	28.5	38.5	21
Degree days			120	182.5	66.5

Table 14. 1. Examples of recommended economic thresholds for wheat insect pests (modified from Government of Saskatchewan, 2016).

Insect	Crop	Growth stage			Notes
Aphids		Seedling	Boot	Dough	
Greenbug [Schizaphis graminum (Rodani)]	Cereals	5 to 15 per stem	10 to 25 per stem	Do not treat	
Birdcherry-oat aphid [Rhopalosiphum padi (Homoptera: Sternoryyncha: Aphididae)]	Cereals	20 per stem	30 per stem	Do not treat	
Russian wheat aphid [Diuraphis noxia (Kurdjumov)]	Cereals	1 aphid on 10% of plants when first node visible	1 aphid on 10% of tillers on flag leaf		
Caterpillars					
Diamondback moth [Plutella xylostella L.]	Canola	100 to 150 per m² in immature fields		200 to 300 per m² in podded canola	Numbers based on stands of 150 to 200 plants per m². In thin stands, EC should be lowered.
Armyworm [Pseudaletia unipuncta (Haworth)]	Cereals, canola	< 5 larvae per m²			

$$\text{Degree day} = \frac{(\text{Max. daily temp.} + \text{Min. daily temp.})}{2} - \text{Developmental threshold value} \quad [14.1]$$

In this equation, the development threshold values are pest specific. The number of degree days that are required for each developmental event are referred to as thermal constants and they differ for the same insect located in different environments. The thermal constants define the number of degree days for the different components in the insect's life cycle, and they may be available for hatching, nymphal molts, pupation, and/or adult emergence.

A degree day represents the accumulation of heat units that are above the known temperature threshold required to produce physiological changes in the insect (*i.e.*, development occurs). If the air temperature is below the threshold, no development occurs, however, if it is above the threshold, then degree days are accumulated. For this reason, the lowest temperature that insect development occurs is referred to as the developmental minimum or developmental threshold value. In addition to a minimum temperature, many insects will also have a threshold for maximum temperatures. These are referred to as developmental optimums, and when they are exceeded, additional degree days are not accumulated (Pedigo and Rice, 2009).

A biofix is often required to start the degree day accumulation. A biofix is simply the first recorded capture or observation of an adult of the species in question.

It is important to note that thermal constants, developmental thresholds (Dev. Thresh) and developmental optimums are species- and location-specific (Table 14.2). Equation [14.1] has several rules, including:

1. No degree days are accumulated if the daily maximum temperature does not exceed the developmental threshold (Pedigo and Rice, 2009). For example, the developmental threshold for the European corn borer is 50°F, so no degree days are accumulated if the maximum and minimum temperatures are 40 and 30°F, respectively.
2. The minimum temperature is set equal to the developmental threshold if the maximum temperature is greater than the developmental threshold, but the minimum temperature is below the developmental threshold (Pedigo and Rice 2009). For the European corn borer (*Ostrinia nubialalis* Hubner), the developmental threshold is 50°F and if maximum temperature is 65°F and the minimum temperature is 40°F, then $50°F$ replaces $40°F$ in the calculation {[(65+50)/2] -50} which results in the accumulation of 7.5 degree days.
3. The maximum temperature is set equal to the developmental optimum if the maximum temperature for a day exceeded this value (Pedigo and Rice, 2009). For the black cutworm (*Agrotis ipsolin* [Hufnagel]), the developmental threshold is 50°F and the developmental optimum is 86°F. For a given day, the maximum temperature was 90°F and the minimum temperature was 70°F. Because the maximum temperature exceeds the developmental optimum, it is replaced with 86°C. Therefore, 26 degree days are accumulated {[(86+70)/2]– 50}.

Estimating Yield Losses

Estimating Yield Loss from Weeds Using a Linear Model

Yield loss estimates are just that, approximations based on the extent of the pest problem and the expected yield. The crop yield is based on crop genetics and interactions between the pest, environment, and management (e.g.,

PROBLEM 14.2.
Estimate the yield loss if the incremental loss per pest is 0.001 bu (weed × acre)$^{-1}$ and the field contains 2000 weeds per acre.

ANSWER:
$$\%YL = I \times D = 0.001 \times 2000 = 2 \text{ bu acre}^{-1}.$$

PROBLEM 14.3.
The weed yellow foxtail [*Setaria pumila* (Poir.) Roem. & Schult] decreases crop yield by 0.1% for each plant present in a square foot (ft^2) area. Calculate the yield reduction if 50 plants ft^2 are uniformly distributed across a field.

ANSWER:
% yield loss = 0.1% × 50 = 5% yield loss
Note that if there were 51 weeds rather than 50, then crop yield loss would be 0.1% greater, or 5.1%.

Table 14. 2. Insect developmental threshold temperatures, developmental optimum, developmental stages, and degree day when the stage is expected to be observed in the field.†

Insect	Developmental threshold (°F)	Developmental optimum (°F)	Developmental stage(s)	DD	Biofix
Black cutworm† [*Agrotis ipsilon* (Hufnagel)]	50	86	egg hatch	310	Trap and/or catch adult
			crop damage	562 to 640	
European corn borer† [*Ostrinia nubilalis* (Hübner)]	50		first spring moth		Trap and/or catch adult
			first eggs	450	
			first summer moth	1400	
			first eggs	1450	
			first egg hatch	1550	
Corn rootworm† (*Diabrotica* sp.)	50		adult beetles	1300	Jan 1
Alfalfa weevil‡ (*Hypera postica* Gyllenhal)	48	87	egg hatch	300	
			first-second instar	301 to 438	
			third-fourth instar	439 to 595	
			pupa-adult	596 to 810	
Corn earworm§ (*Helicoverpa zea* Boddie)	55	92			

† (Delahaut, 2004; Foster, 1986; Murray, 2008; Townsend et. al, 1998).

rainfall and timing, growing degree days [GDD]). For example, corn (Blue River 57H36) planted under weed-free conditions yielded 150 bu acre^{-1} the first year, and 120 bu acre^{-1} in the following year. Both years had identical soil fertility programs and similar rainfall totals. However, during the first month following planting corn growing degree days were higher in year 1 than year 2. Contrast these results to another variety (Blue River 30A57), which yielded 120 bu acre^{-1} in both years.

When estimating yield losses, the crop growth stage and population level, as well as the extent and magnitude of the problem must be known (O'Donovan et al., 2005; Alberta Agriculture and Forestry, 2016). Some pests may devastate the crop, even at low populations, whereas other pests may have very high population densities with little impact to the crop (Table 14.3). If the infestation occurs during a critical crop growth development stage, a low weed density can reduce yields more than if the same density of weeds occurred later in the season. For example, weed presence in corn may severely limit yield if they are not controlled during the weed free period (V1 and V6), however after V12 or 13 weed presence may not reduce yields. Examples with insects typically show similar relationships with greater injury occurring early in the growing season. However, some insects are capable of feeding on the harvestable parts of the crops, which also make them a late season threat.

Pest scientists (weed scientists, entomologists etc.) have conducted numerous experiments to determine yield loss due to different pests at different densities, and sometimes, varying the time when an infestation begins or ends. Values for yield loss can be none (close to unmeasurable or very low) to 100%. In contrast, some plants may retain 50 or 60% of their yield even under very high pest populations, which then becomes the upper limit of the yield loss curve. Knowing the upper boundary of yield loss, helps define yield losses at lower infestation densities.

Table 14. 3. Weed emergence information examples based on growing degree days for the upper Midwest.

Timing†	GDD‡	Weed species	Duration of emergence
Fall or very early before planting	winter annual or early spring	Marestail (*Erigeron canadensis* (L.) Cronquist)	
		White cockle (*Silene latifolia* Poir.)	
		Field pennycress (*Thlaspi arvense* L.)	
		Shepard's purse (*Capsella bursa-pastoris* (L.) Medik.)	
early spring before corn planting	< 150	Giant ragweed (*Ambrosia trifida* L.)	2 to 3 wk
		Common lambsquarters (*Chenopodium album* L.)	4 to 6 wk
		Pennsylvania smartweed (*Polygonum pensylvanicum* L.)	4 to 6 wk
		Common sunflower (*Helianthus annuus* L.)	4 to 6 wk
at corn planting	150 to 300	Woolly cupgrass [*Eriochloa villosa* (Thunb.) Kunth]	2 to 3 wk
		Common ragweed (*Ambrosia artemisiifolia* L.)	4 to 6 wk
		Velvetleaf (*Abutilon theophrasti* Medik)	8 to 10 wk
		Giant foxtail (*Setaria faberi* Herm.)	8 to 10 wk
beginning of soybean planting	250 to 400	Yellow foxtail [Setaria pumila (Poir.) Roem. & Schult.]	4 to 6 wk
		Black nightshade (*Solanum nigrum* L.)	4 to 6 wk
		Common cocklebur (*Xanthium strumarium* L.)	4 to 6 wk
		Wild proso millet (*Panicum miliaceum* L.)	4 to 6 wk
after corn emergence	> 350	Large crabgrass (*Digitaria sanguinalis* (L.) Scop.)	4 to 6 wk
		Fall panicum (*Panicum dichotomiflorum* Michx.)	4 to 6 wk
		Waterhemp (*Amaranthus rudis* Sauer)	8 to 10 wk
		Morningglory (*Ipomoea* spp.)	8 to 10 wk

† Based on http://weeds.cropsci.illinois.edu/extension/Other/WeedEmergePoster.pdf

‡ Base temperature = 48°F

PROBLEM 14.4.

Common ragweed (*Ambrosia artemisiifolia* L.) reduces crop yield by approximately 5% per plant in each ft^2 area. If common ragweed has a density of 25 plants ft^2 in a 10-acre lowland water way, and it has a density of 3 plants ft^2 in a 3-acre upland area, what is the yield loss per acre in each area?

ANSWER:

Upland % yield loss = 5% YL/plant × 3 plants ft^2 = 15% yield loss
Lowland % yield loss = 5%YL/plant ×25 plants ft^2 = 125% yield loss.

This calculation suggests that in the lowland area, a loss of greater than 100% is expected.   Clearly this is not possible, and this illustrates one of the limitations of a linear model.   There is an upper bound to the amount of yield lost due to a pest infestation and the maximum loss is 100%.

Yield losses can be calculated with linear and nonlinear models. If a linear model is used, the assumption is that every pest has a similar impact on yield. The linear model is expressed as,

% yield loss = b + incremental loss of each pest × pest density (or %YL = $I \times D$).

In this model, b is the y-intercept of the equation. In many situations, it is assumed that this value is zero. In Fig. 14.1 the pest density and yield loss are measured. The yield loss is different between the yield at zero pests and the measured yield at each pest density. For example, if the yield at zero pests is 200 bu acre^{-1} and the yield at 100 pests acre^{-1} is 180 bu acre^{-1} then the yield loss and % yield loss are 20 bu acre^{-1} and 10% [(20/200)×100%]. Maximum yield loss is generally less than 100%, because as pest density increases, the pests may begin to compete with themselves (intraspecific interference) and cause somewhat less damage to the crop. At low pest densities, the relationship between pests and damage is often linear.

Yield Losses Due to Weeds Using a Nonlinear Model

To limit the maximum yield loss, Cousens (1985) suggested a hyberbolic model that calculates the maximum yield loss (A) when weed densities approach infinity and an incremental yield loss value (I) when weed density nears zero (Fig. 14.2). The hyperbolic model equation is:

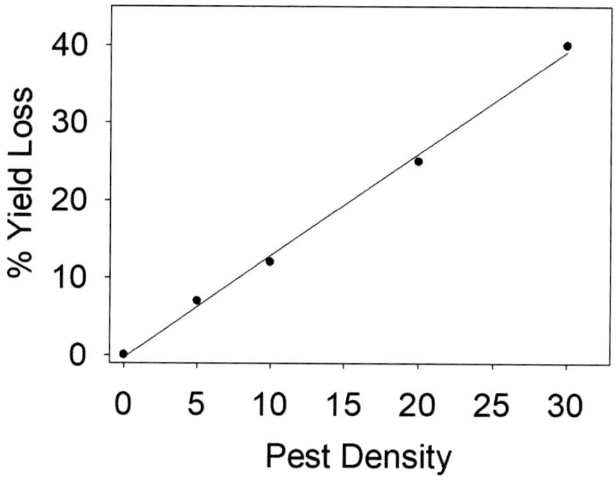

$$\text{Yield loss} = \frac{I \times D}{\left(1 + \frac{I \times d}{A}\right)} \quad [14.2]$$

Fig. 14.1. Example of pest density vs. % yield loss (YL). As density increases, the yield loss increases incrementally. The slope of the line can be calculated, and for each additional pest, the amount of yield lost is equal to the slope.

Where, I is the incremental YL (% per weed/ft^2) for the weed, D is the estimated density of the weed, and A (for asymptote) is the maximum percentage of yield loss at high weed densities (Cousens, 1985). Note that the term in the numerator ($I \times D$) is the same term that was used for the linear regression model discussed above. In this equation, a denominator also is included. As D decreases, the denominator values approaches 1, where yield loss is described by the $I \times D$ term.

Selected I and A values can be found in scientific journals. However, for the vast majority of weeds these values have not been defined. In addition, these values may not be constants and may change with the environment. For example, the incremental yield loss (competitiveness) index (I) for common ragweed in low-moisture soils is 0.5 whereas at

PROBLEM 14.5.

There are 50 yellow foxtail plants ft^2 uniformly distributed across a field. Calculate the yield loss using an I value of 0.1 per weed and a D value 50 plants ft^2. The only other value needed is the A, the maximum yield loss due to high values of yellow foxtail. These values may be obtained from published literature or 'best guesses' can be used. Based on past studies, a reasonable A value is 45% (this means that even at very high densities the maximum yield loss expected is 45%).

ANSWER:
To solve the problem with the above values:

$$\text{Yield loss} = \frac{I \times D}{\left(1 \times \frac{I \times D}{A}\right)} = \frac{0.1 \times 50}{\left(1 + \frac{0.1 \times 50}{45}\right)} = \frac{5}{1.11} = 4.5\%$$

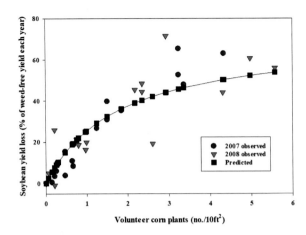

Fig. 14.2. Hyperbolic model example with soybean [*Glycine max* (L.) Merr.] yield loss varying with corn (*Zea mays* L.) density. When density is low (between 0 and 20 plants ft²) there is a near linear relationship with yield loss, whereas at high densities, yield loss reaches a constant or asymptote of about 50% (Modified from Alms et al., 2016).

PROBLEM 14.6.

This problem is similar to 14.4 but here the hyperbolic model is used and different maximum yield loss (*A*) values are included. Common ragweed reduces yield by about 5% per plant in each ft² area. The density of common ragweed is 3 plants per ft² in 3 acres of drier upland areas and in 10 acres of lower, wet areas, there are 25 plants per ft². Literature reports that the maximum yield loss (*A*) is between 45 and 80%. Calculate the yield loss in each area. In this example, two values for *A* are given. An optimist may use the lowest value (hope for the least interference and the greatest yield) whereas a pessimist would probably choose the highest *A* value (i.e., more weed interference and lowest yields). In this example, the average yield maximum yield loss is used, and the yield loss using the highest and lowest *A* values for the low wet areas is shown.

ANSWER:
To calculate the average *A* value:

$$A=\frac{(45+80)}{2}=62.5\%$$

$$\text{Upland yield loss}=\frac{I\times D}{\left(1+\frac{I\times D}{A}\right)}=\frac{3\times5}{\left(1+\frac{3\times5}{62.5}\right)}=\frac{15}{1.24}=12\%$$

$$\text{Lowland yield loss}=\frac{I\times D}{\left(1+\frac{I\times D}{A}\right)}=\frac{25\times5}{\left(1+\frac{25\times5}{62.5}\right)}=\frac{125}{1+2}=41.7\%$$

Using an average maximum yield loss reduced the upland loss from 15% (in Example 14.4) to 12%. However, interpretation in the lowland, while still a large yield loss, was well below 100% yield loss. Now solve the equation using *A* values of 80% and 45% for the lowland weed density.

$$A=80\% \qquad \text{Yield loss}=\frac{I\times D}{\left(1+\frac{I\times d}{A}\right)}=\frac{25\times5}{\left(1+\frac{125}{80}\right)}=\frac{125}{1+1.56}=48.8\%$$

$$A=45\%$$

$$\text{Yield loss}=\frac{I\times D}{\left(1+\frac{I\times D}{A}\right)}=\frac{25\times5}{\left(1+\frac{125}{45}\right)}=\frac{125}{1+2.77}=33\%$$

moderate to high moisture contents this value may be 0.15 (Maryland Cooperative Extension and Illinois Cooperative Extension values) (Table 14.4).

In most cases, weed loss calculations only examine a single species and do not account for weed complexes (Swinton et al., 1994). In many fields, complexes of yellow foxtail, kochia (*Bassia scoparia* (L.) A.J. Scott), common lambsquarters (*Chenopodium album* L.), or other weeds may exist. Florez et al. (1999) proposed that a visual estimate of relative ground cover could be used to estimate yield loss. The value of relative cover (RC) takes the place of density (*D*) in Eq. [2].

$$RC = \% \text{ cover by weeds}/(\% \text{ cover by crop} + \% \text{ cover by weeds}).$$

The yield loss equation is then:

$$YL = RC \times I/(1+((RC*I)/A)$$

The unknowns remain the *A* (maximum yield loss) and *I* (incremental yield loss) values. As above, these values may not be available, but they can be estimated.

Another method is to use a 'competitive load' estimate for yield loss (Gherekhloo et al., 2010). In this method, the weed density for the most competitive weed is added to the relative densities of the other weed species present. The other weeds species are multiplied by the proportion of their incremental yield loss divided by the incremental yield loss of the most competitive weed species. This is the Total Competitive Load (TLC). The equation is:

Problem 14.7.

This problem uses the value of relative cover (RC) in place of the weed density. In a field of cotton, the total ground cover including crop is 60%, weeds are estimated to make of 30% of the cover. The weeds present have an average competitive index (*I*) of 0.4. Maximum yield loss (*A*) is 80%. Estimate the yield loss.

The relative cover (RC) = 30/60 = 50 relative cover

Answer:

$$\text{Yield loss} = \frac{0.4 \times 50}{\left(1 + \dfrac{0.4 \times 50}{80}\right)} \times 100\% = \frac{20}{(1+0.25)} \times 100\% = 16\%$$

Problem 14.8.

Determine the (Economic Injury Level) EIL if the gain threshold is 4 bu/acre and each insect reduces yield 0.25 bu acre^{-1}.

$$EIL = \text{Gain Threshold(GTH)}/YL \text{ per pest}$$
$$EIL = (4 \text{ bu acre}^{-1})/(0.25 \text{ bu acre}^{-1} \text{ per pest}) = 16 \text{ pests per acre}$$

Answer:

From the literature it was possible to determine that each individual of 'insect A' causes yield loss of 0.25 bushels/acre. The EIL for 'insect A' is then 16 insects per unit of area.

Problem 14.9.

For "insect A" the cost of management (*C*) is $10 per acre, the market value of the crop (*V*) is $8 per bushel, the injury caused per pest per acre (*I*) is 48 and the damage per unit of injury (*D*) is 0.05. (*I*) was based on scouting the field and examining one square yard (1 yd^2 = 0.00021 acre), and knowing that each pest causes 0.1% defoliation per acre. The management tactic is 100% effective (*K*=1)

Answer:

$$EIL = \frac{C}{V \times I \times D \times K} = \frac{10}{8 \times 0.25 \times 0.05 \times 1} = 4.87 \text{ insects}$$

The EIL for this problem is five insects per sq yd, which indicates that it will require a population of five "insect A" per yd^2 before observed yield loss is equal to the GTH.

Table 14.4. Competitive Index Factor in high and low moisture environments for various weed species. (Modified from NE-IPM Module #10. Weed management in row crops, original information for low moisture environments from Maryland Cooperative Extension service and high moisture environments from Illinois Cooperative Extension Service).

Weed†	Low moisture	High moisture	Weed	Low moisture	High moisture
Barnyardgrass (*Echinochloa crus-galli* (L.) P. Beauv.)	0.125	0.06	Common ragweed (*Ambrosia artemisiifolia* L.)	0.5	0.15
Crabgrass species [*Digitaria* sp.]	0.1	0.05	Giant ragweed (*Ambrosia trifida* L.)	1.25	0.6
Fall panicum (*Panicum dichotomiflorum* Michx.)	0.125	0.06	Jimsonweed (*Datura stramonium* L.)	0.8	0.3
Giant foxtail, yellow foxtail, green foxtail (*Setaria* spp.)	0.125	0.06	Morning glory species (*Ipomoea* spp.)	0.5	0.25
Johnsongrass [*Sorghum halepense* (L.) Pers.]	0.25	0.125	Pigweed species (*Amaranthus* sp.)	0.5	0.15
Shattercane [*Sorghum bicolor* (L.) Moench subsp. *arundinaceum* (Desv.) de Wet & Harlan]	0.25	0.125	Prickly sida (*Sida spinosa* L.)	0.125	0.05
Black nightshade (*Solanum nigrum* L.)	0.25	0.1	Smartweed (*Polygunum* sp.)	0.5	0.25
Burcucumber (*Sicyos angulatus* L.)	0.5	0.25	Spurred anoda (*Anoda cristata* (L.) Schlecht.)	0.5	0.25
Common cocklebur (*Xanthium strumarium* L.)	1.25	0.6	Velvetleaf (*Abutilon theophrasti* Medik.)	0.8	0.25
Common lambsquarters (*Chenopodium album* L.)	0.5	0.15			

† For most grass species, density is based on tiller number if low densiies present, and plant number at higher densities. For example, green foxtail at low densities may be one plant but have five tillers, whereas at high densities, there may be single plants and no tillers.

$$\text{TLC} = w_i + (b_{cw1}/b_{wi})\, w_1 + (b_{cw2}/b_{wi})\, w_2 + \ldots. + (b_{cwn}/b_{wi})\, w_n$$

where w_i is the density of the most competitive weed, b_{wi} is the *I* value for the most competitive weed, w_1 through w_n are the densities of the other weeds present, and b_{cw1} through b_{cwn} are the *I* values for each weed in the complex. The yield loss equation is:

$$\text{YL} = (b_{wi} * \text{TLC})/(1 + (b_{wi} * \text{TLC})/A).$$

In this case, A is the maximum yield loss for the most competitive species (w_i).

Calculating the Economic Injury level for insects

To determine the economic impact of a pest on a crop the gain threshold (GTH) should be calculated. It defines the management cost relative to expected selling price. The gain threshold in bu acre⁻¹ is calculated with the equation,

$$\text{GTH} = \frac{\text{Management cost (price per acre)}}{\text{Market value (price per bu)}} = \frac{\text{bushels}}{\text{acre}}$$

For example, if the cost of management is \$30 acre⁻¹ and the market value of the crop is \$3 bu⁻¹, then the gain threshold is 10 bu acre⁻¹ to break even. This value suggests that control is necessary at the point where the expected yield loss ≥ 10 bu acre⁻¹. The next value to be calculated is the economic injury level (EIL). The economic injury level represents the lowest number of pests that are capable of causing yield loss that is equivalent to the gain threshold (or insect management costs). The EIL is expressed as the population of insects (or other pests) present per unit of area.

The EIL represents a pest population that will cause economic damage. When the EIL has been reached, management of the pest is justified because of the potential economic loss that is greater than the GTH. The economic injury level can be calculated using the equation:

$$EIL = \frac{C}{V \times I \times D \times K}$$

[14.3]

Where, EIL is the density or intensity of the pest population (e.g., pests per acre or pests per plant), C is the cost of managing the pest per acre (cost per acre), V is the market value per unit of commodity (cost per bushel), I is injury units per pest per unit of commodity (injury per pest activity per acre) (e.g., I can represent the percent defoliation or removed plants caused by each individual insect per acre), D is the damage per unit of injury (yield loss per acre per pest activity) (e.g., D can represent the yield loss in bushels per acre associated with either defoliation or removed plants, and K is the proportionate reduction in the pest's potential to cause injury or damage (i.e., the effectiveness of the management tactic) (Pedigo and Rice, 2009). The value for D is obtained through research and can be found in the related scientific literature. There are pest such as the European corn borer where K is assumed to be 1 or 100%.

For pests such as aphids and other insects with piercing-sucking mouthparts, it is difficult to estimate EIL because the injury units per insect are difficult to quantify. To estimate EIL's for these insects, the slope of regression analyses (b) between the insect population and the yield loss is used to estimate the $I \times D$ value (i.e., $b = I \times D$). In these cases, the EIL formula is:

$$EIL = \frac{C}{V \times I \times D \times K}$$

$$I \times D = b$$

$$EIL = \frac{C}{V \times b \times K}$$

Where, b is the slope of the linear equation between yield (y-axis) and insect (x-axis) and represents yield loss (bu acre^{-1}) per insect. For example, the southern green stink bug (*Nezara viridula* L.) is known to reduce soybean yields later in the season. The cost of management (C) is $9 acre^{-1}, the market value of the crop (V) is $9 bushel^{-1}, and the injury caused per stinkbug per square yard is (b) 0.037 (0.044 m^{-2}). The management tactic is 95% effective (K = 0.95).

The EIL for the southern green stink bug based on these values is 28 per yd^2. This indicates that there must be 28 southern green stink bugs per yd^2 before economic loss is greater than the GTH.

In some instances, b is used for insects that are feeding within the plant. For example, the European corn borer caterpillar causes injury by feeding within corn stalks. Furthermore, the value of b varies by the 10-leaf and 16-leaf growth stage of the corn. The cost of management (C) is $10 acre^{-1}, the market value of the crop (V) is $3 per bushel, and the injury caused per caterpillar is 1.73 at the 10-leaf stage and 1.25 at the 16-leaf stage. 100% management of the caterpillars is not expected, and the proportionate reduction in injury (K) is 80%.

The EIL for the European corn borer is then two caterpillars per plant at the 10-leaf growth stage and three caterpillars per plant at the 16 leaf growth stage.

Calculating the Economic Threshold

The economic threshold is calculated from the expected loss per individual pest. The economic threshold (ET) is a monetary decision, where the cost of treatment is equal to or less than the expected return. These values are most often calculated from previously determined economic injury level (EIL), which is a measure of number of pests needed to meet the gain threshold.

The ET represents a population of insects that trigger management action, with the goal of preventing those populations from reaching the EIL, that is, where the yield being lost is greater than the GTH. To calculate ETs, an EIL is necessary. The two broad categories of ET are fixed and objective. For a fixed ET, a fixed population of pests that is below the EIL is used to determine when action is necessary. These provide a reaction time to prevent populations from reaching the EIL. Objective ETs take into account the EIL and growth rates of the pest (if these are known).

The ET can also be broken into additional categories that include nominal thresholds, simple thresholds, or comprehensive thresholds. Nominal thresholds are based on experience of the decision maker and they are most

PROBLEM 14.10.

For "insect B" the cost of management (C) is $10 acre^{-1}, the market value of the crop (V) is $8 per bushel, the injury caused per pest per acre (I) is 0.25 and the damage per unit of injury (D) is 0.05. However, 100% of the injury caused by the pest can't be reduced, so the proportionate reduction in the potential yield loss (K) is 70% or 0.7.

ANSWER:

$$EIL = \frac{C}{V \times I \times D \times K} = \frac{10}{8 \times 0.25 \times 0.05 \times 0.7} = 143 \text{ insects}$$

The EIL is 143 insects as 100% management of the pest cannot be achieved and some yield loss has occurred.

PROBLEM 14.11.

A sugar beet (*Beta vulgaris* subsp. *vulgaris* L.) grower has two fields, one is 75 acres and has an estimated 2% YL. The second field is 50 acres and has an estimated 30% yield loss. How much can the grower spend in each area to break even?

ANSWER:

Two additional pieces of information are needed. The first is the expected yields in each portion of the field. The second is the selling price of the sugar beets.

75 acre field: After visiting with the grower, the yield history of the 75-acre field is droughty in dry years, has been determined to have a high yield potential of 35 tons per acre in wet years, and a low of 18 tons per acre of raw sugar beet in dry years. In the 75 acres the high and low yields will provide estimates for the greatest and least amounts that can be spent on control. The agronomist should look at the condition of the crop and soil moisture to help determine if the highest, lowest, or a mid-value yield potential should be used for the following calculations. In the example, the high and low yields will be used for the field.

If low yielding, Economic threshold = 18 tons per acre × (0.02) = 0.36 tons lost per acre at $66 per ton, cost of control could be up to (0.36 tons × $66 per ton) = $24 acre^{-1} to break even on control versus return.

If high yielding, economic threshold = 35 tons acre^{-1} × (0.02) = 0.7 tons lost per acre at $66 ton^{-1}. The breakeven point for this higher yield is $46 per acre. The 50 acre field: The 50-acre field has been producing 25 tons per acre consistently. The raw value per ton of sugar beet for fall delivery is $66. The expected loss is 30%, resulting in control costs that could be as high as (25×0.3×66) $495 per acre.

PROBLEM 14.12.

Now, let's examine on how the selling price affects the breakeven point. Before the crop is sold to the sugar processing plant, the price of the sugar beets decreases from $66 per raw ton to $46 per raw ton. What is the breakeven value for the 75 acre field where 18 tons acre^{-1} are expected to be harvested? Yield loss is 2%.

ANSWER:

$$18 \text{ ton} \times 0.02 \times \$46 \text{ ton}^{-1} = \$16.56.$$

widely used thresholds. These thresholds are based on historical observations. Simple thresholds are based off of calculated EILs, while comprehensive thresholds are rarely used due to their complexity (Pedigo and Rice, 2009). Table 14.1 contains examples for insects with calculated economic thresholds at different plant growth stages.

Choosing a Management Tool and Calculating Returns

The input costs associated with pest control vary with the control strategy (hand-labor at $260 per acre vs. mechanical cultivation at $20 per acre). Based on a USDA-ERS report (http://www.ers.usda.gov/media/943070/sb974–8.pdf) for the 2000 crop year, the average amount spent on chemicals (other than fertilizer) for sugar beet production in the Great Plains region was $77.68 acre^{-1} and $88.64 acre^{-1} for Northwest region. In addition, custom operations (assuming 50% of the total operation was for pest control) was an additional $17.50 per acre for the Great Plains and $25.20 in the Northwest. Yield in the Great Plains averaged 20.3 tons per acre and yield in the Northwest was 28.1 tons per acre. Based on these values, the yield loss in the Great Plains and Northwest were 15% and 11%, respectively. The assumptions with the above example is that ALL of the yield loss is recoverable with treatment.

In many cases this is not true, and therefore the effectiveness of the control strategy must be consideration. For example, if only 2.5 out of 3 tons is recovered, then the net effectiveness in 83% (100 × 2.5/3).

Estimation of After Harvest Crop Loss: Tuber Losses to Weevils

Sometimes, crop losses often are estimated after harvest. This can be done by determining yield from an undamaged area and comparing with yield from damaged areas (Sandifolo, 2003). In this example, a percent weight loss of a potato (*Solanum tuberosum* L.) crop to a weevil infestation is estimated. The % weight loss can be calculated by knowing (i) weight of clean tubers (CT); (ii) number of clean tubers (#CT); (iii) weight of damaged tubers (DT); and (iv) number of damaged tubers (#DT). The equation is:

$$\% \text{ weight loss} = \frac{(CT \times \#CT) - (DT \times \#DT)}{CT[CT \times (\#CT + \#DT)]} \times 100\%$$

Summary

In summary, the ability to estimate the impact of pests on crop yield is based on accurate scouting reports and previous research that has defined the relationship between the pest and the yield. Based on the information provided above, these examples illustrate several points, including that realistic estimates for:

1) Yield,

2) Yield loss,

3) Selling price,

4) Price of control

are needed to make decisions that are economically sensible. The lower the yield estimate, the lower the dollar amount lost and the less that can be spent to break even. As yield losses increase, the potential

PROBLEM 14.13.

Based on a sugar beet season-average price of \$31.65 ton^{-1} for the Great Plains and \$37.38 ton^{-1} for the Northwest, calculate the yield loss from pests needed to have a breakeven scenario in each region.

ANSWER:

Great Plains = cost of control = 77.68 + 17.50 = \$95.18
\$95.18/31.65 ton^{-1} = 3 tons acre^{-1} would pay for the chemicals
% YL = 3/20.3 (average yield) = 14.8% yield loss.

If yield losses were less than about 15%, control would be valued more than the amount of yield recovered. In the Northwest control costs = 88.64 + 25.20 = 113.64

113.38/37.38 = 3.04 tons lost per acre would pay for control
% YL = 3.04/28.1 = 10.8% yield loss.

PROBLEM 14.14.

In an area, 1000 clean potato tubers and 500 damaged tubers were harvested. The weight of clean tubers averaged 32 oz., whereas the weight of damaged tubers averaged 26 oz. Calculate the % weight loss.

ANSWER:

$$\% \text{ weight loss} = \frac{(CT \times \#CT) - (DT \times \#DT)}{CT[CT \times (\#CT + \#DT)]} \times 100\%$$

$$\% \text{ weight loss} = \frac{(32 \times 1000) - (26 \times 500)}{[32 \times (1000 + 500)]} \times 100\% = \frac{32000 - 13000}{48000} \times 100 = 39\%$$

amount that can be spent on control measures increase.

Once the yield losses, the effectiveness of the control strategy, and the estimated value of the product at harvest are estimated, economic-based decisions can be developed. However, these decisions do not consider the long-term ramifications of not controlling the pest.

Other suggested references include Deneke (2016), Hunt et al. (2009), VanGessel (2017), and Varenhorst et al. (2016).

ACKNOWLEDGMENTS

Support for this document was provided by South Dakota State University, Precision Farming Systems community in the American Society of Agronomy, International Society of Precision Agriculture, and the USDA-AFRI Higher Education program (2014-04572).

Additional Problems

14.15. Calculate the growing degree days and the total GDD for each case.

	Min	Max		Case 1 Min temp = 50	Case 2 Min temp = 40	Case 3 Biofix base temp = 55
day 1	61	80				
day 2	57	82				insect observed
day 3	45	64				
day 4	38	51				
day 5	44	53				
day 6	56	63				
day 7	60	71				
			total GDD			

14.16. There are 40 barnyard (*Echinochloa crus-galli* (L.) P. Beauv.) grass plants ft^{-2} uniformly distributed across a field. Calculate the yield loss using an *I* value of 0.2 per weed and a *D* value 40 plants ft^{-2}. Based on past studies, a reasonable estimate for the *A* value is 30%.

14.17. Determine the EIL if the gain threshold is 10 bu acre^{-1} and each insect reduces yield 0.15 bu acre^{-1}.

14.18. For "insect A" the cost of management (*C*) is $6 acre^{-1}, the market value of the crop (*V*) is $3.50 per bushel, the injury caused per pest per acre (*I*) is 0.3 and the damage per unit of injury (*D*) is 0.05. However, 100% of the injury caused by the pest can't be reduced, so the proportionate reduction in the potential yield loss (*K*) is 80% or 0.8. Calculate the Economic Injury Threshold for insect A.

14.19. Common lambsquarter is found at an average density of 5 plants ft^{-2} and Russian thistle (*Kali tragus* (L.) Scop.) was found at an average density of 8 plants ft^{-2}. The competitive index (*I*) for common lambsquarter is 0.25 and Russian thistle 0.6. The greatest yield loss expected is 60%. Use the competitive load to calculate the percent yield loss.

14.20. The cost of control is $11.75 acre^{-1}. The estimated yield loss is 4% on field that averages 60 bu acre^{-1}. What does the selling price of the commodity need to be to break even?

REFERENCES

Alberta Agriculture and Forestry. 2016. Economic thresholds for insects attacking cereals and corn. http://www.agric.gov.
    ab.ca (accessed July 2016).

Alms, J., S.A. Clay, D. Vos, and M. Moechnig. 2016. Corn yield loss due to volunteer soybean. Weed Sci. 64:495-500. doi:10.1614/WS-D-16-00004.1

Bridges, D. 1992. Crop losses due to weeds in the United States. Weed Sci. Soc. Amer. Champaign, IL.

Clay, S.A. 2016a. Chapter 39: Selected broadleaf weeds in South Dakota corn fields. In: D.E. Clay, C.G. Carlson, S.A. Clay, and E. Byamukama, editors, iGROW corn: Best management practices. South Dakota State University. Brookings, SD.

Clay, S.A. 2016b. Identification of South Dakota grass and grasslike weeds of importance Chapter 40. In: D.E. Clay, C.G. Carlson, S.A. Clay, and E. Byamukama, editors. iGROW corn: Best management practices. South Dakota State University.

Clay, S.A., D.E. Clay, D. Horvath, J. Pullis, C.G. Carlson, S. Hansen, and G. Reicks. 2009. Corn (Zea mays) responses to competition: Growth alteration vs limiting factor. Agron. J. 101:1522–1529. doi:10.2134/agronj2008.0213x

Cousens, R.D. 1985. An empirical model relating crop yield to weed and crop density and a statistical comparison with other models. J. Agric. Sci. 105:513-521.

Delahaut, K. 2004. Scouting for vegetable pests. University of Wisconsin Madison, College of Agricultural and Life Sciences. Madison, WI. http://www.cias.wisc.edu/wp-content/uploads/2008/07/scouting.pdf (verified 14 June 2017).

Deneke, D. 2016. Chapter 3: IPM solutions to pest management for corn production. In: D.E. Clay, C.G. Carlson, S.A. Clay, and E. Byamukama, editors, iGROW Corn: Best management practices. South Dakota State University. Brookings, SD.

Deneke, D., and P.O. Johnson. 2016. Chapter 3: IPM solutions to pest management for corn production. In: D.E. Clay, C.G. Carlson, S.A. Clay, and E. Byamukama, editors, iGROW corn: Best management practices. South Dakota State University. Brookings, SD.

Egley, G.H., and R.D. Williams. 1990. Decline of weed seeds and seedling emergence over five years as affected by soil disturbance. Weed Sci. 38:504–510.

Florez, J., A.J. Fischer, H. Ramirez, and M.C. Duque. 1999. Predicting rice yield losses caused by multispecies weed competition. Agron. J. 91:87–92. doi:10.2134/agronj1999.00021962009100010014x

Foster, D.E. 1986. Scouting for and managing the alfalfa weevil. Iowa State University Extension. IC-428. http://publications.iowa.gov/19534/ (verified 14 June 2017).

Gherekhloo, J., S. Noroozi, D. Mazaheri, A. Ghanbari, M.R. Ghannadha, R.A. Vidal, and R. de Prado. 2010. Multispecies weed competition and their economic threshold on the wheat crop. Planta Daninha 28:239–246. doi:10.1590/S0100-83582010000200002

Government of Saskatchewan. 2016. Economic thresholds of insect pests. Government of Saskatchewan. http://www.saskatchewan.ca/business/agriculture-natural-resources-and-industry/agribusiness-farmers-and-ranchers/crops-and-irrigation/crop-protection/insects/economic-thresholds-of-insect-pests (accessed 1 July 2016).

Hunt, T.E., R.J. Wright, and G.L. Hein. 2009. Economic thresholds for today's commodity values. Adapted from Proceedings for the University of Nebraska–Lincoln Crop Production Clinics, Lincoln, NE. January 2009. Cropwatch, University of Nebraska-Lincoln. Lincoln, NE. p. 93-96.

Kieckhefer, R.W., and B.H. Kantack. 1980. Losses in yield in spring wheat in South Dakota caused by cereal aphids. J. Econ. Entomol. 73:582–585. doi:10.1093/jee/73.4.582

Murray, M.S., 2008. Using degree days to time treatments for insect pests. Utah Pests Fact Sheet. Utah State University Extension, Logan, UT. https://climate.usurf.usu.edu/includes/pestFactSheets/degree-days08.pdf

O'Donovan, J.T., R.E. Blackshaw, K.N. Harker, G.W. Clayton, and D.C. Maurice. 2005. Field evaluation of regression equations to estimate crop yield losses due to weeds. Can. J. Plant Sci. 85:955–962. doi:10.4141/P05-041

Pedigo, L.P., and M.E. Rice. 2009. Entomology and pest management, 6th ed. Pearson Education Inc., Saddle River, NJ.

Sandifolo, V.S. 2003. Paper 15: Estimation of crop losses due to different causes in root and tuber crops: The case of Malawi. Proceedings of the Expert Consultation on Root Crop Statistics, Volume II: Invited Papers. Harare, Zimbabwe, 3-6 December 2002. Food and Agriculture Organization of the United Nations Statistics Division and regional office for Africa FAO, Rome, Italy.

Strunk, C. L., and E. Byamukama. 2016. Corn diseases in South Dakota and their management. Chapter 47. In: D.E. Clay, C.G. Carlson, S.A. Clay, and E. Byamukama, editors, iGrow Corn: Best management practices. South Dakota State University. Brookings, SD.

Swinton, S.M., D.D. Buhler, F. Forcella, J.L. Gunsolus, and R.P. King. 1994. Estimation of crop yield loss due to interference of multiple weed species. Weed Sci. 42:103–109.

Toole, E.H., and E. Brown. 1946. Final results of the Duvel buried weed experiment. J. Agric. Res. 72:201–210.

Townsend, L., R. Bessin, and D. Johnson. 1998. Predicting insect development using degree days. ENTFACT-123. University of Kentucky. http://entomology.ca.uky.edu/ef123 (verified 19 June 2017).

VanGessel, M. 2017. Weed management in row crops: Application to corn production. NE-IPM Model #10. http://www.northeastipm.org/saremod/weedmgmt.pdf (verified 25 July 2017).[2017 is year accessed]

Varenhorst, A.J., M.W. Dunbar, B. Fuller, and B.W. French. 2016. Chapter 51. Corn insect pests. In: D.E. Clay, C.G. Carlson, S.A. Clay, and E. Byamukama, editors, iGrow corn: Best management practices. South Dakota State University. Brookings, SD.

Determining the Economic Optimum Rate for Second Polynomial Plateau Models

15

Christopher J. Graham, David E. Clay,* and Stephanie A. Bruggeman

Chapter Purpose

The 2nd order polynomial and 2nd order polynomial plateau models can be used to calculate the economic optimum fertilizer and seeding rates using the approach discussed in Chapter 11 (Fausti, 2017). For these models to provide accurate predictions, the equation must be unbiased in the response region where the economic optimum rate is likely to occur. However, solving the polynomial plateau model requires a regression analysis followed by iteration. Iteration is the process of modifying the model parameters in an attempt to obtain a closer approximation of the solution. Some may call iteration a process of trial and error. This chapter provides examples on how to solve these problems.

Key Terms

iteration, bias, genetic capacity, 'defensive hybrids', 'race horse' hybrids, 2nd order polynomial model, 2nd order polynomial plateau model

Mathematical Skills

Using regression analysis, fitting equations using iteration to minimize bias.

Developing a Yield Response Curve

Many different models have been used to define the relationship between agronomic inputs and yield (Cerrato and Blackmer, 1990; Bélanger et al., 2000; Clay et al., 2011). In many situations, yields increase with increasing fertilizer or increasing seeding rates to a maximum value, thereafter yields either remain stable or decrease with additional increases (Fig. 15.1, Nielsen et al., 2016; Reid, 2002). The relationship between inputs and yield is dependent on many factors including the plant's genetic capacity, climatic conditions, soil productivity, and nature of the inputs.

For fertilizer inputs, each nutrient has unique characteristics that directly impact its management. For example, the crop response curve for nitrogen (N) is often different than responses to phosphorus or other nutrients. Based on a unique crop response, locally-based recommendations can be developed (Kim et al., 2013).

Understanding Genetic Capacity

Because cultivars respond differently to N fertilizer and changes in plant population, combining multiple cultivars into a common analysis may produce recommendations that have little meaning (O'Neill et al., 2004; Clay et al., 2009). To minimize these errors, experiments should be grouped into appropriate categories.

For corn, the ability of specific hybrid to respond favorably or negatively to environmental stress conditions are characterized by terms such as *fixed ear*,

South Dakota State University, Brookings, SD 57007. *Corresponding author (david.clay@sdstate.edu)
doi: 10.2134/practicalmath2016.0110

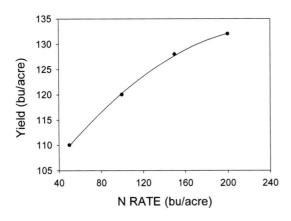 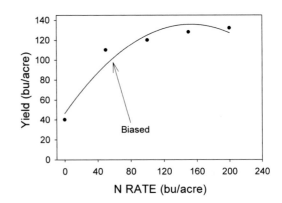

Fig. 15.1. The relationship between N rate and yield as defined by a second order polynomial model. Because the model runs through the data, the model on the left is unbiased. However, in the model on the right, the model under predicts the yield when 40 lb N/acre are applied. When this model is used to predict the economic optimum N rate, the calculated value will not be correct. This is called biased. This bias can be reduced using techniques described below.

flexible ear, racehorse, and *defensive.* In corn hybrids, the term "fixed ear" indicates that the size of the ear is fixed by the plants' genetic capacity, and regardless of the environment, the final ear size is unchanged. The term "flexible ear" indicates that ear size can vary depending on the climatic and soil conditions. Plants with flexible ears have the capacity to reduce the size of the ear with increases in the seeding rate (Clay et al., 2009). "Racehorse" hybrids are those that will take advantage of optimum growing conditions, and may have characteristics of a flexible ear. "Defensive" hybrids are those that produce "good" yields under less than optimum conditions (Kleinjan, 2016).

Plants may also be characterized according to length of the growing period for it to reach maturity. For example, a company may offer 100, 110, and 120-d hybrids. However, the maturity rating does not indicate the exact number of days the plant will take to mature. The rating system provides a *relative* value for assessing the suitability of a hybrid for a given environment. Soybean also has a relative rating system with 'groups' that range from 000 (grown in areas with a short growing season) to X (grown in areas with long growing seasons) Each major division is further subdivided into 10 subdivisions.

The lower the maturity rating, the shorter the growing season needed to reach harvest maturity. Unfortunately, the definition of relative-day maturity differs from company to company, and therefore this definition has value only within a company's product line.

Another category used to define the plant's genetic capacity are *indeterminate* and *determinate* growth. For soybeans, "indeterminate" growth represents continuous vegetative growth after flower initiation, whereas for "determinate" growth, vegetative growth is completed when flowering begins. In the United States, most soybean varieties released for central and northern environments (maturity group < IV) have indeterminate growth, whereas farmers in the Southern United States may seed soybean cultivars with determinate growth characteristics.

Other terms used to describe plant growth characteristics are cool season, warm season, C_3 and C_4, tillering potential, and day length (long-day plants, short-day plants, or day-neutral plants). "Cool season" plants grow best when the temperatures are cool. Examples are wheat, lettuce, spinach, carrots, peas, and radishes. "Warm season" plants grow best when the air and soil temperatures are warm. Examples include corn, tomatoes, and squash. "C_3" and "C_4" refers to the photosynthetic pathway and the reactions used to fix CO_2 into glucose. "Tillering potential" refers to the number of side shoots produced by the main stem of a grass crop like wheat or rice. Plants with high tillering capacity, such as wheat, are less sensitive to seeding rates than plants with low tillering capacity. Photoperiod is the physiological response of the plant to daylength. "Long-day plants" such as peas, barley, and wheat, require less darkness to initiate flowering. In the northern hemisphere, long-day plants flower when the daylengths are increasing in the spring and early summer. "Short-day plants" such as cotton, rice, and soybeans, flower as the nights grow longer toward the end of summer and fall. "Day-neutral plants" such as cucumbers or tomatoes, do not initiate flowering based on day-length but instead by the age of the plant.

Yield Response Models

In many situations, fertilizer and seeding rates can be defined with a mathematical model where the yield increases with inputs up to a maximum value, and thereafter yield remain constant or decreases with increasing inputs (Figure

15.1). Data for these experiments can be obtained by conducting replicated experiments containing multiple N or seeding rates. Additional information for calculating economic optimum rates is available in Chapter 11 (Fausti et al., 2017).

Two statistical approaches that have been widely used to determine agronomic or economic optimum rates are the second order polynomial model and the second order polynomial plateau model.

Derivation of the second order polynomial model

The second order polynomial model is defined by the equation,

$$Yield = a + b \times input + c \times input^2,$$

where a is the y-intercept or yield when zero inputs were applied, b and c are fitted coefficients of the second order polynomial equation, and the variable *input* refers to the amount of fertilizer applied or number of seeds planted per unit area. In some situations, polynomial models will not adequately define the relationship between the outputs and inputs in the region surrounding the economic optimum rate (Cerrato and Blackmer, 1990)(Fig. 15.1). The problem may be recognized by (or a low R^2 value or a large gap between the measured data and the line). Under these conditions consider an alternative model.

Derivation of the second order polynomial plateau model

This identical approach can be used to determine economic optimum N and seeding rates. In the second order polynomial plateau model, the yield response is defined by two curves. The first curve is the second order polynomial equation that defines the quadratic relationship between yield and inputs (fertilizer or seeding rates). This portion of the curve is defined by the equation,

$$Yield = a + b \times input + c \times input^2$$

where input < critical input value. The second curve is the plateau where additional inputs do not increase yield (Fig. 15.2). The plateau portion of the curve is defined by the equation,

$$Yield = plateau\ yield,\ where\ input \geq critical\ input\ value$$

In this model, the intersection between the plateau and polynomial models is calculated with the equation,

$$Yield\ at\ intersection = a - \frac{b^2}{4 \times c}$$

where, a, b, and c are the coefficients in the second order polynomial equation. Once the intersection between these curves is defined, iteration (trial and error) is used to fine-tune the fit between the observed and predicted findings. For example, if a is 101, b is 0.939, and c is -0.00317, then the intersection of the two models occur at 170.5 (101-[0.9392/{4x(-0.00317)}]). To combine the polynomial and plateau equations, two conditions must be achieved. First, the curves are continuous and they meet at some value of x (x_o) and second, they have identical slopes at the intersection. Mathematically, these conditions are defined as,

$$Yield = a + b \times input + c \times input^2$$

$$\frac{dyield}{dinput} = b + 2 \times c \times input$$

where a, b, and c are the coefficients associated with second order polynomial equation, and b+2cinput is the first derivative of the second order polynomial equation. The solution for yield and the plateau model is determined by solving for input, $\left(input_0 = \frac{-b}{2c}\right)$ and substituting this solution into the second order polynomial equation. The resulting equation is $Yield\ at\ intersection\ (d) = a - \frac{b^2}{4 \times c}$, where a, b, and c are coefficients and d is the point on the y axis where the plateau and polynomial equations intersect (Fig. 15.2). For example, if a is 101, b is 0.939, and c is -0.00317, then the intersection of second order polynomial and plateau models occurs at 170.5 [101-(0.9392/(4×-0.00317))]. Cerrato and Blackmer (1990) used this model to improve calculated economic optimum N rates.

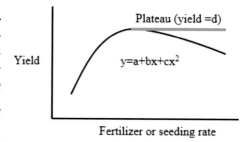

Fig. 15.2. A graphical representation of the second order polynomial plateau model. The value d is defined by the equation, $d = a - \frac{b^2}{4 \times c}$, where a, b, and c are coefficients in the polynomial equation.

PROBLEM 15.1.

What is the equation for the plateau model, if $a = 80$, $b = 1.90$, and $c = -0.0065$?

ANSWER:

For this model, the plateau is the model where yield = 219.

$$d = a - \frac{b^2}{4 \times c} \qquad d = 80 - \frac{1.90^2}{4 \times (-0.0065)} = 219$$

PROBLEM 15.2.

Determine the 2nd order polynomial equation between yield and N rate.

N rate (lb/acre)	0	40	80	120	160	200
Yield (bu/acre)	100	140	170	200	220	221

ANSWER:

$$\text{Yield} = 98.7 + 1.15(\text{N rate}) - 0.00262(\text{N rate})^2$$

(See results in "*Coefficients*" column in last table below).

In Microsoft Excel, this problem was solved using the discussion below.

Step 1: Enter the data.

Step 2. Select Data.

Step 3. Select Data Analysis.

Step 4. Select Regression.

Step 5. Input $C$2:$C$7 into the *y* range and $A$2:$B$7 into the *x* range.

Step 6. Select OK.

This was developed using Analysis ToolPak, following steps 2 through 4 above.

For this program to work, the data must be in form as typed above.

This equation can be derived in Microsoft Excel 2013. However, the equation is determined differently in different version. In Excel 2013, highlight the data, select insert, select recommended charts, click on scatter and select OK. At this point the chart should show, left click on the first point, select *add trendline*, select *polynomial*, select *display equation on chart* and *display R^2 on chart*. The equation can also be derived using regression under Data and Data Analysis. However, to use the Data Analysis program it must be activated. In Microsoft Excel 2013 select *Developer, Add-Ins*, and click on *Analysis ToolPak*, and click on *OK*. Highlight column with yield data for the *Y*-range and the column with fertilizer rate and fert rate^2 for the *x*-range.

Regression Statistics	
Multiple *R*	0.998027
R square	0.996058
Adjusted *R* Square	0.993429
Standard Error	3.911217
Observations	6

ANOVA

	df	SS	MS	F	Significance F
Regression	2	11594.94	5797.47	378.9786	0.000248
Residual	3	45.89286	15.29762		
Total	5	11640.83			

	Coefficients	Standard Error	t Stat	P-value	Lower 95%	Upper 95%	Lower 95.0%	Upper 95.0%
Intercept (a)	98.67857	3.544842	27.83723	0.000102	87.3973	109.9598	87.3973	109.9598
X Variable 1 (b)	1.149554	0.08336	13.79031	0.000825	0.884266	1.414841	0.884266	1.414841
X Variable 2 (c)	-0.00262	0.0004	-6.55566	0.007218	-0.0039	-0.00135	-0.0039	-0.00135

This analysis shows that 99% of the variability was explained by the model. The *y* intercept is 98.68 and the coefficients are 1.15 and -0.00262. The *P*-value (probability) and confidence intervals of these values are also provided. The intercept, *X* variable 1, and X variable 2 are the *a*, *b*, and *c* coefficients used in the equation.

Problem 15.3.

Determine the 2nd order polynomial plateau model that minimizes error, for the following data set.

N rate (lb/acre)	0	30	60	90	120	150	180	210
Yield (bu/acre)	80	130	170	200	220	221	215	210

Step 1. Determine the second order polynomial equation. This data was fit the 2nd order polynomial equation. The resulting model was

$$\text{yield} = 80.5 + 1.8456(\text{N rate}) - 0.00594(\text{N rate})^2$$

Step 2. Determine the intersection between 2nd order polynomial and plateau models, the yield at the intersection is 224 bu/acre ($80.5 - (1.8465^2/(4 \times (-0.00594)))$), and the N rate for a yield at 224 bu/acre is 155 lb N/acre.

Step 3. Determine the predicted values using the 2nd order polynomial and the 2nd order polynomial with plateau, and calculate the difference between the measured and predicted values.
The purpose of the iteration is to minimize the difference between predicted and the measured value as shown in the table above. In this example the difference for the 90 lb N rate is 1.5. If the difference was zero for all of the N rates, the iteration would be complete. Because the difference is relatively large, continue the iteration calculations.

			First iteration			Second iteration		
N rate	N rate²	Measured yield	Yield w/o plateau	Yield with plateau	Predicted– measured	Yield w/o plateau	Yield with plateau	Predicted– measured
0	0	80	80.5	80.5	-0.5	78	78	-2
30	900	130	130.5	130.5238	-0.5238	128	128	-2
60	3600	170	169.95	169.8571	0.1429	167	167	-3
90	8100	200	198.5	198.5	1.5	196	196	-4
120	14400	220	216.45	216.4524	3.5476	214	214	-6
150	22500	221	223.7	223.7143	-2.7143	221	221	0
180	32400	215	220.28	223.8	-8.8	217	221	6
210	44100	210	206.2	223.8	-13.8	204	221	11

Step 4. During the first iteration, decrease the a value from 80.5 to 78. Recalculate the intersection (N rate is 134.4 and yield is 218.8).

Step 5. Determine the difference between the predicted and measured values. The new values for the 90 and 120 lb N rate have switch from a positive value (1.5 and 3.5) to negative values. This would suggest that the corrected a is between 80.5 and 78. Continue the iterations.

Step 6. This process is continued until the selection criteria is achieved. During the iterations, the a, b, and c coefficients are adjusted.

Step 7. The final solution is, polynomial-plateau = $78.5 + 1.9899(\text{N rate}) - 0.0072(\text{N rate})^2$ with a plateau at 216.7 bu/acre and 139 lb N/acre.
There are software packages such as SAS and R that automate the iteration.

Summary

In summary, models can be used to calculate the relationship between inputs and yield. However, the response functions are different for different fertilizers and cultivars. For example, racehorse hybrids may have a different yield response function than defensive hybrids. Software programs such as SAS and R are available to assist in model development (SAS/STAT(R), 2016).

Additional Problems

15.3. Develop a second order polynomial model for the following experiment. For this equation, what is the plateau model?

Seeding rate	Yield
1000 seeds/acre	bu/acre
20	100
25	140
30	160
35	160
40	150

15.4. For the model, yield = 100 + 0.86 (N rate) – 0.004 (N rate)^2, what is the maximum yield predicted by the model? What is the N rate associated with this value?

15.5. What are the benefits of using a second order polynomial plateau model to define the yield response function?

15.6. Would a second order polynomial plateau model improve the fit between the measured and predicted yields for the 24,000 and 27,000 seeding rates for the following experiment? Why? Why not?

plants/acre (1000 seeds)	12	15	18	21	24	27	30	33	36
bu/acre	105	134	153	162	157	175	167	183	188

ACKNOWLEDGMENTS

Support for this document was provided by South Dakota State University, Precision Farming Systems community in the American Society of Agronomy, International Society of Precision Agriculture, and the USDA-AFRI Higher Education Grant (2014-04572).

REFERENCES AND ADDITIONAL READING

Bélanger, G., J.R. Walsh, J.E. Richards, P.H. Milburn, and N. Ziadi. 2000. Comparison of three statistical models describing potato yield response to nitrogen fertilizer. Agron. J. 92:902–908. doi:10.2134/agronj2000.925902x

Cerrato, M.E., and A.M. Blackmer. 1990. Comparison of models for describing yield response to nitrogen fertilizer. Agron. J. 82:138–143. doi:10.2134/agronj1990.00021962008200010030x

Clay, D.E., S.A. Clay, C.G. Carlson, and S. Murrell. 2011. Mathematics and science for agronomists and soil scientists. International Plant Nutrition Institute, Peachtree Corners, GA.

Clay, S.A., D.E. Clay, D.P. Horvath, J. Pullis, C.G. Carlson, S. Hansen, and G. Reicks. 2009. Corn (Zea mays) response to competition: Growth alteration vs limiting factor. Agron. J. 101:1522–1529. doi:10.2134/agronj2008.0213x

Fausti, S., B.J. Erickson, D.R. Clay, and C.G. Carlson. 2017. Chapter 11: Deriving and using an equation to calculate economic optimum fertilizer and seeding rate. In: D.E. Clay, S.A. Clay. S. Bruggeman, editors, Practical mathematics in precision farming. ASA, SSSA, CSSA, Madison, WI.

Kim, K., D.E. Clay, S.A. Clay, G.C. Carlson, and T. Trooien. 2013. Testing corn (Zea mays L.) preseason regional nitrogen recommendation models in South Dakota. Agron. J. 105:1619–1625. doi:10.2134/agronj2013.0166

Nielsen, R.L., J. Lee, J. Hettinga, and J. Camberato. 2016. Yield response of corn to plant population in Indiana. Applied crop production update. Purdue University, West Lafayette, IN.

O'Neill, P.M., J.F. Shanahan, J.S. Schepers, and B. Caldwell. 2004. Agronomic responses of corn hybrids from different eras to deficit and adequate levels of water and nitrogen. Agron. J. 96:1660–1667. doi:10.2134/agronj2004.1660

Reid, J.B. 2002. Yield response to nutrient supply across a wide range of conditions: 1. Model derivation. Field Crops Res. 77:161–171. doi:10.1016/S0378-4290(02)00088-6

SAS/STAT(R). 2016. Example 60.1, Segmented model. SAS Institute Inc. https://support.sas.com/documentation/cdl/en/statug/63033/HTML/default/viewer.htm#statug_nlin_sect033.htm (accessed 21 Sept. 2016).

A Site-Specific Fertilizer Program Assessment Using Soil and Nutrient Removal Benchmarks

16

David E. Clay

Chapter Purpose

Accurate soil nutrient benchmarks are needed to assess the impact of precision farming on soil heath. For site-specific soil fertility assessments, archived soil test results, yield monitor data sets, and benchmark values for grain soil nutrient concentrations can be used to determine nutrient removal and additions at targeted locations across a field. This information can be combined with soil organic matter assessment to evaluate changes in soil health. This chapter provides examples on how to use yield monitor data and soil nutrient concentrations for developing P and K budgets, and conducting organic C and N soil health assessments.

Key Terms

Benchmark, fertilizer efficiency, carbon sequestration, soil health.

Mathematical Skills

Comparison of measured values with appropriate soil nutrient benchmarks. To assess if nutrient additions fall short, meet, or exceed nutrient removal.

Soil Nutrient Assessments

In precision farming, it is important to conduct periodic assessments of the fertilizer program. These assessments can be conducted based on archived soil test results, fertilizer application rates, and yield monitor data (Murrell, 2008). For fertilizer assessments, the soil samples should be collected at the same relative location, at the same relative time (spring vs. fall), and analyzed using the same method by the same laboratory. Following a fixed protocol is important because the reported values are influenced by sampling protocol and laboratory procedures (Vaughan, 1999; Clay et al., 2002; Murrell, 2008). For example, N, P, and K soil test results are often lower following harvest than prior to planting in the spring.

When assessing a fertilizer program, it is important to consider the laboratory accuracy (Vaughan, 1999), which is evaluated by comparing the laboratory results of standards (samples with known values) with the results from other laboratories. Laboratory certification is provided by the Soil Science Society of America North American Proficiency Testing (NAPT) program and a list of certified laboratories is available at http://www.naptprogram.org/. Certification insures that the laboratory has the appropriate systems in place to generate reliable data. Ask your laboratory if it participates in the certification program.

South Dakota State University, Department of Agronomy, Horticulture, and Plant Science, Box 2207A, Brookings, SD 57007-2201.*Corresponding author (david.clay@sdstate.edu).

doi:10.2134/practicalmath2017.0028

Nutrient assessments and budgets are based on comparing changes in the soil test value with fertilizer additions and the amount of nutrient removed by the harvested crop(s). Factors that influence the reliability of these calculations include:

1) Soil test values are often lower in the fall following harvest than the spring
2) Field-moist samples often have lower K concentrations than dried and ground samples
3) Soil variability
4) The soil redox (reduction–oxidation reaction) status, which can change some soil nutrient oxidation states (e.g., Fe), affecting solubility, and/or extraction efficiency.

Yield Measurement

Developing nutrient budgets requires yield measurements. Generally, yield information is obtained from a calibrated yield monitor system. Yield monitor accuracy depends on a yield, grain moisture, and temperature sensors. Information for preparing a yield monitor and associated sensors for the field is available in Franzen and Humburg (2016). The accuracy of the yield monitor data depends on model calibration and an ability to accurately identify the combine location and speed, which is generally measured with differentially corrected global positioning systems (GPS). When determining nutrient budget, erroneous data should be removed from the data set. For example, as the combine slows down, the reported yields increase. Slowing the combine from 4 to 3 mph (1.788 to 1.341 m s^{-1}) can result in a reported yield increase from 200 to 267 bu acre^{-1}. This increase is not real and is an artifact of the grain flowing through the system past the sensor. Increasing the speed has an opposite impact. These artificially high and low points must be removed from the data set because they artificially affect yield values. The process of removing these data is often called "cleaning" the data set.

Determine P and K Budgets

Routine soil testing assesses temporal soil nutrient status, helps determine if a nutrient addition is needed, and provides a suggested level of application for the cropping season. Nutrient budgets determine the balance between nutrient addition and removal over a longer time period, often a cropping rotation. The first step in developing a budget is to collect representative soil samples for a field, in either the fall or spring, but not both, at the beginning of the time period of interest. Additional information on soil sampling methodology is available in Clay et al. (2002, 2007, 2016). Once these samples are collected, they should be stored, shipped to the laboratory, and analyzed using appropriate methods, followed by archiving the information. These data provide the benchmark or reference point against which future soil sampling results will be compared. In addition, the results can be used to set up a nutrient plan for the current and future cropping seasons. For example, if the P concentrations are low, a single high rate of P fertilizer may be applied the first year as P (unlike N) does not leach and can be available for several seasons. If the soil pH is low, applying liming materials may be appropriate.

The second step of estimating a nutrient budget for a crop rotation is to calculate nutrient removal and additions. Nutrient removal can be estimated by multiplying the measured yields times the average nutrient levels in grain and tissue samples (Table 16.1). Nutrient removal must consider all the crops in the rotation, whereas additions must consider all additions over the entire rotation.

At the end of the cropping period, the beginning nutrient concentrations need to be compared to the end concentrations. This will help determine if the nutrient management plan was appropriate or if changes are needed. To get appropriate data for comparison, the same sampling sites and protocols should be followed at the end of the time period and the nutrient status from the first and last sampling compared. These results will help determine the performance of the nutrient plan for the field and rotations. If nutrient concentrations become depleted or enriched, changes to the plan should be developed to better meet the crop needs.

Soil testing laboratories and fertilizer companies may report nutrients differently. For example, in the United States fertilizer grade is reported as the percent of N, P_2O_5, and K_2O, whereas many other counties report fertilizer grade as the percentage of N, P, and K. In addition, many laboratories report NO_3–N, P, and K values. To understand, these values the conversions from one form to another is often required. To convert P to P_2O_5, divide P by 0.436 and to convert K to K_2O divide K by 0.83. In addition, it may be necessary to convert the laboratory reported values from a dry weight basis to a wet weight basis (chapters 12, 13, 17) or, conversely, adjust the moisture content of the harvested crop to obtain values on the same reference point.

Nutrient Inputs

Nutrient inputs are the total amounts of nutrients added by the fertilizer or manure applications (Carlson et al., 2016, Table 10.1). If stover from the previous crop is baled and removed from the field, this is considered a removal, and the amount of nutrients should be calculated and subtracted from additions. If the crop residue remains on the field, it is not considered a removal *nor* an addition, so the nutrients in the residue need not be calculated. Every quantity of fertilizer sold has an associated grade, providing information on the %N, %P_2O_5, and %K_2O.

Nutrient Budgets

The soils yield potential can be reduced if analysis suggests that nutrient mining has occurred (outputs > inputs) (Problem 16.5). If this occurs it may be necessary to increase the fertilizer rate. The potential impact of increasing or decreasing fertilizer rates can be tested by placing side-by-side fertilizer strips (areas with no fertilizer or with different rates of fertilizer) in the field that are harvested using a yield monitor. Yields can then be compared to help decide if and where fertility adjustments are needed. Of course, this method has a few disadvantages including that

Table 16.1. Estimates of nutrient removal of N, P_2O_5 (phosphorus pentoxide), K_2O (potassium oxide), Mg, and S by major annual crops grown in the U.S. Great Plains (Murrell, 2008).

Crop	Plant Part	Unit	N	P_2O_5	K_2O	Mg	S
Corn	Grain	Lb bu^{-1}	0.67	0.35	0.25	0.09	0.08
	Stover	Lb ton^{-1}	16	5.8	40	5	2.6
Soybean	Grain	Lb bu^{-1}	3.3	0.73	1.2	0.21	0.18
	Stover	Lb ton^{-1}	40	8.8	37	8.1	6.2
Wheat	Grain	Lb bu^{-1}	1.2–1.5	0.48–0.57	1.50.29–0.33	0.15	0.1
	Straw	Lb mt^{-1}	15	3.7	29	2	5.4
Oat	Grain	Lb bu^{-1}	0.77	0.28	0.19	0.04	0.07
	Straw	Lb ton^{-1}	12	6.3	37	4	4.5
Sorghum	Grain	Lb bu^{-1}	0.66	0.39	0.27	0.06	0.06
	Straw	Lb ton^{-1}	28	8.3	42	6.1	5.9

PROBLEM 16.1.

Determine the amount of N and P_2O_5 harvested in a 250 bu acre^{-1} corn crop. All the stover is baled and removed after grain harvest. This calculation assumes a harvest index [HI= dry grain/(grain + stover)] = 0.50. A harvest index of 0.5 indicates that for each unit of grain there is an equal amount of stover produced. A harvest index of 0.4 means the ratio between harvested grain and stover is 0.66. Note: (i) A bushel of corn weights 56 lbs, and (ii) the harvest index uses (grain/grain + stover). Therefore, a 250 bu acre^{-1} corn crop produces [(250 bu acre^{-1})×(56 lbs dry grain bu^{-1} grain)×(1 lb stover per lb grain)] =14,000 lbs of dry stover. In Table 16.1, 1 ton of stover contains 16 lb of N and 5.8 lbs of P_2O_5 and each bushel of corn contains 0.67 lb of N and 0.35 lbs of P_2O_5.

ANSWER:

N and P_2O_5 in the grain + stover

$$\text{Corn N} = \frac{250 \text{ bu}}{\text{acre}} \times \frac{0.67 \text{ lb N}}{\text{bu}} + \frac{14,000 \text{ lbs stover}}{\text{acre}} \times \frac{\text{ton}}{2000 \text{ lbs}} \times \frac{16 \text{ lbs}}{1 \text{ ton}} = \frac{279.5 \text{ lbs N}}{\text{acre}}$$

$$\text{Corn } P_2O_5 = \frac{250 \text{ bu}}{\text{acre}} \times \frac{0.35 \text{ lb } P_2O_5}{\text{bu}} + \frac{14,000 \text{ lbs stover}}{\text{acre}} \times \frac{\text{ton}}{2000 \text{ lbs}} \times \frac{5.8 \text{lbs}}{1 \text{ ton}} = \frac{128.1 \text{ lbs}}{\text{acre}}$$

Note: Check your answer by cancelling units and determining if the value is reasonable.

PROBLEM 16.2

Calculate the amount of P_2O_5 and K_2O removed by a two year rotation that produced a 50 bu acre^{-1} soybean crop and 175 bu acre^{-1} corn crop. In this calculation ONLY the grain is removed from the field.

ANSWER:

Pounds of P_2O_5 per acre removed by a 50 bu acre^{-1} soybean crop $= \dfrac{50 \text{ bu}}{\text{acre}} \times \dfrac{0.73 \text{ lbs } P_2O_5}{\text{bu}} = \dfrac{36.5 \text{ lbs } P_2O_5}{\text{acre}}$

Pounds of P_2O_5/acre removed by a 175 bu acre^{-1} corn crop $= \dfrac{175 \text{ bu}}{\text{acre}} \times \dfrac{0.35 \text{ lbs } P_2O_5}{\text{bu}} = \dfrac{61.3 \text{ lbs } P_2O_5}{\text{acre}}$

Total removal is 36.5 + 61.3 = 98 pounds of P_2O_5 per acre
Pounds of K_2O acre^{-1} removed by a 50 bu acre^{-1} soybean crop $= \dfrac{50 \text{ bu}}{\text{acre}} \times \dfrac{1.2 \text{ lbs } K_2O}{\text{bu}} = \dfrac{60 \text{ lbs } K_2O}{\text{acre}}$

Pounds of K_2O/acre removed by a 175 bu acre^{-1} corn crop $= \dfrac{175 \text{ bu}}{\text{acre}} \times \dfrac{0.25 \text{ lbs } K_2O}{\text{bu}} = \dfrac{43.8 \text{ lbs } K_2O}{\text{acre}}$

Total K_2O per acre removed is 104 lb K_2O per acre (60+43.8).

PROBLEM 16.3.

Calculate the total amount of P_2O_5 added to a soil if 60 gal acre^{-1} of ammonium polyphosphate (10-34-0, density = 11.7 lb gal^{-1}) are applied annually for four years.

ANSWER:

$$4 \text{ years} \times \dfrac{60 \text{ gal}}{\text{acre}} \times \dfrac{11.7 \text{ lbs}}{1 \text{ gal}} \times \dfrac{34 \text{ lbs } P_2O_5}{100 \text{ lbs fert}} = 955 \text{ lbs } P_2O_5 \Big/ \text{acre}$$

PROBLEM 16.4.

How much total K_2O has been applied if 100 lb of potassium chloride (KCl) (0-0-62) has been applied for three years?

ANSWER:

$$3 \text{ years} \times \dfrac{100 \text{ lbs}}{\text{acre}} \times \dfrac{62 \text{ lbs } K_2O}{100 \text{ lbs fert}} = 186 \text{ lbs } K_2O \Big/ \text{acre}$$

PROBLEM 16.5A.

Determine the amount of P_2O_5 that has been applied if 120 lb acre^{-1} of monoammonium phosphate (MAP, 11-54-0; Carlson et al., 2016) is applied every other year over a 6-year period.

ANSWER:

6 yrs/2 × 120 lb acre^{-1} = 360 lb of MAP per acre (applied every other year over the six-year period)

PROBLEM 16.5B.

If 360 pound of MAP per acre is applied, how much P_2O_5 was applied per acre during the 6 years?

ANSWER:

360 lb MAP × 0.54 lb per lb of MAP = 194.4 lb of P_2O_5.

PROBLEM 16.5C.

If 120 lb P_2O_5 are removed annually, how much total P_2O_5 was removed during the 6-yr period?

ANSWER:

total P_2O_5 removed = 120 lb yr^{-1} × 6 = 720 lb

PROBLEM 16.5D.

What is the total P_2O_5 removal or gain during the 6 years?

ANSWER:

P_2O_5 added	194.4 lb
P_2O_5 removed	720 lb
Change	-525.6 lb of P_2O_5.

PROBLEM 16.6A.

Determine the K_2O budget after two years if removal is 150 lb K_2O per acre per year and fertilizer additions are 100 lb K_2O per acre per year.

ANSWER:

Removal = 150 lb K_2O acre^{-1} × 2 yr = 300 lb
Addition = 100 lb K_2O acre^{-1} × 2 yr = 200 lb
Nutrient budget = -300 + 200 = -100 K_2O acre^{-1}

These results suggest that mining has occurred.

PROBLEM 16.6B.

Determine the P_2O_5 budget if removal over two years is 160 lb P_2O_5 per acre and additions are 90 lb P_2O_5 acre^{-1}.

ANSWER:

Removal = 160 lb × 2 = 320 lb P_2O_5 acre^{-1}
Addition = 90 ×2 = 180 lb P_2O_5 acre^{-1}
Nutrient budget = -320 +180 = -140 lb acre^{-1}

These results suggest that mining has occurred.

PROBLEM 16.6C.

If over this time period, the soil test P level decreased from 16 to 14 ppm (15 to 14 mg kg^{-1}), what is your recommendation? How about if the soil test P level dropped from 100 to 98 ppm (100 to 98 mg kg^{-1})?

ANSWER:

Whether to increase or decrease the amount of fertilizer applied depends on the expected yield response (see Fig. 15.1 for a yield response curve).

adjustments are always made in future time and there is a degree of uncertainty because yields can vary from one year to the next. This method does allow for comparison at similar field positions. The width of the strips should be the width of the combine head and should be well-marked so that that mixing of treatments can be avoided.

In the past, fertilizer recommendations were based on the plants economic responses to specific nutrients. Agronomists are now asked to consider both production and environmental goals simultaneously. Achieving these goals require that fertilizer Best Management Practices become aligned with the 4-R program (Fixen, 2007). The 4-R program is the application of fertilizers using the right source, at the right rate, at the right time, and at the right place. This basic concept is designed to increase or sustain yields and soil health while minimizing the impact of agriculture on the environment. Worldwide, research is being conducted to achieve these goals simultaneously.

Assessing Changes in Soil Organic C and N

In many agricultural systems, the amount of N fertilizer purchased is an important annual management decision. Many agronomists view N as a short term investment because N not taken up by the crop can be lost within a year to leaching or denitrification. Even though it is viewed as a short term investment, the relative amount of N remaining in the soil from one year to the next may be relatively large (Clay et al., 1990). For example, Sebilo et al. (2013) reported that from 1982 to 2012, about 63% of the ^{15}N labeled fertilizer that was applied in 1982 was taken up by the plant, about 10% of the applied N was leached out of the soil, and about 14% remained in the soil. These results suggest that nitrogen fertilizer has short- and long-term impacts on productivity. The short-term impact of applying N fertilizer is related to providing the N needed for the growing plant, and it may be reported as apparent nutrient recover efficiency [(nutrient uptake fertilized– nutrient uptake unfertilized)/quaintly of applied nutrient] are generally higher for N (at or below 50%), than P (10%) or K (40%) (Baligar et al., 2007).

Understanding the fate of N in agricultural systems is important, as too little N limits plant growth, and excess N (over and above growth requirements) can have an adverse impacts on the environment. Plants obtain N from multiple sources including fertilizer, N fixation, soil organic carbon and manure. Increasing the amount of organic N contained in the soil, as suggested by Sebilo et al. (2013) and Clay et al. (1990), has the potential to improve soil resilience (i.e., the ability to resist or recover from disturbances), and increase both yield and N mineralization potentials.

Long-term changes in productivity may be assessed by calculating changes in soil organic matter (SOM, Combs and Nathan, 1998) or soil organic N (SON). Because SOM and SON are related (Malo et al., 2005) in many soils, SON can be estimated from the SOM value. This relationship is defined by the ratio between the soil organic C and soil organic N (SOC/SON). The SOC/SON ratio is defined by the amount of C in the soil divided by the amount organic N in the soil. For example, if a soil contains 10,000 lb organic C and 1000 lb of organic N, then the C/N ratio is 10 (10,000/1,000).

Worldwide, the C to N ratios have been reported to range from 9 to 18 (Kirkby et al., 2011). However, within a geographical area and cropping system, research suggests that the ratios may be relatively constant. For example, in South Dakota, Malo et al. (2005) reported that over a 75-yr period (1921 to 1996) tillage reduced soil organic C and had a nonsignificant impact on the soil C to N ratio, which averaged 9.5. In central Europe, Anderson and Domsch (1984) reported that the N%, C%, and C to N ratios were, 0.143 ± 0.038, 1.35 ± 0.127, and 10.79 ± 0.323, respectively.

Many commercial soil testing laboratories use the loss on ignition method determine soil organic matter (SOM) (Combs and Nathan, 1998), not soil organic C. SOM value can be converted to SOC using the equation:

$$SOC = SOM \times 0.58.$$

After the SOC is determined, SON can be estimated using an appropriate C/N ratio. This method is based on the assumption that for a given soil and environment, stable ratios exist, although this assumption may be valid in some soils but not others. To estimate soil organic N multiply SOC by an appropriate C to N ratio to estimate soil organic N.

$$\text{C/N ratio} = \frac{\text{lbs SOC}/\text{acre}}{\text{lbs SON}/\text{acre}} = \frac{\text{lbs SOC}}{\text{lbs SON}}, \quad \frac{1}{\text{SOC}} = \frac{\text{lbs SON}}{\text{lbs SOC}}$$

$$\frac{\text{pounds SON}}{\text{acre}} = \frac{\text{pounds SOC}}{\text{acre}} \times \frac{\text{pounds SON}/\text{acre}}{\text{pound SOC}/\text{acre}}$$

For example if SOC = 40,000 lb/acre then SON = 4,000 lb/acre is the C/N ratio is 10.

The N that is contained in the soil organic matter can be made available to plants through microbial mineralization. In this process, organic N is mineralized to ammonium which subsequently can be converted to nitrate (NO_3^-) through the nitrification process.

$$\text{Organic N} \rightarrow NH_4^+ \rightarrow NO_3^-$$

Both ammonium (NH_4^+) and nitrate can be taken up by plants, however nitrate (NO_3^-) can be denitrified under oxygen limited conditions moved out of the root zone with percolating water. The movement of nitrate from surface to subsurface soils is called leaching. The amount of organic N that can be mineralized annually is highly variable and depends on many factors including tillage, soil organic matter content, and climatic conditions (Clay et al., 1990). Generally, increasing soil organic N increases the N mineralization potential.

Soil organic matter assessments provide information about the soil productivity and N mineralization potential. Soils with higher organic matter increase both available water and the soils N mineralization potential (Clay et al., 2014). Soils with a higher N mineralization may require less N fertilizer to optimize yields.

Summary

In summary, soil fertility and organic matter assessments are an important step in creating locally based fertilizer recommendations. These assessments can be conducted by tracking changes in soil nutrient and organic matter levels and by estimating nutrient removal. If removal is greater than additions, then soil test values can decrease. If additions are greater than removal then the soil test values may increase and the amount of fertilizer applied may be decreased. If the soil test level is much greater than the critical level, it is unlikely that adding additional fertilizer will increase yields (Marschner, 1995). The critical nutrient level is often defined as the soil nutrient concentration where the yield is 90% of the maximum yield.

PROBLEM 16.7A.

The soil organic matter in the surface 6 inches increased from 1.25 to 1.4% over 25 years. How much additional C and N are stored in the soil organic matter? Assume the surface 6 in acre^{-1} weighs 2,000,000 lb, the organic matter to organic C is 0.58, and the C/N ratio is 10:1.

ANSWER:

Soil organic matter increased 0.15% (1.4-1.25) over 25 years.
Rewrite 0.15% as 0.0015 lb SOM per lb soil

$$\frac{0.0015 \text{ lbs SOM}}{\text{lbs soil}} \times \frac{2,000,000 \text{ lbs soil}}{\text{acre}} = 3000 \text{ lbs SOM}\Big/\text{acre}$$

Assume 1 lb of SOM contains 0.58 lb SOC

$$\frac{3000 \text{ lbs SOM}}{\text{acre}} \times \frac{0.58 \text{ lbs C}}{1 \text{ lbs SOM}} = 1740 \text{ lbs SOC}\Big/\text{acre}$$

Assume the C/N ratio is 10:1

$$\frac{1740 \text{ lbs SOC}}{\text{acre}} \times \frac{1 \text{ lbs SON}}{10 \text{ lbs SOC}} = 174 \text{ lbs SON}\Big/\text{acre}$$

PROBLEM 16.7B.

If 2% of the organic N is mineralized (organic N $\rightarrow$ NH_4–N) annually, how much N is available if the soil contains 1.4% soil organic matter? In this calculation, SON is soil organic N and SOC is soil organic carbon.

ANSWER:

Assumes the C/N ratio is 10:1, and converts 1.4% to 0.014 lb SOM per lb soil

$$\frac{2{,}000{,}000 \text{ lbs soil}}{\text{acre}} \times \frac{0.014 \text{ lb SOM}}{1 \text{ lbs soil}} \times \frac{0.58 \text{ lbs SOC}}{1 \text{ lb SOM}} \times \frac{1 \text{ lbs SON}}{10 \text{ lbs SOC}} \times \frac{0.02 \text{ lbs mineralized N}}{1 \text{ lbs SON}}$$

$$= 32.5 \text{ lbs SON} \Big/ \text{acre}$$

Check you answer by cancelling units and making sure the answer is reasonable.

Additional Problems

16.8. Soil samples are collected from two landscape positions (summit and footslope) in the spring of 2008 and fall of 2011. In the summit area, soil test P has increased from 20 to 25 ppm (20 to 25 mg kg⁻¹), whereas in the footslope area, soil test P has decreased from 20 to 19 ppm (20 to 19 mg kg⁻¹). Over this time period, DAP, was applied annually at a rate of 100 lb acre⁻¹ and the footslope yields were 200, 220, and 210 bu acre⁻¹. In the summit, yields were 110, 120, and 140 bu acre⁻¹. Develop P_2O_5 and K_2O budgets for both landscape positions. Given these budgets would you change your recommendations? In these calculations, ppm is µg P_2O_5 per gram of soil or m P_2O_5 per kg of soil. In this solution, only consider the change in the amount of extractable P. Decreases suggest that removal is greater than additions.

16.9. If P removal is 300 lb P acre⁻¹ and P additions are 150 lb P acre⁻¹ and soil test P has decreased from 17 to 15 ppm (17 to 15 mg kg⁻¹), what is your recommendation? Why?

16.10. The soil C to N ratio is 10:1. How much N is contained in the surface soil if it contains 30,000 lb C acre⁻¹?

16.11. The soil organic matter has increased from 2 to 2.2%, estimate the change in organic matter, the change in soil organic N, and soil organic carbon for the surface 6 inches if the bulk density is 1.25 g cm³. In this calculation assume that the C/N ratio is 10.

16.12. The soil organic matter in the surface 6 inches has decreased from 6% in 1910 to 3% in 1980. How much N and C was mined from the soil? In this calculation assume that the C/N ratio is 10.

16.13. Over a five-year period a field produced two 200 bu acre⁻¹ corn crops, two 70 bu acre⁻¹ wheat crops, and a 120 bu acre⁻¹ oat crop. How much N, K_2O, P_2O_5, and sulfur were removed over this time period?

ACKNOWLEDGMENTS

Support for this document was provided by South Dakota State University, Precision Farming Systems community in the American Society of Agronomy, International Plant Nutrition Institute, International Society of Precision Agriculture, and the USDA-AFRI Higher Education Grant (2014-04572).

REFERENCES

Anderson, T.H., and K.H. Domsch. 1989. Ratios of microbial biomass carbon to total organic carbon in arable soils. Soil Biol. Biochem. 21:471–479. doi:10.1016/0038-0717(89)90117-X

Baligar, V.C., N.K. Fageria, and Z.L. He. 2007. Nutrient use efficiency in plants. Commun. Soil Sci. Plant Anal. 32:921–950. doi:10.1081/CSS-100104098

Carlson, C.G., D.E. Clay, and C.L. Reese. 2016. Chapter 28: Common fertilizers used in corn production. In: D.E. Clay, C.G. Carlson, S.A. Clay, and E. Byamukama, editors, iGROW Corn: Best management practices. South Dakota State University. Brookings, SD.

Clay, D.E., G. Reicks, J. Chang, T. Kharel, and S.A.H. Bruggeman. 2016. Assessing a fertilizer program: Short and long-term approaches. In: A. Chatterjee and D. Clay, editors, Soil fertility management in agroecosystems. ASA, CSSA, SSSA. Madison, WI.

Clay, D.E., G. Carlson, J. Chang, G. Reicks, S.A. Clay, and C. Reese. 2017. Calculating soil organic turnover at different landscape positions in precision conservation. In: J. Delgado, S. Sassenrath, and T. Mueller, editor, Agronomy Monograph 59, Precision Conservation: Geospatial Techniques and Natural Resource Conservation. ASA, Madison WI.

Clay, D.E., S.A. Clay, K.D. Reitsma, B.H. Dunn, A.J. Smart, C.G. Carlson, D. Horvath, and J.L. Stone. 2014. Does the conversion of grasslands to row crop production in semi-arid areas threaten global food security? Glob. Food Secur. 3:22–30. doi:10.1016/j.gfs.2013.12.002

Clay, D.E., N.R. Kitchen, C.G. Carlson, J.L. Kleinjan, and W.A. Tjentland. 2002. Collecting representative soil samples for N and P fertilizer recommendations. SSMG-38 Site-Specific Management Guidelines. Potash and Phosphate Institute. http://www.ipni.net/publication/ssmg.nsf/0/0528E52FEFC58040852579E50077CBF3/$FILE/SSMG-38.pdf (verified 27 June 2017).

Clay, D.E., N. Kitchen, J. Kleinjan, and C.G. Carlson. 2007. Using historical management areas to reduce soil nutrient sampling error. GIS in Agriculture, CRC Press. Boca Raton, FL.

Clay, D.E., G.L. Malzer, and J.L. Anderson. 1990. Tillage and dicyandiamide influence on nitrogen fertilization immobilization, remineralization, and utilization by maize (Zea mays L.). Biol. Fertil. Soils 9:220–225. doi:10.1007/BF00336229

Combs, S.M., and M.V. Nathan. 1998. Recommended soil organic matter tests. In: J.R. Brown, editor, Recommended chemical soil test procedures for the North Central Region. NCR Publ. no. 221 (revised). Missouri Agricultural Experiment Station SB 1001. NCR-13 Committee. p. 53–58.

Fixen, P. 2007. Can we define a global framework within which fertilizer best management practices can be adapted to local conditions? In: Fertilizer best management practices. International Fertilizer Industry Association, Paris, France.

Franzen, A., and D. Humburg. 2016. Chapter 50: Calibrating yield monitors. In: D.E. Clay, C.G. Carlson, S.A. Clay, and E. Byamukama, editors, iGROW corn: Best management practices. South Dakota State University, Brookings, SD.

Kirkby, C.A., J.A. Kirkegaard, A.E. Richardson, L.J. Wade, C. Blanchard, and G. Batton. 2011. Stable organic matter: A comparison of C:N:P:S ratios in Australia and other world soils. Geoderma 163:197–208. doi:10.1016/j.geoderma.2011.04.010

Malo, D.D., T.E. Schumacher, and J.J. Doolittle. 2005. Long-term cultivation impacts on selected soil properties in the northern Great Plains. Soil Tillage Res. 81:277–291. doi:10.1016/j.still.2004.09.015

Marschner, H. 1995. Mineral nutrition of higher plants. 2nd ed. Academic Press, Orlando, FL.

Murrell, T.S. 2008. Measuring nutrient budget and calculating nutrient budgets. In: S. Logsdon, D.E. Clay, D. Moore, and T. Tsegaya, editors, Soil science: Step-by-step field analysis. SSSA, Madison, WI. p. 159–182. doi:10.2136/2008.soilsciencestepbystep.c13

Sebilo, M., B. Mayer, B. Nicolardot, G. Pinay, and A. Mariotti. 2013. Long-term fate of nitrate fertilizer in agricultural soils. Proc. Natl. Acad. Sci. USA 110:18185–18189. doi:10.1073/pnas.1305372110

Vaughan, B. 1999. How to determine an accurate soil testing laboratory. Online. Site Specific Management Guidelines No. 4 (SSMG 4). Potash and Phosphate Institute, Norcross, GA.

Understanding Grain Moisture Percentage and Nutrient Contents for Precision Grain Management

17

Cheryl L. Reese* and C. Gregg Carlson

Chapter Purpose

Crops at different landscape positions mature at different times. Harvesting crops following maturity at an appropriate moisture content can reduce harvest losses, grain quality, and drying costs. To assess if targeted harvesting is a feasible operation, it is important to understand how grain moisture percentage affects selling prices. To calculate the value of the crop, grain yields must be converted to a standard moisture content. This chapter discusses how to convert grain at one moisture content to another. Moisture is calculated differently for different products. For example, soil, wood, and nutrient analysis are generally based on a dry weight basis (0% water content). However, grain, silage, hay, and compost, are reported on the wet weight basis, with each product having a different standard moisture content. Manure can be reported based on either the dry or wet weight basis. Wheat protein is often based on a 12% moisture standard, whereas wheat weight (per bushel) is based on a 13.5% moisture standard. This chapter discusses and provides examples on how to calculate grain moisture contents and determine the nutrient concentration in the harvested grain. When converting yields from one moisture content to another it is important to check your answer. This is done by cancelling units and checking the answer to see if it makes sense.

Key Terms

Grain moisture percentage, bushel, moisture shrinkage, protein content, soil nutrient concentration, storage.

Mathematical Skills

Determining the weight of grain at different moisture percentages, convert dry weight to wet weights, convert N% to protein.

Grain Moisture History and Storage Requirements

Historically, grain has been sold by volume, and in the United States the standard volume was a bushel (2150.42 in³). Over time, this convention has changed from grain being sold on a volume basis to being sold on a weight basis at a specified moisture. For corn, a standard bushel is defined as 56 lb at 15.5% moisture content. Different grains have different standard weights and moisture contents (Table 17.1). However, depending on the buyer's requirement, the moisture

Department of Agronomy, Horticulture, and Plant Science, South Dakota State University, Brookings, SD 57007. *Corresponding author (cheryl.reese@sdstate.edu)

doi:10.2134/practicalmath2017.0030

content can vary, and grain delivered at a moisture content that is different than the specified value may receive a discount. This discount is associated with the change in volume and weight during drying, a term called shrinkage. The length of time that grain can be safely stored is indirectly related to the moisture content (Table 17.2) and the ambient air temperature. According to Carlson (2016) rules of thumb for grain storage are provided below.

1. Determine the desired moisture content of the grain, and the short and long-term storage requirements.
2. Clean all equipment that will be in direct contact the grain. Bin and equipment should be cleaned enough so that evidence from the last use is eliminated (Ess et al., 2005).
3. Minimize the number of broken kernels placed into the grain bin. Broken kernels can be minimized by correcting adjusting the combine and drying the grain at an appropriate temperature. For example, stress cracking can occur by high temperature drying and broken kernels and other trash can be removed by screening (Hanna, 2008).
4. Grain with high moisture content (> 22%) needs to be dried prior to storage. Information on drying grain is available in Behlen Grain Systems (2012) and Hellevang (1998).
5. If corn will be stored for 6 to 12 months, the moisture content should be reduced to 14%, and if storage is a year or longer, the moisture content should be 13%.
6. Periodically, at least every two weeks, monitor the grain bin and electronic monitoring devices, if problems are detected, immediately resolve them, waiting will worsen the problem or make the problem uncontrollable. The two important factors to monitor are temperature and moisture, and these sensors do malfunction. In many situations, problems can be minimized by selecting a good service provider, and checking the system when the bin is empty.

Grain Moisture

Grain moisture content information can be obtained by measuring the moisture content of the grain in the bin, combine, field, or truck or by using data provided by a yield monitor. In many fields, grain with relatively low moisture percentages are harvested from high elevational areas, whereas high moisture grain is harvested from low elevational areas. The grain moisture content can be lowered by delaying harvest or drying the grain in a dryer. Drying the grain to too low a moisture results in a higher propensity to shatter and shrink, which results in discounts at the elevator.

Table 17.1. The weight and moisture content of standard bushel of different annual crops.

Crop	Weight bushel	Specified moisture	lb at 0% moisture
Corn	56	15.5%	47.32
Soybean	60	13%	52.2
Wheat	60	13.5%	51.9
Barley	48	14.5	41.04
Oat	32	14%	27.52
Rye	56	14%	48.16

Table 17.2. Approximate storage time of grains as influenced by moisture content and temperature (°F). (Carlson, 2016).

% Moisture content	Temperature (°F)					
	30	40	50	60	70	80
			—days—			
14					200	140
15				240	125	70
16			230	120	70	40
17		280	130	75	45	20
18		200	90	50	30	15
19		140	70	35	20	10
20		90	50	25	14	7
22	190	60	30	15	8	3
24	130	40	15	10	6	2
26	90	35	12	8	5	2
28	70	30	10	7	4	2
30	60	25	5	8	3	1

In many combines, grain moisture can be measured by a moisture sensor located on the clean grain elevator or fountain auger is used to measure grain moisture (Franzen and Humburg, 2016). This sensor provides critical information needed to process the yield monitor data or determine if drying is necessary. The sensor should be calibrated at period intervals following protocols outlined by Franzen and Humburg (2016).

When yield is measured by a yield monitor, the value measured is related to mass and moisture content. To understand yield, the agronomist or engineer must understand how to convert the grain yield at one moisture content to the grain yield at another moisture content. Grain is reported on the wet weight basis, and the moisture content is calculated with the equation,

$$\text{Grain moisture percentage} = 100\% \times \frac{\text{water weight}}{\text{water weight} + \text{dry grain}}$$

The amount of water in wet grain is determined by the equation:

$$\text{Water weight} = \text{grain moisture \%} \times \text{wet weight of the grain}$$

For example, 56 lbs of grain at 15.5% moisture contains 8.68 lbs of water (56 × 0.155). Therefore, the amount of dry grain at this weight and moisture content is 47.32 lbs (56–8.58 lbs). When grain is dried to a specified moisture content, water is lost to the atmosphere. This loss is called the moisture shrinkage and it is calculated with the equation,

$$\%\text{Moisture shrinkage} = 100 \times \frac{\text{original moisture content \%} - \text{final moisture\%}}{100 - \text{final moisture \%}}$$

The amount of grain that remains after drying is calculated with the equation,

$$\text{grain weight}_{final} = \frac{\text{grain weight}_{initial} \times (100 - \%\,\text{Initial moisture})}{(100 - \%\,\text{final moisture})}$$

PROBLEM 17.1.
Convert 50,000 lb of corn at 15.5% moisture to bushels.

ANSWER:
A bushel of corn at 15.5% moisture weighs 56 pounds.

$$50,000 \text{ lb corn} \times \frac{1 \text{bu}}{56 \text{ lb corn}} = 893 \text{ bu}$$

However, in many situations, it is assumed that a bushel of corn weights 56 lbs, regardless of the moisture content.

PROBLEM 17.2.
Convert 50,000 lb of wheat at 13.5% moisture to bu.

ANSWER:
A bushel of wheat weighs 60 lb at 13.5% moisture

$$50,000 \text{ lb wheat} \times \frac{1 \text{ bushel}}{60 \text{ lb wheat}} = 833 \text{ bu wheat}$$

This calculation is based on a bushel of wheat at 13.5% moisture weighing 60 lb (Table 17.1). In many situations, a bushel is based on weight and does not consider the moisture content. However, to make valid comparison, yields need to be converted to a standard moisture contents, which are provided in Table 17.1.

PROBLEM 17.4.
Convert 50,000 lb of oats at 14% moisture to bushels)

ANSWER:
A bushel of oats weighs 32 pounds at 14% moisture (Table 17.1)

$$50,000 \text{ lb oats} \times \frac{1 \text{bu}}{32 \text{ lb oats}} = 1562 \text{ bu oats}$$

PROBLEM 17.6.
If a sample contains 2.5% N (dry weight basis) and you harvested 150 lb of grain (wet weight) at 25% moisture, how much N was contained in the harvested grain?

ANSWER:
Calculate moister in the sample: Moisture lb = 0.25 × 150 lb = 37.5 lb.
Calculate dry grain: Dry grain = 150 lb - 37.5 lb = 125.5 lb at 15.5% moisture
Determine N in grain: N = (0.025 lb N/lb dry grain) × 125.5 lb dry grain = 2.81 lb of N

Adjusting Wheat Protein Percentage to 12% Moisture

For wheat sold in the United States, a standard bushel contains 13.5% moisture, while the standard protein content is based on 12% moisture. If the sample was submitted for total N analysis at a laboratory, the value that is returned to you might be 2.1% on a dry weight basis. This value needs to be converted to protein (multiply by 6.25) at 12% moisture. A two-step process is used for this conversion. First, the percent dry weight is determined by the equation,

$$\% \text{Dry weight} = 100\% \times \frac{\left(\text{Wet weight} - \text{Dry weight}\right)}{\text{Dry weight}}$$

To convert nutrient concentration at a dry weight to the wet weight use the following equation,

$$\text{nutrient at wet weight basis} = \text{nutrient dry weight basis} \times \frac{\left(100 - \%\text{moisture}\right)}{100}$$

Measuring Grain Moisture Content

Step-by-step directions for measuring grain and hay moisture contents using sensors, ovens, and microwaves are available in Anderson and Grant (1993). For accurate estimates, product specific protocols should be followed. Anderson and Grant (1993) report that:

1. Different protocols must be followed for different grains,
2. Oven drying requires one to three days, and
3. Calibrated grain moisture sensors should provide acceptable accuracy levels if protocols are followed, and
4. Information for calibrating yield monitor moisture sensors are available in Franzen and Humburg (2016).

PROBLEM 17.7.

You have 1000 kg or wheat at 12% moisture. How many kg of wheat would you have at 13.5% moisture?

ANSWER:

Calculate grain at 13.5% moisture: $= 1000 \times \dfrac{\left(100 - 12\right)}{\left(100 - 13.5\right)} = 1017 \, \text{kg}$

PROBLEM 17.8.

The N percentage of dry wheat is 2.1%. What is its protein content at 12% moisture? Note: In some situations, protein premium and discounts are based on wheat at 12% moisture and N is converted to protein by multiplying %N by 6.25.

ANSWER:

Calculate protein at 0% moisture: $\text{Protein}\% = \text{N}\% \times 6.25 = 13.13\%$

Calculate protein at 12% moisture:

$$\%\text{protein}_{12\%} = \%\text{protein}_{dry} \times \frac{\left(100 - \%\text{moisture}\right)}{100} = 13.13\% \times \frac{100 - 12}{100} = 11.55\% \text{ protein}$$

PROBLEM 17.9.

How many bushels at 15.5% moisture are contained in 1000 lb of corn grain at 22% moisture content?

ANSWER:

Calculate weight of corn at 15.5% moisture: $= 1000 \times \left(\left(100 - 22\right) \middle/ \left(100 - 15.5\right)\right) = 923$

Calculate bu acre^{-1}: Bushels $= 923 \, \text{lb} \times \dfrac{1 \, \text{bu}}{56 \, \text{lb}} = 16.5 \, \text{bu corn}$

Summary

In summary, grain moisture content of harvested grain can vary across the landscape, and this variability should be considered when deciding when to harvest. The grain moisture content can influence the drying costs as well as discounts when delivered to a buying agent. When calculating the amount of harvested grain, the moisture content should be considered. These calculations should be checked by canceling units and if the calculated value makes sense. To minimize storage losses, grain must be stored at an appropriate moisture content. This chapter provides examples on how to convert grain moisture percentage from one value to another and provides references for

Additional Problems

17.10. You deliver 50,000 lb of grain to the elevator. The moisture content of the grain is 18%. How many pounds of grain at 15.5% moisture did you deliver?

17.11. You deliver 150,000 lb of grain at 13% moisture to and elevator. How much water should be added to increase the moisture content to 15.5%?

17.12. A bushel of corn at 15.5% moisture weighs 56 lb. How much water does this bushel contain?

17.13. The dry weight N content of grain is 1.2%, if the grain is at 15.5% moisture content, what is its N% on a wet weight basis?

17.14. You harvest 300 bu corn acre⁻¹ at 15.5% moisture. You send a sample to a laboratory to determine how much N and P were removed. On a dry weight basis the N content is 1.1% and the P content is 0.28%. How much N, P, and P_2O_5 were removed?

measuring grain moisture contents. Additional information is available at Hellevang and Wilcke, 2013, Hurburg hand Elmore, 2009, and McKenzie and Fossen, 1995.

ACKNOWLEDGEMENTS

Support for this document was provided by South Dakota State University, Precision Farming Systems community in the American Society of Agronomy, International Society of Precision Agriculture, and the USDA-AFRI Higher Education Grant (2014-04572).

REFERENCES AND ADDITIONAL INFORMATION

Anderson, B., and R. Grant. 1993. G93-1168 Moisture testing of grain, hay, and silage. Paper 1312. University of Nebraska Extension, Lincoln, NE.

Behlen Grain Systems, 2012. Grain bin operation manual. Behlen Grain Systems, Columbus, NE.

Carlson, C.G. 2016. Chapter 53: Corn storage and drying. In: D.E. Clay, C.G. Carlson, S.A. Clay, and E. Byamukama, editors, iGrow corn: Best management practices. South Dakota State University. Brookings, SD.

Ess, D.R., N.A. Fleck, and D.E. Maier. 2005. Where grain hides in a combine. Purdue Extension GQ-49-W. West Lafayette, IN.

Franzen, A., and D. Humburg. 2016. Chapter 50: Calibrating yield monitors. In: D.E. Clay, C.G. Carlson, S.A. Clay, and E. Byamukama, editors, iGROW corn: Best management practices. South Dakota State University. Brookings, SD.

Hellevang, K.J., and W.F. Wilcke. 2013. Corn drying and storage. NDSU Extension, AE-1119. Available at: https://www.ag.ndsu.edu/graindrying. (accessed 5 November 2015).

Hellevang, K.J., 1998. Temporary grain storage. AE-84 NDSU Extension Service. Fargo, ND.

Hurburgh, C., and R. Elmore. 2009. 2009 Corn quality issues. Integrated corn management. Iowa State University Extension and outreach. Ames, IA.

McKenzie, B.A., and L. Van Fossen. 1995. Managing dry grain in storage. Agricultural engineers digest-20, Purdue University Cooperative Extension Service, West Lafayette, IN

Calculating the Impacts of Agriculture on the Environment and How Precision Farming Can Reduce These Consequences

18

Clay Robinson*

Chapter Purpose

During the 20th century, the unintended impact of food and fiber production on the environment became increasingly apparent. For example, during the 1930's, the wind transported 850 million tons of topsoil from areas of the United States Great Plains to the Midwest, New England, and even the Atlantic Ocean. Water erosion had similar impacts on long-term sustainability as soil sediments from plowed, unvegetated fields filled streams, lakes, and reservoirs and contributed to eutrophication and algal blooms. During the 1990s, science showed that an unintended consequence of tile drainage from Corn Belt States has been anoxia in the Gulf of Mexico and the creation of a Dead Zone (Alexander et al., 2008). This chapter discusses tools available to estimate effects of tillage on residues and erosion rates, methods to determine nitrogen and phosphorus loads to drainage ditches, streams, rivers and lakes, and conceptual models behind greenhouse gas (GHG) estimates. Problems on how to estimate the impact of agriculture to various environmental parameters are provided.

Key Terms

Wind erosion, tile drainage, anoxia, carbon footprints, greenhouse gas emissions (GHG).

Mathematical Skills

Determining erosion losses, carbon footprints, estimating N loads, GHG emissions, carbon footprint.

Erosion and Residue Estimates

Historically, tillage implements that worked on small-scale operations with animal-powered plows left some vegetation intact. However, when animals were replaced by tractors, large areas of bare soil, with little or no vegetation left on the soil surface, was the end result (Clay et al., 2017). The bare soil was easily carried away by wind and water. Tillage and crop management systems have improved greatly since the Dust Bowl era of the 1930's, and precision agriculture will continue making progress to limit wind and water erosion. There are three types of water erosion: sheet, rill, and gully. Sheet involves the uniform loss of thin layers of soil due to raindrop impact and overland flow, and rill erosion begins as water begins to flow in small rivulets or channels. Gully erosion is the removal of soil along drainage lines by surface water runoff with channels

Dept. of Agriculture, Illinois State University, Normal, IL 61790-5020.*Corresponding author (c.drdirt.robinson@gmail.com)
doi:10.2134/practicalmath2017.0108

Table 18.1. Values of the topographic factor, LS (Adapted from Table 3, USDA Science and Education Administration, 1981).

%	Slope length											
	ft											
slope	25	50	75	100	150	200	300	400	500	600	800	1000
0.200	0.060	0.069	0.075	0.080	0.086	0.092	0.099	0.105	0.110	0.114	0.121	0.126
0.500	0.073	0.083	0.090	0.096	0.104	0.110	0.119	0.126	0.132	0.137	0.145	0.152
0.800	0.086	0.098	0.107	0.113	0.123	0.130	0.141	0.149	0.156	0.162	0.171	0.179
2.000	0.133	0.163	0.185	0.201	0.227	0.248	0.280	0.305	0.326	0.344	0.376	0.402
3.000	0.190	0.233	0.264	0.287	0.325	0.354	0.400	0.437	0.466	0.492	0.536	0.573
4.000	0.230	0.303	0.357	0.400	0.471	0.528	0.621	0.697	0.762	0.820	0.920	1.010
5.000	0.268	0.379	0.464	0.536	0.656	0.758	0.928	1.070	1.200	1.310	1.520	1.680

wider and deeper than can be obliterated by tillage. The Universal Soil Loss Equation (USLE) is an empirical (statistical) model developed to estimate average annual sheet and rill erosion (USDA Science and Education Administration, 1981). This tool can be used to provide estimates on the potential impact of adopting precision farming and conservation techniques. The USLE is a good model to understand the basic factors affecting erosion. It was developed to estimate the annual sheet and rill erosion from a single slope; it does not account for gully, wind, or tillage erosion. The current version used by the NRCS for conservation planning is RUSLE2 (Revised Universal Soil Loss Equation, Version 2) (USDA-ARS, 2008, Renard et al., 1994).

Universal Soil Loss Equation

Many different combinations of rotations, production levels, rotations, and tillage methods are possible. The Universal Soil Loss Equation was developed to estimate the impacts of different types of tillage and cropping practices on soil loss (Equation 18.1)

$$A = R \cdot K \cdot LS \cdot C \cdot P \text{ [Eq 18.1]}$$

where, A is the average annual soil loss, tons acre^{-1} yr^{-1}, R is a local rainfall and runoff factor, k is the soil erodibility factor, LS is the topographic factor (slope length and gradient), C is the cover and management factor, and P is the support management practice factor. The rainfall and runoff factor (R) considers the mean rainfall, timing and other factors, and provides the erosivity index, which is determined by the amount of rainfall and peak intensity sustained over an extended period of time. The soil erodibility factor (k) is determined using the percent sand, silt, organic matter, structure, and permeability, and represents the susceptibility of the soil to erosion, as affected by the rate of runoff. The topographic factor (LS) is determined by the slope gradient (steepness) and length. The cover and management factor (C) is the ratio of soil loss of a management system relative to the erosion from a clean-tilled surface. The support practice (P) is the ratio of soil loss with additional management practices such as contouring, strip-cropping, and terracing compared with straight row up- and down-slope farming. When there is not a support practice in place, there is no reduction in erosion prediction ($P = 1$). The K, LS, C, and P tables were developed from more than 20 years of erosion trials; Tables 18.1, 18.2, and 18.3 provide excerpts to use in the calculations below.

For any given site, the values for R, LS (Table 18.1), and K (Table 18.2) are constant, while the values for C (Table 18.2) and P vary with practices (Table 18.3). In this equation, a high value indicates increased erosion. For example, in Table 18.1, increasing the length of the slope increases the LS value. In a second example, removing the crop residues and tilling the field in a continuous corn system has a C value of 0.6, whereas using no-tillage in continuous corn with 30% residue cover after planting has a C value of 0.22 (Table 18.2). In a third example, reducing meadow and increasing row crops increases the P value (Table 18.3). There are many on-line calculators that simplify these calculations.

Estimating Residue Cover

As observed in the USLE, cover and management affect the amount of water erosion from a field. The primary factors affecting residue cover are the crop, residue management (e.g., leaving all residue or baling), and tillage practices. A teaching tool was developed to demonstrate the effects of crop type and tillage implements on residue cover (Kok and Thien, 1994).

Table 18.2. Soil characteristics, slope gradient and length, soil erodibility, and local rainfall and runoff factors for three locations, and cover and management factors for two systems (USDA Science and Education Administration, 1981).

Location		Slope		K	R
		Length			
		ft	%		
Southern Iowa	Silt loam, 4% organic matter, fine granular structure, Moderate permeability	200	4	0.35	180
Central Nebraska	Loam, 2% organic matter, coarse granular structure Rapid permeability	600	2	0.32	110
Southeastern Missouri	Clay loam, 1% organic matter, blocky structure Slow permeabillity	300	5	0.42	240

Crop and tillage practices	C
Corn after Corn, with residues removed, using conventional tillage with a moldboard plow in the spring before planting	0.60
Corn after Corn, no-till, 30% cover after planting	0.22
Corn after Corn, no-till, 70% residue cover after planting	0.07

Table 18.3. Values of support management practice factor (P) within the Universal Soil Loss Equation. The management practices in Column A are for a four-year rotation that consists of a row crop, small grain, and two years of a meadow (typically alfalfa). The management practices in Column B are for a four-year rotation that consists of two years of row crops, one year of winter grain, and one year of a meadow. The management practices in Column C include alternating contour strips (of the defined width) of row crops and small grains. (Adapted from Table 14, USDA Science and Education Administration, 1981).

% slope	Crop rotations			Strip width	Maximum length
	A	B	C		
			P-values		
1-2	0.30	0.45	0.60	130	800
2-5	0.25	0.38	0.50	122	600
6-8	0.25	0.38	0.50	100	400
9-12	0.30	0.45	0.60	80	240
13-16	0.35	0.52	0.74	80	160
17-20	0.40	0.60	0.80	60	120
21-25	0.45	0.69	0.90	50	100

Table 18.4. Residue indices and conversion factors (modified from Kok and Thien, 1994).

Crop	Residue index	
	lb residue bu^{-1}	α
Wheat	100	0.0006
Corn	60	0.00038
Grain sorghum	60	0.00034
Oats	55	0.00106
Soybean	45	0.00058
Sunflowers	1.5 lb lb^{-1}	0.00024

PROBLEM 18.1.

Determine the soil loss in Central Nebraska for corn after corn with a conventional tillage system and no conservation practice. Using Table 18.1, for a 2.0% slope, 600-ft long, the length by slope (LS) factor is 0.344.

ANSWER:

Use these factors:

$$R = 110; K = 0.32; LS = 0.344; C = 0.6; P = 1$$
$$A = R \cdot K \cdot LS \cdot C \cdot P = 110 \cdot 0.32 \cdot 0.344 \cdot 0.60 \cdot 1 = 7.3 \text{ tons acre}^{-1} \text{ yr}^{-1}$$

Table 18.2 indicates incorporating contour strip cropping, and alternating corn and wheat in 130-ft strips, changes the P-factor to 0.60.

$$A = R \cdot K \cdot LS \cdot C \cdot P = 110 \cdot 0.32 \cdot 0.344 \cdot 0.60 \cdot 0.60 = 4.4 \text{ tons acre}^{-1} \text{ yr}^{-1}$$

PROBLEM 18.2.

Determine the soil loss in Southeastern Missouri for corn after corn with a conventional tillage system and no conservation practice. Using Table 18.1, for a 5.0% slope, 300-ft long, the LS factor is 0.928. In this system, the R value is 240 and the K factor is 0.42.

ANSWER:

$$R = 240; K = 0.42; LS = 0.928; C = 0.6; P = 1.$$
$$A = R \cdot K \cdot LS \cdot C \cdot P = 240 \cdot 0.42 \cdot 0.928 \cdot 0.60 \cdot 1 = 56.1 \text{ tons acre}^{-1} \text{ yr}^{-1}$$

Switching to no-till with 30% residue cover after planting changes the C factor to 0.22.

$$R = 240; K = 0.42; LS = 0.928; C = 0.22; P = 1.$$
$$A = R \cdot K \cdot LS \cdot C \cdot P = 240 \cdot 0.42 \cdot 0.928 \cdot 0.22 \cdot 1 = 20.6 \text{ tons acre}^{-1} \text{ yr}^{-1}$$

To reduce the erosion further, implement a four-year corn–corn–wheat–meadow rotation on the contour. Use Table 18.2 to find the P factor.

$$R = 240; K = 0.42; LS = 0.928; C = 0.22; P = 0.38.$$
$$A = R \cdot K \cdot LS \cdot C \cdot P = 240 \cdot 0.42 \cdot 0.928 \cdot 0.22 \cdot 0.38 = 7.8 \text{ tons acre}^{-1} \text{ yr}^{-1}$$

Increasing the residue from 30% to 70% cover would change the C factor to 0.07 and would reduce soil loss even further.

PROBLEM 18.3.

Estimate cover in the summit and lower back slope positions if the yields were 100 and 200 bu acre^{-1} corn, respectively.

ANSWER:

Corn produces 60 lb residue bu^{-1} yield (Table 18.4)

Summit:

$$\text{Residue} = \frac{100 \text{ bu}}{\text{acre}} \times \frac{60 \text{ lb}}{1 \text{ bu}}$$

$$\text{Residue} = \frac{100 \text{ bu}}{\text{acre}} \times \frac{60 \text{ lb}}{1 \text{ bu}} = \frac{6000 \text{ lb}}{\text{acre}}$$

$$\% \text{cover} = 100 \times \left(1 - e^{-0.00038 \times 12000}\right) = 89.8\% \text{ cover}$$

Footslope:

$$\text{Residue} = \frac{200 \text{ bu}}{\text{acre}} \times \frac{60 \text{ lb}}{1 \text{ bu}} = \frac{12000 \text{ lb}}{\text{acre}}$$

$$\% \text{ cover} = 100 \times \left(1 - e^{-0.00038 \times 12000}\right) = 99.0\% \text{ cover}$$

The amount of biomass produced by different grain crops varies by crop and grain yield (Table 18.4). For example, 50 bu soybean, 37.5 bu corn, and 22.5 bu wheat all produce 2250 lb residue. Yield increases directly with biomass. The amount of residue cover on the soil surface, if any is left, is described by an exponential function,

% COVER = 100 • (1 - e$^{-\alpha \cdot [\text{residue}]}$) [18.2]

where α is the conversion factor from Table 18.4 and residue is in lb acre^{-1}.

Every tillage operation buries residues, and the amount of residue buried varies by tillage implement (Table 18.5). In addition, because of their fragile nature, crops such as soybeans and sunflowers lose more residue with each tillage pass than many other crops. Residues also decompose during a fallow or overwintering period. The percent residue cover is determined as the product of the previous cover amount and the factor associated with the tillage operation Eq. [18.3].

Remaining cover (%) = Previous cover (%) × % maintained [18.3]

Nitrogen and Phosphorus

Just like in production fields, nutrients stimulate organism growth in streams, rivers, lakes and oceans. The 4Rs Nutrient Management Stewardship practices help reduce the amount of nutrients that end up in streams, rivers, lakes, and the ocean (Illinois Council on Best Management Practices, 2015). The 4R management practices include using the **R**ight source at the **R**ight rate at the **R**ight time in the **R**ight place. The amount of nutrients that are transported out of production fields can be staggering. Figure 18.1 shows a graph of daily flow and nitrogen concentration in the Mississippi River over time at Clinton, IA. Because the flow is quite high, even during dry periods, small concentrations of a chemical in the water can lead to high total amounts of nutrients being present (Fig. 18.2). Table 18.6 shows concentration and flow data for a hypothetical watershed.

The daily load is calculated as the product of the daily discharge and the concentration,

$$\text{Load} = \text{discharge} \cdot \text{concentration} \quad [18.4]$$

Table 18.5. Crop residue maintained by implements (Adapted from Table 1, Kok and Thien, 1994). Fragile residues refer to soybean and sunflower residues. For use in Eq. [18.3].

Action	% residue maintained per trip	
	nonfragile	fragile
Moldboard plow, 5- to 7-in depth	20	5
Moldboard plow, > 8-in depth	10	5
Chisel, twisted points	50	35
Chisel, straight points	75	55
Chisel-disk tandem	50	30
Tandem disk, regular blades	50	30
Tandem disk, large blades	30	20
Disk plow	35	15
One way disk, 24 to 36 in blades	50	30
Subsoiler	60	55
Field cultivator, 12- to 20-in sweep	85	65
Field cultivator, 6- to 12-in sweep	75	55
Mulch treader, spike tooth	70	65
Blades, ≥ 36 inches	90	85
Sweeps, 24 to 36 in	90	80
Rodweeder, plain	90	85
Harrow	80	60
Fertilizer appl., injection type	80	60
No-till slot planter	90	85
Row-crop planter	80	60
Low-till drill	80	75
Furrow drill	80	50
Conventional drill, disk openers	95	90
Overwintering	90	70

PROBLEM 18.4.

A field of wheat in a wheat–corn–fallow cropping system averaged 50 bu acre^{-1}. How many pounds of residue are expected on the field after wheat harvest?

ANSWER:

Wheat produces 100 lb residue bu^{-1} yield (Table 18.4).

$$\text{Residue} = \frac{100 \text{ bu}}{\text{acre}} \times \frac{100 \text{ lb}}{1 \text{ bu}} = \frac{10{,}000 \text{ lb}}{\text{acre}}$$

What is the percent residue cover before plowing the field? The wheat residue conversion factor, $\alpha = 0.00060$

$$\% \text{ COVER} = 100(1 - e^{-\alpha \cdot (\text{residue})}) = 99.75\%$$

After harvest in July, the farmer used a tandem disk (regular blades). How much residue remained on the soil surface (lb acre^{-1}) before going into winter? Tandem disks leave 50% of the residue on the surface.

$$\text{Remaining cover (\%)} = 99\% \times 50\% = 49.5\%$$

In the next spring, what is the remaining cover? Decomposition removes 10% of the wheat residues, leaving 41% of the soil covered.

$$\text{Remaining cover (\%)} = 49.5\% \cdot 90\% = 44.6\%$$

Agriculture and Soil Organic Carbon

Prior to 1850, wood was the primary energy source because the technology to mine and produce fossil fuels on a large scale was not available. Since the beginning of the Industrial Revolution, consumption of coal, petroleum, and natural gas has increased dramatically (Fig. 18.3). Agriculture is an energy-intensive enterprise (Clay and Shanahan, 2011; Clay et al., 2012). Energy is required to plow fields, produce fertilizers and pesticides, plant and harvest crops, and dry grain. Further, it takes a great deal of fuel to transport all the equipment, inputs, and outputs from manufacturer to wholesaler to retailer to farm, and from farm to market after harvest.

Essentially all the energy currently used in agricultural production comes from fossil fuels. Environmental concerns associated with fossil fuels include that they are a finite resource, which release greenhouse gases into the atmosphere, including carbon dioxide (CO_2) and methane

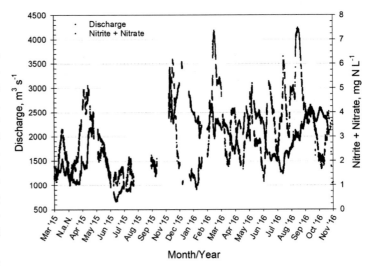

Fig. 18.1. Water volume discharge and nitrite plus nitrate concentration in the Mississippi River at Clinton, IA (USGS, 2017). Note this chart reports the nitrite plus nitrate values as mg N L^{-1}, which is identical to ppm N.

PROBLEM 18.5.

Use Eq. 18.4 to determine the total discharge of nitrate and nitrate–N in one hour (1 hr) if the discharge is 100,000 ft^3 s^{-1} and the concentration is 3 mg L^{-1}.

Conversions: 1 mg L^{-1} = 1 ppm   1 ft^3 = 7.48 gal

Answer:

$$1\ hr \times \frac{3600\ s}{1\ hr} \times \frac{100{,}000\ ft^3}{1\ s} \times \frac{7.48\ gal}{1\ ft^3} \times \frac{8.344\ lb}{1\ gal} \times \frac{3\ lb\ N}{1{,}000{,}000\ lb\ water}$$

Now cancel the units and complete the calculation.

$$1\ hr \times \frac{3600\ s}{1\ hr} \times \frac{100{,}000\ ft^3}{1\ s} \times \frac{7.48\ gal}{1\ ft^3} \times \frac{8.344\ lb}{1\ gal} \times \frac{3\ lb\ N}{1{,}000{,}000\ lb\ water} = 67{,}406\ lb\ N$$

PROBLEM 18.6.

Use the data in Table 18.6 to determine the daily total N load on 6 April.

ANSWER:

$$Load = discharge \times concentration = \frac{863\ m^3}{h} \times \frac{1.5\ mg\ N}{L}$$

The discharge is given in m^3 h^{-1} and ft^3 h^{-1}. These units do not cancel to give the load in mass/day (g d^{-1}). Converting from mg h^{-1} to g d^{-1} is straightforward, but volume is currently shown as m^3 and L, and must be converted into the same units.

1 d = 24 h, 1 g = 1000 mg, 1 ml = 1 cm^3, 1 m = 100 cm, 1 L = 1000 mL, 1 m^3 = 1000 L

Based on these values, $1\ m^3 = 1{,}000{,}000\ mL \times \dfrac{1\ L}{1000\ mL} = 1{,}000\ L$

Now substitute these into the equation.

$$Load = \frac{863\ m^3}{hr} \times \frac{1.5\ mg\ N}{L} \times \frac{1000\ L}{m^3} \times \frac{1\ g}{1000\ mg} \times \frac{24\ hr}{day} = 31{,}068\ g\ N\ per\ day$$

(CH₄). Greenhouse gases allow shortwave radiation (light) to enter the atmosphere and reach the earth's surface, but do not allow longwave radiation (heat) to escape, trapping it in the atmosphere. As a consequence, this warms the atmosphere and influences the earth's climate system. Since the beginning of the Industrial Revolution, the atmospheric CO_2 concentration has increased almost 150% and CH_4 by 250%, while CO_2 emissions increased almost 40 times (Fig. 18.4).

Concerns over increasing atmospheric CO_2 levels have resulted in developing programs to:

1. Estimate the carbon footprint of a farm operation. A carbon footprint is the amount of greenhouse gas emissions produced by a product from cradle (creation) to grave (disposal).
2. Calculate the carbon footprint of fertilizers, plant protection chemicals, and fuel, and
3. Determine the amount of carbon sequestered in soil

The breakdown of organic materials in soil has been another source of CO_2 to the atmosphere. Undisturbed soils with native vegetation varied widely in organic carbon concentrations, from about 1 to 8%, depending on climatic factors and species of native vegetation present, but historic inversion-tillage methods and cropping management practices resulted in a loss of about half the organic carbon (measured as the carbon atoms in organic matter) from the surface 15 cm (six-inch plow layer) of almost 160 million ha (400 million acres) of U.S. cropland. As a conservative estimate, if the average organic carbon loss was 1%, the resulting amount of CO_2 released into the atmosphere would be 24 billion Mg (52 trillion lb), equivalent to about two-thirds of the cumulative CO_2 emissions since 1850. Research has demonstrated that decreasing tillage, incorporating high-residue and/or perennial crops into a rotation, and planting areas to permanent grass increases the amount of organic carbon in the soil (Clay et al., 2015; Robinson et al., 1994; Robinson et al., 1996).

The growing concern over the environmental effects of carbon is manifested in the number of online "carbon footprint" calculators. A carbon footprint is a way to express the net energy usage of a system, whether that system is a person, farm, company, industry, or other entity. The primary inputs into a carbon footprint are energy usage and other items that may reduce energy usage such as recycling, using energy-efficient appliances or low-emission vehicles, or using tactics that may sequester carbon. COMET-Farm (http://cometfarm.

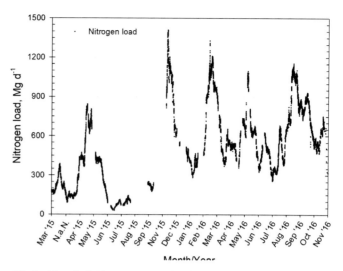

Fig. 18.2. Total daily nitrogen load as nitrite plus nitrate in the Mississippi River at Clinton, IA. (1 Mg = 1 metric ton = 1 long ton = 2200 lb).

Table 18.6. Water discharge and total N and P concentrations in the water from a watershed.

Date	Water discharge m³ h⁻¹	Water discharge ft³ h⁻¹	N concentration mg L⁻¹	P concentration mg L⁻¹
6 April	863	30,442	1.5	0.2
15 April	3367	118,821	2.3	0.3
23 April	466	16,430	2.7	0.4
27 April	3852	135,931	3.3	0.7
4 May	4553	160,676	2.8	0.6

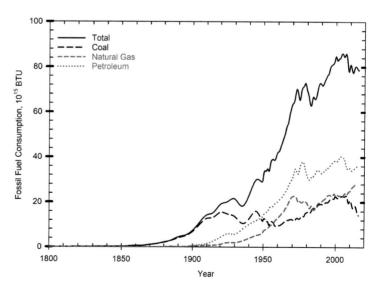

Fig. 18.3. Fossil fuel consumption in the United States since the beginning of the Industrial Revolution (United States Energy Information Administration, 2010, 2017).

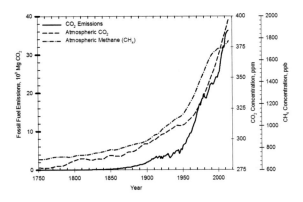

Fig. 18.4. Fossil fuel emissions (Boden et al., 2017) and atmospheric concentrations of carbon dioxide and methane (MacFarling Meure et al., 2006; Carbon Dioxide Information Analysis Center, http://cdiac.ornl.gov/GCP/).

nrel.colostate.edu/, USDA-NRCS, 2017) is one program that is used to estimate the carbon footprint of an operation based on farm management practices.

Case Study to Assess Carbon Footprint

Though oversimplified, the following case study demonstrates the concept of determining a carbon footprint for a conventional tillage operation and the resultant change when implementing a no-till system with a prescription nitrogen application. This management change should increase soil organic carbon.

First, the amount of energy and carbon to produce fertilizer will be examined. The first step in producing all nitrogen fertilizer is to produce anhydrous ammonia from atmospheric nitrogen with the Haber–Bosch process using natural gas as the fuel. All other nitrogen fertilizers are produced from ammonia, and require additional processing and energy. Based on the data in Tables 18.7 and 18.8, 1 gallon #2 diesel will produce 8.5 lb of anhydrous ammonia (NH_3) and 2.7 pounds of urea [$(NH_2)_2CO$].

Most on-farm equipment such as tractors and combines use diesel fuel. The amount of energy in fuel is reported in British Thermal Units (BTU). The following tables identify the energy in natural gas and diesel and their CO_2 emissions (Table 18–7; United States Energy Information Administration, 2016; Downs and Hansen, 1998), the amount of natural gas required to manufacture various nitrogen fertilizers (Table 18–8; PotashCorp, 2013), and the gallons of diesel required per acre to conduct several field operations (Table 18–9; Downs and Hansen, 1998).

Case Study: Carbon Footprint from Switching from Plowed to No-tillage

This case will consider a before and after situation involving a producer farming 1000 acres. Previously the farmer used a conventional, inversion tillage, continuous corn system. Nitrogen was applied as anhydrous ammonia in the fall, with supplemental N and other fertilizers prior to planting in the spring, and a sidedressed application of nitrogen midseason The producer then switched to a variable-rate fertilizer application, continuous corn, no-till

Table 18.7. Energy and carbon dioxide emissions from natural gas and diesel (adapted from **USEIA, 2016;** Downs and Hansen, 1998).

Source	Unit	Energy	CO_2 equivalent
		×10⁶ BTU †	kg CO_2
Natural gas	1000 cu. ft.	1.000	53.2
No. 2 diesel	1 gal	0.138	10.2 to 13.7

† 10⁶ BTU = 1000,000 BTU

Table 18.8. Natural gas requirement to manufacture select nitrogen fertilizers using natural gas (adapted from PotashCorp, 2013).

Fertilizer	Grade	Energy to produce	
		1 ton	1 Mg
		× 10⁶ BTU	×10⁶ BTU
Anhydrous ammonia	82-0-0	32.5	35.8
Ammonium nitrate	34-0-0	49.7	54.7
Urea	45-0-0	56.5	62.2
UAN	32-0-0	46.2	50.8

system and incorporated cover crops. Incorporating cover crops and precision ag fertilizer application decreased the total amount of nitrogen used. The operations, dates of the operations, and average fertilizer rates are provided in Table 18.10. Nitrogen information is available in Table 18.11. This example demonstrates how to determine the carbon footprint and the decrease in carbon footprint due to the change in management system. Determining the partial carbon footprint requires several calculations including to:

1. Determine the amount of fertilizer used, the energy requirement to produce the fertilizer, and the associated CO_2 emissions;
2. Determine the amount of diesel used in each system and associated CO_2 emissions;
3. Determine the total amount of emissions from each system;
4. Determine the amount of carbon sequestered in the soil after the conversion to no-till.

These calculations are demonstrated below.

Partial Carbon Footprints: Fertilizer and Fuel Consumption

Determining the amount of fertilizer used requires converting the pounds of nitrogen into the amount of fertilizer needed for each source: anhydrous ammonia, urea, and UAN. Anhydrous ammonia is 82–0-0, each pound is 82% nitrogen. To find the amount of fertilizer required, divide the recommended nitrogen rate by the percentage of nitrogen in the fertilizer. Multiply by 1000 acres to find the total amount of fertilizer required.

$$\frac{100\,lb\,N}{acre}\times\frac{anhydrous\,ammonia}{82\%\,N}=\frac{100\,lb\,anhydrous\,ammonia}{0.82}\frac{anhydrous\,ammonia}{acre}=122\,lb\,anhydrous\,ammonia\,per\,acre$$

$$\frac{122\,lb\,anhydrous\,ammonia}{acre}\times1,000\,acre\,field^{-1}=122,000\,lb\,anhydrous\,ammonia\,per\,field$$

Follow the same process with the other fertilizers using the grades provided in Table 18.8.

PROBLEM 18.7.

Determine the pounds of anhydrous ammonia (AA) (82-0-0) produced per gal diesel.

ANSWER:

$$\frac{2000\,lb\,AA}{ton}\times\frac{0.82\,lb\,N}{1\,lb\,AA}\times\frac{1\,ton}{32.5\times10^6\,BTU}\times\frac{0.138\times10^6\,BTU}{gallon\,\#2\,diesel}=\frac{6.96\,lb\,N}{gallon\,\#2\,diesel}$$

Data from tables 18.7 and 18.8.

Table 18. 9. Diesel fuel consumption for selected field operations (adapted from Downs and Hansen, 1998).

Operation	Diesel fuel consumption	
	gal acre^{-1}	L ha^{-1}
Fertilizer, Sidedress, Solution	0.20	0.31
Fertilizer, surface broadcast	0.25	0.38
Fertilizer, knife injection, 30-in centers	0.65	1.00
Sprayer, pre-emergence	0.10	0.15
Chisel, sweep shovel	1.10	1.69
Disk, tandem secondary operation	0.65	1.00
Field cultivator, 12-in sweeps	0.60	0.92
No-till drill	0.35	0.54
Planter, double-disk opener	0.50	0.77
Planter, double-disk opener with fluted coulter	0.35	0.54
Harvest, leave 50% standing stubble	1.60	2.46

Table 18.10. Field operations for a conventional tillage system to no-till with precision fertilizer application (data for Case Study 1).

Date	System and operation	
	Conventional	No-till system
16 September		No-till drill, plant cover crops
1 November	Chisel, sweep shovel	
5 November	100 lb N acre⁻¹ as anhydrous ammonia, knife injection, 30-in centers	
10 April	Sprayer, preemergent herbicide	Sprayer: herbicide, terminate cover crop
16 April	75 lb N acre⁻¹ as urea, surface broadcast	
20 April		120 lb N acre⁻¹ as UAN, banded
25 April	Disk, tandem secondary operation	
30 April	Field cultivator, 12-in sweeps	
1 May	Planter, double-disk opener	Planter, double-disk opener, fluted coulter
Jun 10	50 lb N acre⁻¹ as UAN, Sidedress	50 lb N acre⁻¹ as UAN, Sidedress
Jul 1	Sprayer, fungicide application	Sprayer, fungicide application
Oct 20	Harvest, leave 50% standing stubble	Harvest, leave 50% standing stubble

Table 18.11. Nitrogen amount and sources.

Date	System and operation	
	Conventional	No-till system
5 November	100 lb N acre⁻¹ as anhydrous ammonia, knife injection, 30-in centers	
16 April	75 lb N acre⁻¹ as urea, surface broadcast	
20 April		120 lb N acre⁻¹ as UAN, banded
10 June	50 lb N acre⁻¹ as UAN, Sidedress	50 lb N acre⁻¹ as UAN, Sidedress

Use the data in Table 18-8 to find the energy required to produce the fertilizer, then the data in Table 18-7 to determine the CO_2 emissions. The energy requirement is given in BTU units per ton of fertilizer, so first the pounds of fertilizer must be converted to tons. Because of assumptions leading to the energy requirement calculation, round CO_2 emissions to the nearest 5000 kg.

Tons fertilizer:

$$\frac{122{,}000 \text{ lb anhydrous ammonia}}{\text{field}} \times \frac{1 \text{ ton}}{2{,}000 \text{ lb}} = 61.0 \text{ tons anhydrous ammonia per field}$$

Energy requirement:

$$\frac{61.0 \text{ tons anhydrous ammonia}}{\text{field}} \times \frac{32.5 \times 10^6 \text{ BTU natural gas}}{\text{tons anhydrous ammonia}} = \frac{1{,}983 \times 10^6 \text{ BTU natural gas}}{\text{field}}$$

CO_2 emissions:

$$\frac{1{,}983 \times 10^6 \text{ BTU natural gas}}{\text{field}} \times \frac{53.2 \text{ kg } CO_2}{10^6 \text{ BTU natural gas}} = 105{,}000 \text{ kg } CO_2 \text{ per field}$$

Follow the same process with each of the fertilizers and complete Table 18.12. The tons of fertilizer and total emissions are given so that you may check your work.

Even though less total nitrogen is applied, the emissions from nitrogen production in the prescription fertilizer application increased almost 20% relative to the conventional system. This increase was attributed to the use of UAN. To determine the diesel usage, multiply the rate of fuel use (gal acre⁻¹) for each operation by the number of

acres (1000 acre). For a chisel operation, $\frac{1.10 \text{ gal}}{\text{acre}} \times 1{,}000 \text{ acre} = 1{,}100 \text{ gal}$. Do the same for all the other operations in

both systems to complete Tables 18.13 and 18.14.

Table 18.12. Fertilizer amounts, energy requirement, and CO$_2$ emissions. Complete the worksheet using data and examples provided above.

Nitrogen	Fertilizer source	Amount	Amount per 1000-acre field			
			Fertilizer		Natural gas	Emissions
			lb	tons		
lb acre^{-1}		lb acre^{-1}			10^6 BTU	kg CO$_2$
Conventional system						
100	Anhydrous ammonia	122	122,000	61.0	1983	105,000
75	Urea			83.5		
50	UAN			78.0		
225					Total	545,000
No-till, prescription fertilizer						
120	UAN			187.5		
50	UAN			78.0		
170					Total	650,000

Table 18.13. Diesel usage in the conventional system.

Date	Conventional	Diesel	Diesel
		gal acre^{-1}	gal
1 November	Chisel, sweep shovel	1.10	1100
5 November	75 lb N acre^{-1} as anhydrous ammonia, knife injection, 30-in centers	0.65	
10 April	Sprayer, preemergent herbicide	0.10	
16 April	75 lb N acre^{-1} as urea, surface broadcast	0.25	
25 April	Disk, tandem secondary operation	0.65	
30 April	Field cultivator, 12-in sweeps	0.60	
1 May	Planter, double-disk opener	0.50	
10 June	50 lb N acre^{-1} as UAN, Sidedress	0.20	
1 July	Sprayer, fungicide application	0.10	
20 October	Harvest, leave 50% standing stubble	1.60	
		Total	5750

Table 18.14. Diesel usage in the no-till system.

Date	No-till system	Diesel	Diesel
		gal acre^{-1}	gal
16 September	No-till drill planter	0.35	
10 April	Sprayer: herbicide, terminate cover crop	0.10	
20 April	120 lb N acre^{-1} as UAN, banded	0.10	
1 May	Planter, double-disk opener, fluted coulter	0.35	
10 June	50 lb N acre^{-1} as UAN, Sidedress	0.20	
1 July	Sprayer, fungicide application	0.10	
20 October	Harvest, leave 50% standing stubble	1.60	
		Total	2800

Table 18.15. Total annual CO₂ emissions from each tillage, cropping, and fertilizer management system.

System	Diesel emissions	Fertilizer emissions	System total
		kg CO₂	
Conventional system	70,000	545,000	615,000
No-till, prescription fertilizer system	35,000	650,000	685,000

Then, sum the column to find the total quantity of diesel used. The total diesel consumption is provided so that you may check your work.

Adopting no-tillage reduced the total diesel consumption by almost half. The data in Table 18.7 show combustion of 1 gal diesel releases 10.2 to 13.7 kg CO_2. For the sake of this example, select 12 kg CO_2 to determine the emissions from each system. The calculation is straightforward, the product of the gallons of diesel and the emissions per gallon. Remember to round CO_2 emissions to the nearest 5000 kg.

$$\frac{12 \text{ kg CO}_2}{\text{gallon diesel}} \times 5{,}570 \text{ gallon diesel} = 66{,}840 \text{ kg CO}_2$$

Next sum the amount of CO_2 from fertilizer manufacture and diesel consumption to determine the total emissions from both systems and complete Table 18.15.

The calculations up to this point suggest that change from conventional tillage to no-till increased the carbon emissions by about 12%. This increase was attributed to the use of UAN as opposed to anhydrous ammonia. The footprint could be reduced by changing the N source. However, these calculations do not consider the amount of carbon sequestered in the soil (Clay et al., 2015)

Sequestered Carbon

No-till systems have been shown to increase organic carbon levels (Clay et al., 2015). For partial C footprints, the amount of carbon sequestered must be subtracted from the carbon emissions (Clay et al., 2012). The next series of calculations demonstrate how to determine the carbon sequestration if the conversion from conventional tillage to no-till resulted in a 0.25% increase in the soil organic carbon levels in the surface 12 inches over a 5-yr period. This calculation is separated into: (i) determining the net carbon footprint for the no-till system, (ii) adjusting the CO_2 emissions for the carbon sequestered; and (iii) comparing the resulting carbon footprint for each system. Use the following steps to determine the carbon sequestered. Because of the assumptions made in these estimates, round the C sequestered to the nearest 10,000 lb (approximately 5000 kg).

1. Determine the mass of one acre-foot of soil.
2. Determine the mass of carbon sequestered with a 0.25% soil organic carbon increase.
3. Determine the difference in carbon sequestered and carbon emissions.
4. Compare the carbon footprint between the two systems.

Step 1. Determine the mass of one acre-foot of soil. An acre–furrow slice represents a volume of soil that is one acre of surface area by about 6.67 inches deep, and that has a mass of about 2000,000 lb. To determine the mass of one acre-foot, divide by 6.67 to find the mass of one acre-in of soil, then multiply by 12 inches. Round to the nearest 10,000 lb.

Weight of one acre-inch of soil:

$$\frac{2{,}000{,}000 \text{ lb}}{\text{acre-furrow slice}} = \frac{2{,}000{,}000 \text{ lb}}{1 \text{ acre} \times 6.7 \text{ in}} = \frac{298{,}500 \text{ lb}}{1 \text{ acre} \times 1 \text{ in}}$$

Weight of 1 acre-foot of soil:

$$\frac{300{,}000 \text{ lb}}{1 \text{ acre} \times 1 \text{ in}} \times \frac{12 \text{ in}}{1 \text{ ft}} = \frac{3{,}600{,}000 \text{ lb}}{1 \text{ acre} \times 1 \text{ ft}}$$

Step 2. Determine the mass of carbon sequestered in a 0.25% C (soil organic carbon) increase. Multiply the mass of 1 acre-ft of soil by the percentage increase in organic carbon. Remember that percent means per 100. It helps to remember this when determining how to "move the decimal", forgetting the meaning can result in a math error.

$$0.25\% \; C = \frac{0.25 \; SOC}{100} = 0.0025 \; C$$

Increase in soil organic carbon:
$$\frac{3,600,000 \; lb}{1 \; acre \times 1 \; ft} \times \frac{0.25 \; C}{100} = \frac{3,600,000 \; lb}{1 \; acre \times 1 \; ft} \times 0.0025 \; C = \frac{9,000 \; lb \; C}{1 \; acre \times 1 \; ft}$$

Amount of sequestered carbon in 1000 acres:
$$\frac{9,000 \; lb \; C}{1 \; acre \times 1 \; foot} \times \frac{1,000 \; acre}{field} \times 1 \; ft = 9,000,000 \; lb \; C \; per \; field$$

Step 3. Determine the difference in carbon sequestered and carbon emitted. Notice that emissions are reported in kg CO_2 field[-1] while sequestration is identified in lb C field[-1]. The units must be the same, so two conversions will be necessary: mass and chemical composition. The order of conversions does not matter. For the sake of this example, convert CO_2 to C since the goal is to determine the carbon footprint, then convert lb to kg.

To convert carbon dioxide to carbon, use the molecular weights to determine the fraction of C in CO_2: C = 12, O = 16
Convert C to CO_2:
$$\frac{C}{CO_2} \rightarrow \frac{C}{C + O \times 2} \rightarrow \frac{12}{12 + 16 \times 2} = \frac{12 \; C}{44 \; CO_2} = \frac{0.2727 \; C}{CO_2}$$

Using the fraction form helps prevent making errors in unit conversions. Multiply CO_2 by 0.2727 to convert to C.
Total annual C emissions from no-till system:
$$\frac{685,000 \; kg \; CO_2}{1 \; field \times 1 \; yr} \times \frac{12 \; C}{44 \; CO_2} = \frac{185,000 \; kg \; C}{1 \; field \times 1 \; yr}$$

Total five-year C emissions in no-till system:
$$\frac{185,000 \; kg \; C}{1 \; field \times 1 \; yr} \times 5 \; yr = 925,000 \; kg \; C \; per \; field$$

Do the same for the conventional system and enter both values into Table 18–16.
Next, convert lb to kg for the total carbon sequestered, and enter the value into Table 18–16.

$$\frac{9,000,000 \; lb \; C}{field} \times \frac{1 \; kg}{2.2 \; lb} = 4,090,000 \; kg \; C \; per \; field \quad total \; C \; sequestered$$

Step 4. Compare the carbon footprint between the two systems.
Complete the calculations for the carbon footprint using Table 18.16. Carbon sequestration will be represented as a negative value because carbon is being removed from the atmosphere and stored in the soil.
Total C sequestered: For the carbon sequestered in the no-till system, add across the rows.

$$925,000 \; kg \; C - 4,090,000 \; kg \; C = -3,165,000 \; kg \; C$$

To determine if no-tillage increased carbon sequestration, subtract the second row from the first.

$$925,000 \; kg \; C - 850,000 \; kg \; C = 75,000 \; kg \; C$$

Do the same for the other two rows and columns to complete Table 18–16. As a check for the calculations, the table is organized such that the sum of the last row equals the difference of the last column. This overall difference in carbon footprint is provided to check your answers.

Table 18.16. Total emissions and sequestration during five years for a 1000-acre field from each tillage, cropping, and fertilizer management system.

System	Emissions	Sequestration†	System total
		—kg C—	
No-till, prescription fertilizer system	925,000	-4,090,000	
Conventional system	850,000	0	
Difference in C emissions			-4,015,000

†Negative value indicates C removed from the atmosphere and stored in the soil.

Summary

Agriculture is an energy-intensive activity that strongly effects the environment. This chapter provided an introduction for methods to identify the potential effects of agricultural management systems on erosion, nutrient loading, and carbon emissions and sequestration. In addition, this chapter provides details on how to determine a partial carbon footprint.

Additional Problems

18.8. Calculate how many pounds of UAN are produced per gallon of No. 2 Diesel.

18.9. Calculate how many pounds of ammonium nitrate (NH_4NO_3) are produced per gallon of No. 2 Diesel.

18.10. Determine how much organic C in lb acre^{-1} is contained in the surface 12 inches of a soil that contains 3% soil organic C? Assume that each foot of soil contains 4 million lb.

18.11. What is the expected residue cover for a 40 bu acre^{-1} winter wheat crop?

18.12. How will changing from a chisel straight point to chisel–disk tandem influence residue cover?

18.13. A producer decides to knife inject manure (slurry) rather than surface broadcast 100 lb of N acre^{-1} applied as urea? Assume the manure contains 26 lb N per 1000 gallon. How much manure needs to be injected to get 100 lb of N? What is the energy cost in diesel fuel for the fertilizer plus application for each type of application? Assume that the N in the manure would be equivalent to the energy cost of anhydrous ammonia.

ACKNOWLEDGMENTS

Support for this document was provided by the Precision Farming Systems community in the American Society of Agronomy, International Society of Precision Agriculture, and the USDA-AFRI Higher Education Grant (2014-04572).

REFERENCES AND ADDITIONAL INFORMATION

Alexander, R.B., R.A. Smith, G.E. Schwarz, E.W. Boyer, J.V. Nolan, and J.W. Brakebill. 2008. Differences in phosphorus and nitrogen delivery to the Gulf of Mexico from the Mississippi River Basin. Environ. Sci. Technol. 42:822–830. doi:10.1021/es0716103

Boden, T.A., G. Marland, and R.J. Andres. 2017. Global, Regional, and National Fossil-Fuel CO2 Emissions. Carbon Dioxide Information Analysis Center, Oak Ridge National Laboratory, U.S. Department of Energy, Oak Ridge, TN. doi 10.3334/CDIAC/00001_V2017.

Clay, D.E., and J. Shanahan. 2011. GIS applications in agriculture: Nutrient management for improved energy efficiency. CEC Press, Taylor and Francis, New York.

Clay, D.E., G. Reicks, C.G. Carlson, J. Moriles-Miller, J.J. Stone, and S.A. Clay. 2015. Tillage and corn residue harvesting impacts surface and subsurface carbon sequestration. J. Environ. Qual. 44:803–809. doi:10.2134/jeq2014.07.0322

Clay, D.E., T.M. DeSutter, S.A. Clay, and C. Reese. 2017. From plows, horses, and harnesses to precision technologies in the north American Great Plains. Oxford Research Encyclopedia of Environmental Science. DOI: doi:10.1093/acrefore/9780199389414.013.196.

Clay, D.E., J. Chang, S.A. Clay, J.J. Stone, R.H. Gelderman, C.G. Carlson, K. Reitsma, M. Jones, L. Janssen, and T. Schumacher. 2012. Corn yields and no-tillage affects carbon sequestration and carbon footprint. Agron. J. 104:763–777. doi:10.2134/agronj2011.0353

Downs, H.W., and R.W. Hansen. 1998. Estimating farm fuel requirements. Farm and Ranch Series: Equipment, Fact Sheet No. 5.006. Colorado State University Extension. Fort Collins, CO. http://www.waterandenergyprogress.org/library/05006.pdf (accessed 1 June 2017).

Illinois Council on Best Management Practices. 2015. Nitrogen management: The 4Rs. http://illinoiscbmp.org/Practices/Nitrogen-Management/ (Accessed 25 June 2017).

Kok, H., and J. Thien. 1994. RES-N-TILL: Crop residue conservation and tillage management software. J. Soil Water Conserv. 49:551–553.

MacFarling Meure, C., D. Etheridge, C. Trudinger, P. Steele,R. Langenfelds, T. van Ommen, A. Smith, and J. Elkins. 2006. The Law Dome CO_2, CH_4 and N_2O Ice Core Records Extended to 2000 years BP. Geophysical Research Letters, 33(14): L14810 doi:10.1029/2006GL026152

PotashCorp. 2013. Overview of PotashCorp and its industry. PotashCorp. http://www.potashcorp.com/overview/resources/nitrogen (verified 10 July 2017).

Renard, K.G., J.M. Laflen, G.R. Foster, and D.K. McCool. 1994. The revised soil loss equation. In: R. Lal, editor, Soil Erosion, Research Methods. 2nd ed. Soil and Water Conservation Society, Ankeny, IA. p. 105–124.

Robinson, C.A., R.M. Cruse, and M. Ghaffarzadeh. 1996. Cropping system and nitrogen effects on Mollisol organic carbon. Soil Sci. Soc. Am. J. 60:264–269. doi:10.2136/sssaj1996.03615995006000010040x

Robinson, C.A., R.M. Cruse, and K.A. Kohler. 1994. Soil management. In: J.L. Hatfield and D.L. Karlen, editors, Sustainable agriculture systems. Lewis Publishers, Boca Raton, FL. p. 109–134.

USDA Science and Education Administration. 1981. Predicting rainfall erosion losses: A guide to conservation planning. Agriculture Handbook No. 537. U.S. Gov. Print. Office. Washington, DC.

USDA-NRCS. 2017. COMET-Farm. USDA-NRCS, Colorado State University. http://cometfarm.nrel.colostate.edu/Home (verified 10 July 2017).

USDA-ARS. 2008. Draft User reference guide: Revised Universal Soil Loss Equation, Version 2 (RUSLE2). USDA-ARS, Washington, DC.

United States Energy Information Administration. 2010. Annual Energy Review 2009. DOE/EIA-0384(2009). US Energy Information Administration, Washington, DC.

United States Energy Information Administration. 2016. Frequently asked questions. U.S. Energy Information Administration. https://www.eia.gov/tools/faqs/faq.php?id=73&t=11 (verified 10 July 2017).

United States Energy Information Administration. 2017. May 2017 Monthly energy review. DOE/EIA-0035(2017/5). US Energy Information Administration. https://www.eia.gov/totalenergy/data/monthly/pdf/mer.pdf (verified 10 July 2017).

USGS. 2017. National Water Information System: Web Interface. USGS. https://nwis.waterdata.usgs.gov (Accessed 25 June 2017).

Index